Underwater Acoustic Modeling

Underwater Acoustic Modeling

Principles, techniques and applications

Second edition

Paul C. Etter

E & FN SPON
An Imprint of Chapman & Hall
London · Glasgow · Weinheim · New York · Tokyo · Melbourne · Madras

Published by E & FN Spon, an imprint of Chapman & Hall, 2–6 Boundary Row, London SE1 8HN, UK

Chapman & Hall, 2–6 Boundary Row, London SE1 8HN, UK

Blackie Academic & Professional, Wester Cleddens Road, Bishopbriggs, Glasgow G64 2NZ, UK

Chapman & Hall GmbH, Pappelallee 3, 69469 Weinheim, Germany

Chapman & Hall USA, 115 Fifth Avenue, New York, NY 10003, USA

Chapman & Hall Japan, ITP-Japan, Kyowa Building, 3F, 2-2-1 Hirakawacho, Chiyoda-ku, Tokyo 102, Japan

Chapman & Hall Australia, 102 Dodds Street, South Melbourne, Victoria 3205, Australia

Chapman & Hall India, R. Seshadri, 32 Second Main Road, CIT East, Madras 600 035, India

First edition 1991

Second edition 1996

Typeset in 10/12 Palatino by Thomson Press (India) Ltd., New Delhi
Printed in Great Britain at the University Press, Cambridge

ISBN 0 419 20190 4

A catalogue record for this book is available from the British Library

∞ Printed on acid-free text paper, manufactured in accordance with ANSI/NISO Z39.48-1992 (Permanence of Paper).

To my wife Alice and my sons Gregory and Andrew

Contents

Preface

The subject of underwater acoustic modeling deals with the translation of our physical understanding of sound in the sea into mathematical formulas solvable by computers. This book divides the subject of underwater acoustic modeling into three fundamental aspects: the physical **principles** used to formulate underwater acoustic models; the mathematical **techniques** used to translate these principles into computer models; and modeling **applications** in sonar technology and oceanographic research.

The material presented here emphasizes aspects of the ocean as an acoustic medium. It shows mathematicians and physical scientists how to use this information to model the behavior of sound in a spatially complex and temporally variable ocean. This approach diminishes the need for discussions of engineering issues such as transducers, arrays and targets. Aspects of hardware design and modeling in underwater acoustics are discussed in other excellent texts.

Recent developments in underwater acoustic modeling have been influenced by changes in global geopolitics. These changes are evidenced by strategic shifts in military priorities as well as by efforts to transfer defense technologies to non-defense applications.

The strategic shift in emphasis from deep-water to shallow-water naval operations has focused attention on improving sonar performance in coastal regions. These near-shore regions, which are sometimes referred to as the littoral zone, are characterized by complicated and highly variable acoustic environments. Such difficult environments challenge the abilities of those sonar models intended for use in deep-water scenarios. This situation has prompted further development of underwater acoustic models suitable for forecasting and analyzing sonar performance in shallow-water areas.

The policy of defense conversion has encouraged the transfer of sonar modeling technology to non-defense applications. Much of this transfer has benefited the growing field of environmental acoustics, which seeks to expand exploration of the oceans through acoustic sensing. Such technology conversion is exemplified by the utilization of naval underwater acousic models as both prognostic and diagnostic tools in sophisticated experiments employing inverse acoustic sensing of the oceans.

These rapid developments in modeling have created a need for a second edition. The intent is to update recent advances in underwater acoustic modeling and to emphasize new applications in oceanographic research. This edition also reflects a broader international interest in the development and application of underwater acoustic models. The coming years promise to be challenging in terms of defining research directions, whether for defense or industry, and this edition should provide technology planners with a useful baseline.

The original organization of material into eleven chapters has served well and therefore remains unchanged. When required, new material has been arranged into additional subsections.

Comments from users of the first edition have evidenced appeal from acousticians, as well as oceanographers, who have enthusiastically endorsed this book as both a practical tool and an instructional aid. In this latter regard, several academic institutions have utilized this book as an adjunct text for graduate-level courses in applied mathematics and ocean sciences.

This edition has benefited from a continuation of my short courses which, since 1993, have been offered through the Applied Technology Institute of Clarksville, Maryland (USA). Continued exposure to the insightful questions posed by my students has provided me with the opportunity to further refine my presentation.

Despite the appearance of several new books in the field of ocean acoustics, this book remains unique in its treatment and coverage of underwater acoustic modeling. It is a pleasure to note that the first edition has been recognized as an authoritative compendium of state-of-the-art models and is often cited as the standard reference.

Paul C. Etter
Rockville, Maryland
USA

Preface to the first edition

The subject of underwater acoustic modeling deals with the translation of our physical understanding of sound in the sea into mathematical formulas solvable by computers. These models are useful in a variety of research and operational applications including undersea defense and marine seismology. There has been a phenomenal growth in both the number and types of models developed over the past several decades. This growth reflects the widespread use of models for the solution of practical problems as well as the considerable advances made in our computational abilities.

The primary motivation for the development of underwater acoustic models is defense related. Researchers involved in Anti-Submarine Warfare (ASW) and associated undersea defense disciplines use models to interpret and forecast acoustic conditions in the sea in support of sonar design and sonar operation. Consequently, the emphasis in this book is placed on those models that are particularly useful in solving sonar performance problems.

Users and potential users of models are commonly ill acquainted with model formulations. As a result, the capabilities and limitations of the models are poorly understood and the models are often improperly used. Moreover, the sheer number of available models complicates the process of model selection.

This book is intended for those who have a fundamental understanding of underwater acoustics but who are not familiar with the various aspects of modeling. Sufficient mathematical derivations are included to demonstrate model formulations, and guidelines are provided to assist in the selection and proper application of these models. Comprehensive summaries identify the available models and associated documentation.

The material in this book is organized into eleven chapters. The order of presentation follows the structure suggested by a hierarchical method of sonar model construction. Chapter 1 introduces the types of underwater acoustic models, provides a framework for the consistent classification of modeling techniques, and defines the terminology common to modeling work. Aspects of oceanography essential to an understanding of acoustic phenomena are presented in Chapter 2. Chapters 3 through 9 address the observations and models dealing with propagation, noise and reverberation in the sea. In Chapter 10, the information from Chapters 3–9 is integrated into sonar performance models. Finally, Chapter 11 describes the process of model evaluation.

Since 1982, I have developed and taught a series of intensive short courses for the Technology Service Corporation of Silver Spring, Maryland (USA). Earlier versions of this course were taught in collaboration with Professor Robert J. Urick of the Catholic University of America. Professor Urick would discuss underwater acoustic measurements while I would review the related modeling techniques. As the course evolved into one in which I became the sole instructor, I borrowed heavily from Professor Urick's several books (with permission) in order to preserve the continuity of the course material. The success of this course encouraged me to publish my class notes as a book.

Many notable books have been published in the field of underwater acoustics. None, however, has dealt exclusively with modern developments in modeling, although some have addressed aspects of propagation modeling. This book is unique in that it treats the entire spectrum of underwater acoustic models including environmental, propagation, noise, reverberation and sonar performance models.

I have intentionally preserved the notation, terminology and formalism used by those researchers whose work I have cited. I have also intentionally emphasized aspects of oceanography since my experience has indicated that many acousticians have little appreciation for the complex role played by the ocean as an acoustic medium. Conversely, oceanographers frequently fail to appreciate the great potential of underwater acoustics as a remote sensing technique.

Paul C. Etter

Acknowledgments

The students who have attended my short courses have provided both a receptive and critical audience for much of the material contained in this book. Many of my colleagues have provided suggestions and useful insights: Dr Aubrey L. Anderson, Dr Stanley A. Chin-Bing, Dr Richard B. Evans, Dr Robert W. Farwell, Dr Richard P. Flanagan, Dr Robert L. Martin, Dr Frederick D. Tappert and Dr Henry Weinberg. Robert S. Winokur provided administrative guidance in the early stages of my work in underwater acoustic modeling.

Professor Robert J. Urick has been a source of great encouragement and has graciously allowed me to liberally borrow material from his several books. Professor John D. Cochrane of Texas A & M University inspired the scholarly discipline that facilitated creation of this book.

1

Introduction

1.1 BACKGROUND

The field of underwater acoustics has been extensively developed over the past four decades in response to practical needs originating within both the naval sonar and the marine seismology communities. The emphasis in this book is focused on those underwater acoustic models developed for sonar (versus seismic) applications.

Since 1960 considerable work has been expended on the development of models with which to analyze data collected in field experiments. These models have then been used to forecast acoustic conditions for planning at-sea experiments, designing optimized sonar system and predicting sonar performance at sea. Modeling has become the chief mechanism by which researchers and analysts can simulate sonar performance in laboratory conditions. Modeling provides an efficient means by which to parametrically investigate the performance of hypothetical sonar designs under varied environmental conditions as well as to estimate the performance of existing sonars in different ocean areas and seasons.

A distinction is made between physical models and mathematical models, both of which are addressed in this book. Physical models pertain to theoretical or conceptual representations of the physical processes occurring within the ocean; the term 'analytical model' is sometimes used synonymously. Mathematical models include both empirical models (those based on observations) and numerical models (those based on mathematical representations of the governing physics). The subject of analog modeling, which is defined here as controlled acoustic experimentation in water tanks employing appropriate oceanic scaling factors, is not addressed in this book. Reviews of acoustic analog modeling have been presented by Barkhatov (1968) and Zornig (1979).

The physical models underlying the numerical models have been well known for some time. The transition to operational computer models, however, has been hampered by several factors: limitations in computer capabilities; inefficient mathematical methods; and inadequate oceanographic and acoustic data with which to initialize and evaluate models.

This book addresses three broad types of underwater acoustic models: **environmental models**, **basic acoustic models** and **sonar performance models.**

The first category, environmental models, includes empirical algorithms which are used to quantify the boundary conditions (surface and bottom) and volumetric effects of the ocean environment. Such models include sound speed, absorption coefficients, surface and bottom reflection losses, and surface, bottom and volume backscattering strengths.

The second category, basic acoustic models, comprises propagation (transmission loss), noise and reverberation models. This category is the primary focus of attention in this book.

The third category, sonar performance models, is composed of environmental models, basic acoustic models and appropriate signal-processing models. Sonar performance models are organized to solve specific sonar applications problems such as submarine detection, mine hunting, torpedo homing and bathymetric sounding.

Figure 1.1 illustrates the relationships among these three broad categories of models. As the applications become more and more system specific, that is, as one progresses from environmental models toward sonar performance models, the respective models become less universal in application. This is a consequence of the fact that system-specific characteristics embedded in the higher level models (for example, signal-processing models) restrict their utility to particular sonar systems. Thus, while one propagation model may enjoy a wide variety of applications, any particular sonar performance models is, by design, limited to a relatively small class of well-defined sonar problems.

The wide breadth of material covered in this book precludes exhaustive discussions of all existing underwater acoustic models. Accordingly, only selected models considered to be representative of each of the three broad categories will be explored in more detail. In addition, comprehensive summaries identifying existing basic acoustic models and sonar performance models are presented. Notable environmental models are identified and discussed in appropriate sections throughout this book.

Modeling applications will generally fall into one of two basic areas: research or operational. Research-oriented applications are conducted in laboratory environments where accuracy is important and computer time is not a critical factor. Examples of research applications include sonar system design and field experiment planning. Operationally oriented applications are conducted as field activities, including fleet operations at sea and sonar system training ashore. Operational applications generally require rapid execution, often under demanding conditions; moreover, modeling accuracy may be subordinate to processing speed.

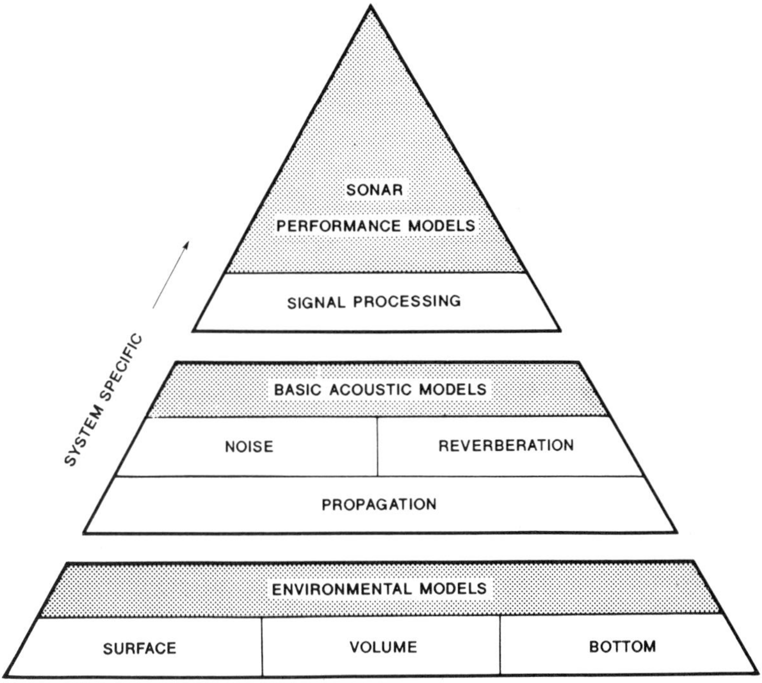

Fig. 1.1 Generalized relationships among environmental models, basic acoustic models and sonar performance models.

1.2 MEASUREMENTS AND PREDICTION

The scientific discipline of underwater acoustics is transitioning from a stage of observation to a stage of understanding and prediction. This transition is not always smooth – direct observations are limited, the resulting prediction tools (models) are not perfected, and much refinement is still needed.

Experimental measurements in the physical sciences are generally expensive due to instrumentation and facility operation costs. In the case of oceanographic and underwater acoustic data collection, this is particularly true because of the high costs of platform operation (ships, aircraft, submarines). Consequently, in the field of underwater acoustics, much use is made of what few field measurements already exist. Notable large-scale field programs that have been successfully conducted in the past include AMOS (Acoustic, Meterological and Oceanographic Survey) and LRAPP (Long Range Acoustic Propagation Project).

Modeling has been used extensively to advance scientific understanding without expending inordinate resources on additional field observations. The balance between observations and modeling, however, is very

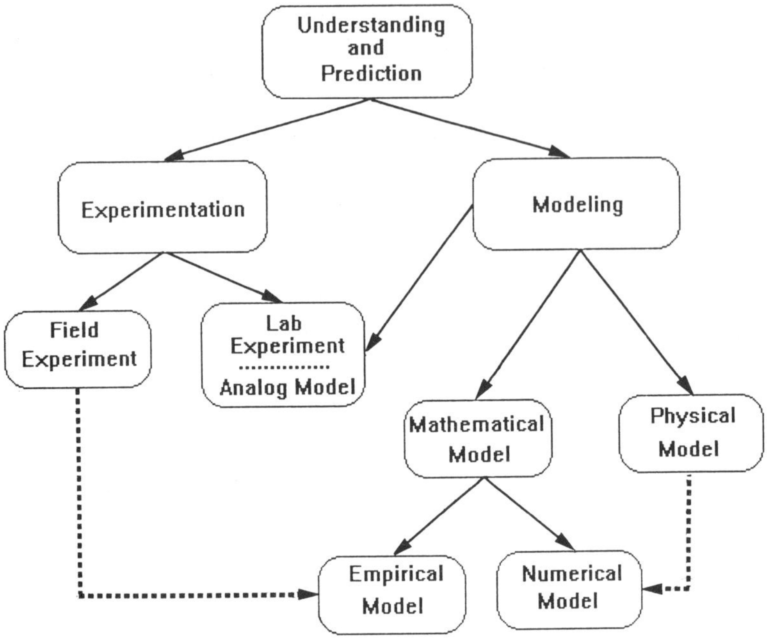

Fig. 1.2 Schematic relationship between experimentation and modeling.

delicate. Experimenters agree that modeling may help to build intuition or refine calculations, but they argue further that only field observations give rise to genuine discovery. Accordingly, many researchers find mathematical models most useful once the available observations have been analyzed on the basis of simple physical models.

The relationship between experimentation and modeling (in the furtherance of understanding and prediction) is depicted schematically in Fig. 1.2. Here, physical models form the basis for numerical models while experimental observations form the basis for empirical models. Moreover, analog modeling is represented as a form of laboratory (versus field) experimentation.

Scientists are becoming more aware of the connection between physical processes and computation and many now find it useful to view the world in computational terms. Consequently, computer simulation is sometimes viewed as a third form of science, halfway between theory and experiment. Furthermore, understanding can be enhanced through the use of advanced computer graphics to convert large volumes of data into vivid and comprehensible patterns.

Because of national security concerns, some existing data sets are limited in accessibility. Also, because of the wide range of acoustic frequencies, ocean areas and geometries of interest to researchers, it is virtually impossible to accommodate all observational requirements within normal fiscal constraints. To make matters worse, acoustic data are sometimes collected at sea without the supporting oceanographic data; thus, models cannot always replicate the observed acoustic results because they lack the necessary input parameters for initialization.

Satellites, together with other remote-sensing techniques, provide a useful adjunct to the prediction of underwater acoustic conditions. Specifically, many dynamic features of the ocean affect the behavior of sound, and knowledge of their location can improve the prediction of sonar performance. Although satellite-borne sensors detect only surface (or near-surface) features of the ocean such as thermal contrast, color or surface roughness, these surface expressions can generally be associated with dynamic oceanographic features below the surface, particularly when comprehensive climatological databases already exist with which to establish such associations. Thus, for example, satellite imagery can be used to provide timely and accurate position information on variable features such as ocean fronts and eddies which are known to significantly affect the propagation of acoustic signals in the sea.

Tactical oceanographic data collection has been augmented by drifting buoys, which use satellite relays to transmit data to mobile or stationary receiving stations, and by Autonomous Underwater Vehicles (AUVs) to access remote ocean areas such as shallow-water and under-ice regions (Brutzman *et al.*, 1992; Selsor, 1993; Dantzler *et al.*, 1993).

The problem of operational sonar prediction embrances many disciplines, one of which is modeling. Such modern operational applications involve not only underwater acoustic models but also oceanographic models (Etter, 1989). The coupling of these two types of models provides a valuable set of prediction tools to the force commanders by enabling them to respond to the changing environmental conditions that affect their sonar performance. The remote-sensing data now available to naval forces afloat can be used in conjunction with oceanographic models to accurately forecast the locations and characteristics of dynamic ocean features (e.g. Robinson, 1992). This information can then be input to the appropriate acoustic models to assess the resultant impacts on sonar performance. These sonar systems can then be optimized for performance in each region of operation at any given time of the year.

1.3 DEVELOPMENTS IN MODELING

A goal of science is to develop the means for reliable prediction to guide decision and action (Ziman, 1978). This is accomplished by finding algorithmic compressions of observations and physical laws. Physical laws are statements about classes of phenomena, and initial conditions are statements about particular systems. Thus, it is the solutions to the equations, and not the equations themselves, that provide a mathematical description of the physical phenomena. In constructing and refining mathematical theories, we rely heavily on models. At its conception, a model provides the framework for a mathematical interpretation of new phenomena.

In its most elemental form, a model is intended to generalize and abstract. A perfect model is one that perfectly represents reality. In practice, however, such a perfect model would defeat its purpose: it would be as complex as the problem it is attempting to represent. Thus, modeling in the physical sciences is normally reduced to many, more easily managed, components.

With the advent of digital computers, modeling in the physical sciences has advanced dramatically. Improvements in computer capabilities over the past several decades have permitted researchers to incorporate more complexity into their models, sometimes with little or no penalty in run time or computer costs (e.g. Hodges, 1987; Runyan, 1991). Frequently, models become data limited. This means that observational data are lacking in sufficient quantity or quality with which to support model initialization and model evaluation.

As modeling techniques have proliferated within the underwater acoustics community, it has become increasingly difficult to take stock of the various models already in existence before launching a new effort to develop yet more models. Moreover, analysts confronted with sonar performance problems have had difficulty in determining what models existed and, of those, which were best for their particular situation. This book had its genesis in just such a dilemma. A small study was sponsored by the US Navy in 1978 (Etter and Flum, 1978) to review the availability of numerical models of underwater acoustic propagation, noise and reverberation as well as the availability of databases with which to support model development and operation. Results of this work, and extensions thereto (Etter and Flum, 1980; Etter *et al.*, 1984), have subsequently been presented at meetings of the Acoustical Society of America (Etter and Flum, 1979; Etter, 1987b) and have been updated periodically in literature review articles (Etter, 1981, 1984, 1987a, 1990). An enhanced version of the first review article (Etter, 1981) was included as Chapter 3 in a book by Urick (1982). Collectively, this work later evolved into a series of lectures and culminated in the writing of this book.

The technical literature cited in this book includes many unpublished reports (so-called 'gray' literature) since no other sources of technical information were available.

Weston and Rowlands (1979) have reviewed the development of models with application to underwater acoustic propagation over the period 1963–1978. DiNapoli and Deavenport (1979) have provided a highly mathematical examination of a select number of propagation models. Brekhovskikh and Lysanov (1991) have presented a comprehensive

Russian perspective on underwater acoustics with a limited treatment of modeling. Jensen *et al.* (1994) have provided a comprehensive review of recent theoretical developments in ocean acoustic propagation modeling.

1.4 INVERSE ACOUSTIC SENSING OF THE OCEANS

Inverse sensing involves the extraction of information from direct measurements of physical properties. Inverse-sensing techniques that employ acoustics have been used in several subdisciplines of geophysics including seismology, meteorology and oceanography. Seismologists have used tomographic techniques to infer the bulk properties of the lithosphere (e.g. Menke, 1989). Atmospheric scientists have employed naturally generated, low-frequency sound (microbaroms) to probe the upper layers of the atmosphere in an inverse fashion (Donn and Rind, 1971). In oceanography, inverse acoustic data provide estimates of spatially integrated, and temporally averaged, oceanic conditions that are not readily available from a traditional constellation of point sensors (e.g. Bennett, 1992).

Traditionally, direct (versus inverse) acoustic sensing of the oceans has been used for sonar applications. This emphasis stems principally from the military significance of sonars in the detection of submarines. Despite the restrictiveness of military security, an extensive body of relevant research has accumulated in the open literature, and much of the recent literature addresses the development and refinement of numerical codes that model the oceans as an acoustic medium. These models are utilized either in a prognostic mode (e.g. sonar performance prediction) or in a diagnostic mode (e.g. sonar design). This situation has stimulated the formation of a new subdiscipline of acoustics known as computational ocean acoustics. Related developments have been documented by Merklinger (1987), Lee *et al.* (1990a, 1990b, 1990c, 1993) and Lau *et al.* (1993).

In the commercial sector, recent applications of direct acoustic-sensing methods include: Acoustic Doppler Current Profilers (ADCPs) for measuring currents; compact sonars for obstacle location and avoidance by Autonomous Underwater Vehicles (AUVs) (e.g. Brutzman *et al.*, 1992); fish-finding devices; underwater communications systems for divers; fathometers for bathymetric sounding and navigation; and side-scanning sonars for topographic mapping of the sea-floor relief. A recent innovation entails acoustic monitoring of marine environmental pollution; for example, point-source and non-point-source pollution studies now use acoustic backscatter and Doppler current meter measurements.

Inverse acoustic sensing of the oceans utilizes one of three natural phenomena: propagation, noise or reverberation. These techniques and their applications are briefly summarized below.

Acoustic propagation characteristics in the deep oceans are determined largely by the refractive properties of the water column and, to a lesser extent, by the surface and bottom boundary conditions. Propagation measurements can be used to infer bulk properties of the water column such as temperature, sound speed, density and currents. In shallow-ocean areas, where propagation characteristics can be strongly affected by the bottom boundary, propagation measurements can be used to infer properties of the sea floor such as composition and scattering characteristics. Inverse acoustic-sensing methods utilizing the propagation characteristics of the oceans include tomography and matched field processing (see Chapter 5).

The ambient noise field in the oceans is described by the spectral, spatial and temporal characteristics of sound generated by both natural and industrial sources. Measurements of these characteristics can provide useful information regarding the nature of the noise sources themselves as well as physical features within the oceans. Examples of inverse applications of the noise field include object imaging ('acoustic daylight'), wind speed determination and rainfall measurements (see Chapter 6).

The reverberation field in the oceans is the product of acoustic scattering by the surface and bottom boundaries, and by inhomogeneities within the oceans. The utility of the reverberation field as an inverse sensing technique is analogous to that of the ambient noise field. For example, the reverberation field can be inverted to image the sea floor (see Chapter 8).

Table 1.1 summarizes selected inverse ocean acoustic-sensing techniques according to the natural phenomenon utilized. The specific techniques identified in Table 1.1 will be discussed in appropriate sections throughout this book.

Table 1.1 Summary of inverse ocean acoustic-sensing techniques

Propagation	*Noise*	*Reverberation*
• Tomography –density field (eddies, currents) –bulk temperature (climate monitoring)	• Surface noise –rainfall rates –wind speeds	• Field inversion –sea-floor imaging
• Matched field processing –source localization –marine environment reconstruction	• Acoustic daylight –object imaging	
• Bottom interaction –sea-floor composition –sea-floor scattering characteristics		

Inverse acoustic-sensing techniques presently constitute adjuncts to direct measurement methods. However, the application of inverse acoustic-sensing techniques to dynamical studies of the oceans' boundaries and interior show great promise for three reasons. First, such data can be used to establish comparative baselines for other remote sensors, such as satellites, by providing synoptic portraitures of the interior oceans together with concurrent groundtruth data at the sea surface. Second, inverse acoustic-sensing techniques often afford useful insights into a broad class of oceanic phenomena since their successful employment relies heavily on the use of numerical models to understand the role of the oceans as an acoustic medium. Third, inverse data provide estimates of spatially integrated and temporally averaged oceanic conditions that are not readily available from traditional oceanographic sensors.

In furtherance of technology transfers from defense to commercial applications, proposals have been advanced to convert existing undersea surveillance assets into a National Acoustic Observatory. Possible research uses include (e.g. Amato, 1993; Carlson, 1994):

- stock and migration monitoring of large marine mammals
- remote ocean observations
- seismic and volcanic monitoring
- acoustic telemetry
- fisheries enforcement.

2

Acoustical oceanography

2.1 BACKGROUND

Acoustical oceanography describes the role of the ocean as an acoustic medium. It relates oceanic properties to the behavior of underwater acoustic propagation, noise and reverberation. Acoustical oceanography crosses four other branches of oceanography: physical, chemical, geological and biological oceanography.

The single most important acoustical variable in the ocean is sound speed. The distribution of sound speed in the ocean influences all other acoustic phenomena. The sound speed field in turn is determined by the density (or temperature and salinity) distribution in the ocean. Advection of the underwater sound field by water currents is also important. Refraction of sound by fronts, eddies and other dynamic features can distort signals. Knowledge of the state of the sea surface as well as the composition and topography of the sea floor is important for specification of boundary conditions. Biological organisms contribute to the noise field and also scatter underwater sound signals.

The balance of this chapter addresses (1) physical and chemical properties, (2) sound speed, (3) boundaries, (4) dynamic features and (5) biologics.

A number of books and published papers already exist on these subjects and appropriate citations will be made to them. Notable text and reference books of general nature include those by Apel (1987), Gill (1982) Clay and Medwin (1977), Neumann and Pierson (1966), Peixoto and Oort (1992), Pickard and Emery (1990) and Sverdrup *et al.* (1942).

2.2 PHYSICAL AND CHEMICAL PROPERTIES

Temperature is basic to any physical description of the oceans. It is the easiest and therefore the most common type of oceanographic measurement made. The exchange of heat between the ocean and the atmosphere depends strongly on temperature. The density and resulting stratification of the ocean depend largely on temperature. The speed of sound in the upper layers of the ocean is most strongly dependent on temperature. Temperature further influences the kinds and rates of chemical reactions occurring in the ocean. The distribution of nutrients and other biologically important substances depends on temperature and the resulting density stratification.

Sea water is a binary fluid in that it consists of various salts in water. Salinity is a term used to measure the quantity of salts dissolved in sea water and is expressed in units of parts per thousand (‰ or ppt). The precise definition of salinity is complicated. Fofonoff (1985) has reviewed the development of the modern salinity scale and the equation of state for sea water. The **Practical Salinity Scale 1978** was introduced to rectify shortcomings associated with the traditional chlorinity–conductivity relationship used to establish salinity (Lewis, 1980; Perkin and Lewis, 1980; Culkin and Ridout, 1989). In the new scale, the existing link between chlorinity and salinity was broken in favor of a definitive salinity–conductivity relationship. The new practical standard is IAPSO (International Association for the Physical Sciences of the Ocean) Standard Seawater, produced and calibrated by the IAPSO Standard Seawater Service. Salinity is now a dimensionless quantity (psu, or practical salinity unit) because the algorithms in the new scale were adjusted to eliminate the ‰ (ppt) used in previous scales. The presence of salts affects a number of oceanic parameters including compressibility, sound speed, refractive index, thermal expansion, freezing point and temperature of maximum density.

The density of sea water is related to temperature, salinity and pressure (which is nearly linearly proportional to depth) through the equation of state (e.g. Fofonoff, 1985). Density provides a measure of the hydrostatic stability in the ocean. Specifically, a stable water column is one in which density increases monotonically with depth.

Sea water is compressible, although less so than pure water. The compressibility of sea water can be expressed by the coefficient of compressibility, which relates fractional changes in water volume to the corresponding changes in pressure (e.g. Apel, 1987).

Compressibility of sea water is an important factor in several applications: the precise determination of the density of sea water, particularly at great depths; the computation of adiabatic temperature changes in the ocean (in an adiabatic process, compression results in warming, while expansion results in cooling); and most importantly for us, the computation of sound speed in sea water.

The speed of sound (c) in sea water is related to the isothermal compressibility (K) as

$$c = \sqrt{\frac{\gamma}{K\rho}} \qquad (2.1)$$

where γ is the ratio of specific heats at constant pressure and constant volume, and ρ is the density of sea water. The isothermal compressibility is easier to measure experimentally than is the adiabatic compressibility.

2.2.1 Temperature distribution

The distribution of temperature at the surface of the oceans is zonal in nature, with isotherms (lines of constant temperature) oriented in an east–west pattern. The annual mean temperature distribution shown in Fig. 2.1 illustrates this general zonal gradation. This pattern is due largely to the zonal distribution of the solar energy received at the sea surface. Specific exceptions to this pattern occur in regions of upwelling (where colder water from below is brought to the surface through action of the winds), and in the vicinity of major current systems such as the Gulf Stream (where the temperature field is distorted). The relatively low equatorial and tropical sea-surface temperatures in the eastern Pacific and Atlantic Oceans are generally ascribed to upwelling. The more meridional trend of the isotherms off the northeast coast of the United States, for example, is evidence of the Gulf Stream current system. Examining only annual averages can sometimes be misleading. The monsoon circulation in the Indian Ocean, for example, makes interpretation of an annual mean temperature field questionable.

The temperature field in the ocean exhibits a high degree of stratification with depth. Since the isotherms are nearly parallel to the horizontal plane, this type of structure is referred to as horizontal stratification. This is

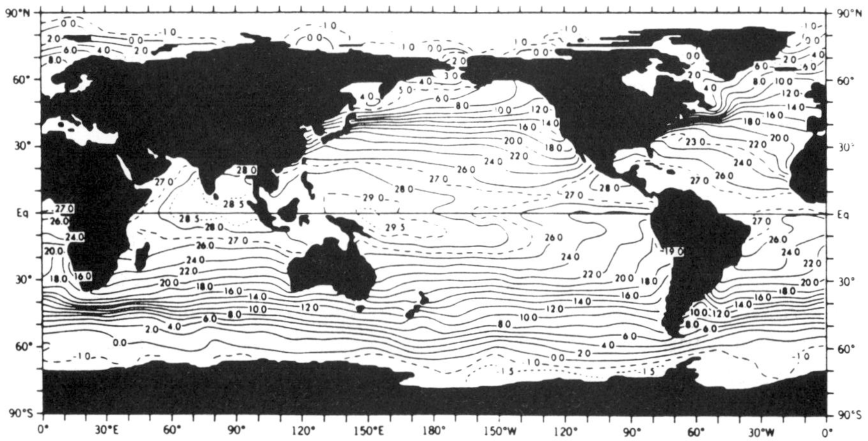

Fig. 2.1 Annual mean temperature (°C) at the sea surface (Levitus, 1982). The distribution of surface temperatures shows a strong latitudinal dependence, due largely to the zonal distribution of solar energy received at the sea surface.

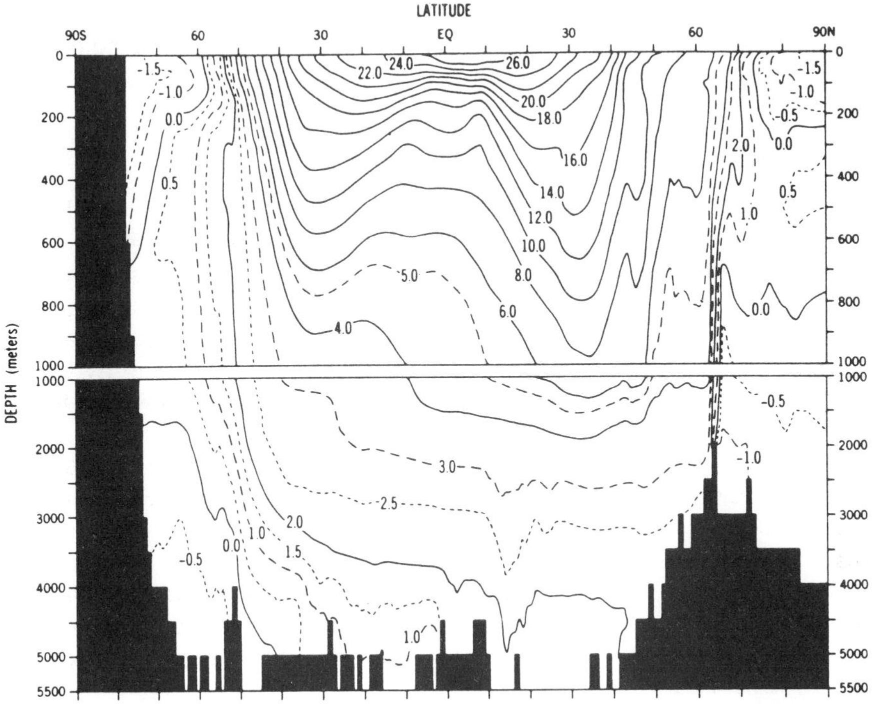

Fig. 2.2 Annual mean Atlantic Ocean zonal average (by one-degree squares) of temperature (°C) as a function of depth (Levitus, 1982). (Note break in the depth scale at 1000 m.)

evidenced in Fig. 2.2, which presents zonal averages of temperature in the Atlantic Ocean by one-degree latitude belts. These zonal averages do not include the Mediterranean Sea, the Baltic Sea or Hudson Bay.

2.2.2 Salinity distribution

The distribution of salinity at the surface of the ocean is shown in Fig. 2.3, and notable features have been summarized by Levitus (1982). Specifically, subtropical salinity maxima associated with the excess of evaporation over precipitation appear in all the individual oceans. Subpolar regions exhibit low salinities associated with the excess of precipitation over evaporation. Low-salinity tongues associated with runoff from the major river systems, such as the Amazon, are also apparent.

Unlike the temperature fields, salinity does not exhibit a consistent stratification with depth in the Atlantic Ocean (Fig. 2.4). These patterns reflect the complex movements of water throughout the oceans.

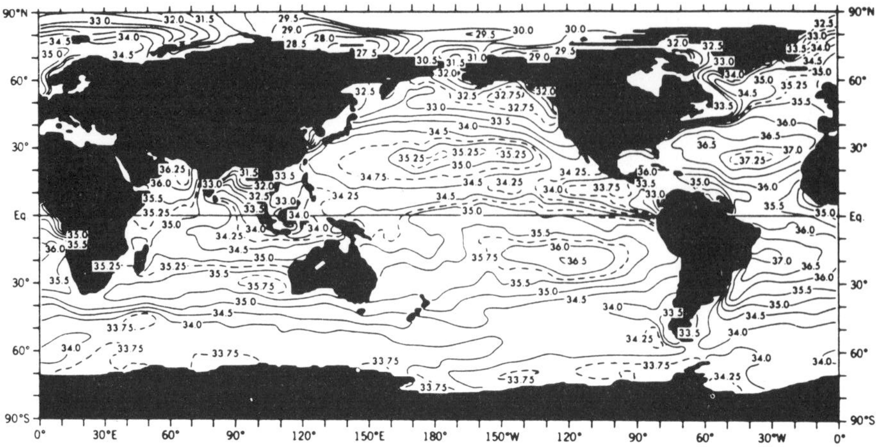

Fig. 2.3 Annual mean salinity (ppt) at the sea surface (Levitus, 1982).

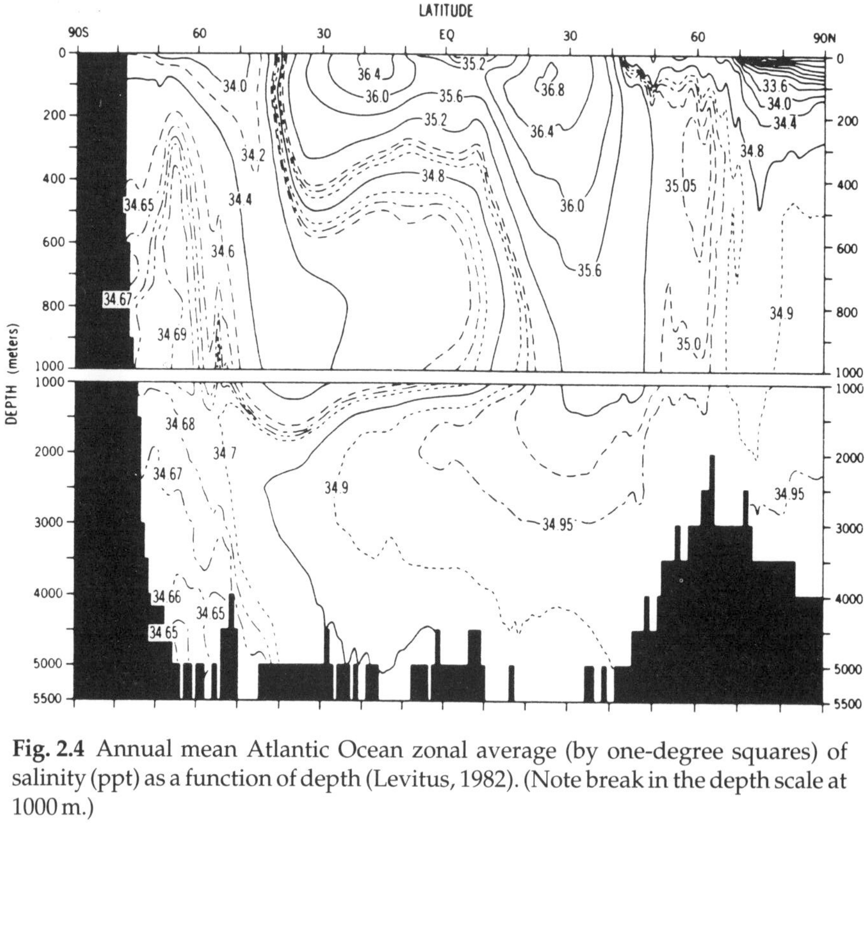

Fig. 2.4 Annual mean Atlantic Ocean zonal average (by one-degree squares) of salinity (ppt) as a function of depth (Levitus, 1982). (Note break in the depth scale at 1000 m.)

2.2.3 Water masses

Discriptions of sea-water characteristics and motions are facilitated by using the concept of water masses. This concept is analogous to that employed by meteorologists to describe air masses in weather patterns. Air masses are identified by characteristic combinations of air temperature and moisture content. These characteristics allow meteorologists to identify the past history (or source regions) of the various air masses. Examples of air masses include continental polar (cold, dry air formed over high-latitude land areas) and maritime tropical (warm, moist air formed over equatorial ocean areas).

Oceanographers (e.g. Sverdrup *et al.*, 1942) have convincingly demonstrated that certain characteristic combinations of water temperature and salinity are associated with water masses formed in particular regions of the world's oceans. These water masses then spread both vertically and laterally and occupy depth ranges of the water column consistent with their density and are distinguishable from one another when plotted on a graph of temperature versus salinity, referred to as a T–S diagram. The T–S relations of the principal water masses of the Atlantic Ocean are presented in Fig. 2.5 (Naval Oceanographic Office, 1972). Approximate

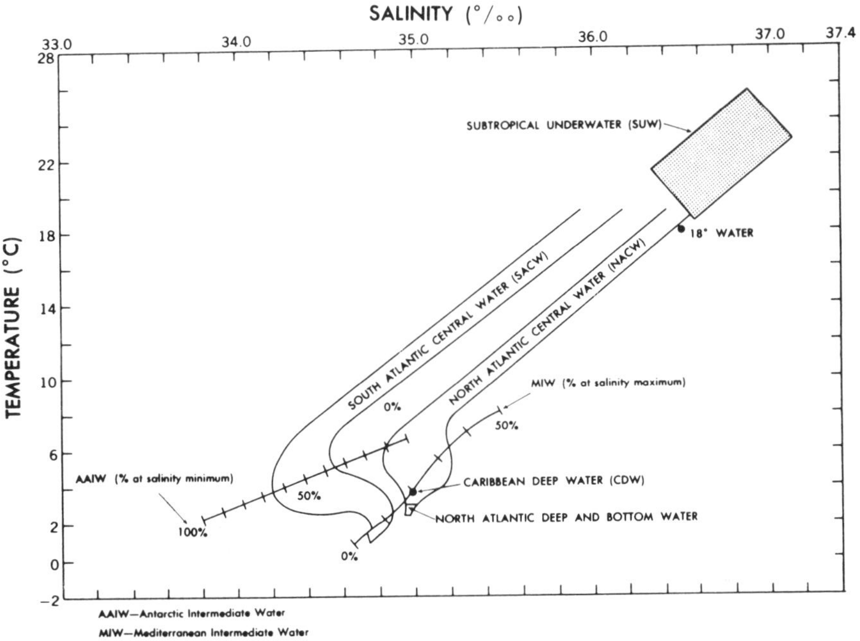

Fig. 2.5 Temperature–salinity (T–S) diagrams of the major water masses of the Atlantic Ocean (Naval Oceanographic Office, 1972).

water depths corresponding to the occurrence of particular T–S combinations are also indicated. Emery and Meincke (1986) have provided an updated review and summary of the global water masses.

The movement of Antarctic Intermediate Water (AAIW), for example, can now be identified in Fig. 2.4. Specifically, the movement of AAIW is evidenced by a low-salinity tongue (34.0–34.6 ppt) extending downward in the latitude belt 60°S to 70°S, then northward at a depth of about 700–800 m, and finally upward near the equator.

The distribution of sound speed in the ocean can be related to the local water mass structure. Knowledge of the water mass structure, then, can greatly enhance our understanding of the large-scale spatial and temporal variability of the sound speed field in the ocean.

2.3 SOUND SPEED

The speed of sound in sea water is an oceanographic variable that determines the behavior of sound propagation in the ocean. Sound speed varies as a function of water temperature, salinity and pressure (or depth). The speed of sound in sea water increases with an increase in any of these three parameters. The term 'sound velocity' is an older term that is used synonymously with 'sound speed'.

2.3.1 Calculation and measurements

Many empirical relationships have been developed over the years for calculating sound speed using values of water temperature, salinity and pressure (or depth). Frequently used formulas include those of Wilson (1960), Leroy (1969), Frye and Pugh (1971), Del Grosso (1974), Medwin (1975), Chen and Millero (1977), Lovett (1978), Coppens (1981) and Mackenzie (1981). Each formula has its own limits for temperature, salinity and pressure (or depth). Calculations outside of the specified limits may be in error. Recent tomographic measurements of sound speed in the oceans have implications for very-long-range ducted propagation (Spiesberger and Metzger, 1991b). Specifically, previous algorithms derived from laboratory measurements have been found to overpredict sound speeds due to pressure effects at great depths (Dushaw *et al.*, 1993). This matter is still the subject of investigation.

For convenience, the formula developed by Mackenzie (1981) is presented here:

$$\begin{aligned} c = {} & 1448.96 + 4.591T - 5.304 \times 10^{-2} T^2 + 2.374 \times 10^{-4} T^3 \\ & + 1.340(S - 35) + 1.630 \times 10^{-2} D + 1.675 \times 10^{-7} D^2 \\ & - 1.025 \times 10^{-2} T(S - 35) - 7.139 \times 10^{-13} TD^3 \end{aligned} \qquad (2.2)$$

where c is the speed of sound in sea water (in meters per second), T is the water temperature (in degrees Celsius), S is the salinity (in parts per thousand) and D is the depth (in meters). This equation is valid over the following ranges: $0^\circ C \leqslant T \leqslant 30^\circ C$; $30\ \text{ppt} \leqslant S \leqslant 40\ \text{ppt}$; $0\ \text{m} \leqslant D \leqslant 8000\ \text{m}$. Mackenzie also discussed the relationship between pressure and depth in the ocean.

Two practical devices commonly used to measure sound speed as a function of depth in the ocean are the bathythermograph (BT) and the velocimeter. The bathythermograph, which is oftan an expendable device (designated XBT), actually measures temperature as a function of depth. Sound speed can then be calculated using one of the available empirical formulas, often on the assumption that salinity is a constant, or nearly so. This assumption is justified by the observation that the typical range of salinities in the open ocean is usually small and that the corresponding impact on sound speed is negligible from a practical standpoint. In coastal areas, and near rivers or ice, this assumption is not generally valid.

The second device, a velocimeter, measures sound speed directly in terms of the travel time of sound over a constant-length fixed path. An expendable version of this instrument is also available (designated XSV).

The velocimeter would seem to be preferred instrument for obtaining measurements of sound speed in support of naval operations. From a broader scientific perspective, however, information on the temperature distribution in the ocean directly supports many other naval applications including ocean dynamics modeling, air–sea interaction studies and various marine biological investigations, to name a few. The demonstrated accuracy of the various empirical formulas (when used within the specified limits) is sufficient for most operational applications. In terms of expendable sensors, measurements of water temperature using the bathythermograph represent a better scientific investment. Nevertheless, some naval operations in coastal areas or in ice-covered regions may be better supported by the velocimeter. Increasing emphasis on naval operations in the littoral zone has placed greater importance on measuring, in real time, coastal processes that are characterized by extreme temporal and spatial variations.

Precise scientific measurements are normally accomplished using expensive, but recoverable, instrumentation. Such instruments are typically deployed from surface ships which maintain station while the instrument package is lowered and then raised in what is termed a hydrographic cast. The instrument package can contain an assortment of sensors for measuring water temperature, salinity (or conductivity), pressure, sound speed, currents and dissolved oxygen, among others.

2.3.2 Sound speed distribution

Typical North Atlantic winter and summer profiles of sound speed versus depth are shown in Fig. 2.6. These profiles represent a region of the North Atlantic Ocean located near 23°N and 70°W (Naval Oceanographic Office, 1972). Temperature–salinity (T–S) diagrams for both winter and summer seasons, based on actual measurements, are also presented to show their relationships to the sound speed profiles. Since the T–S diagrams indicate the ocean depths corresponding to the measured temperature–salinity pairs, the individual temperature and salinity profiles for both winter and summer can be reconstructed. Principal water masses are noted on the sound speed profiles. The so-called 18° Water marks a change in the sound speed gradient at a depth of about 300 m, and Mediterranean intermediate water (MIW) occupies the region of the water column near the sound speed minimum (at about 1200 m).

The profiles in Fig. 2.6 are representative of the sound speed profiles encountered in many tropical and subtropical ocean areas. Such profiles may be divided into arbitrary layers, each having different characteristics and occurrence (Fig. 2.7). Just below the sea surface is the sonic layer where the speed of sound is influenced by local changes in heating, cooling and wind action. The base of the sonic layer is defined as the sonic layer depth (SLD), which is associated with the near-surface maximum in sound speed. This surface layer is usually associated with a well-mixed layer of isothermal water. Oceanographers refer to this well-mixed region as the mixed layer; the base of this layer is then termed the mixed layer depth (MLD).

Below the mixed layer lies the thermocline, a region of the water column in which the temperature decreases rapidly with depth. This region of the water column is also characterized by a negative sound speed gradient (i.e. sound speed decreases with depth).

Below the thermocline and extending to the sea floor is the deep isothermal layer. This layer has a nearly constant temperature in which the speed of sound increases with depth because of the effect of pressure on sound speed. In this region, the sound speed profile becomes linear with a positive gradient of about 0.017 (m/s)/m or 0.017/s. Between the negative sound speed gradient of the thermocline and the positive gradient of the deep isothermal layer is a sound speed minimum. The depth corresponding to this sound speed minimum is referred to as the sound channel axis. At high latitudes, the deep isothermal layer extends nearly to the sea surface; that is, the sound channel axis shoals as one approaches the polar regions. This behavior is vividly demonstrated in Fig. 2.8; the top panel presents contours of the depth of the minimum sound speed (or channel) axis (m) and the bottom panel presents the sound speed (m/s) on this axial surface (Munk and Forbes, 1989). At low latitudes, the depth of the sound channel axis is typically near 1000 m; at high latitudes, the axis is

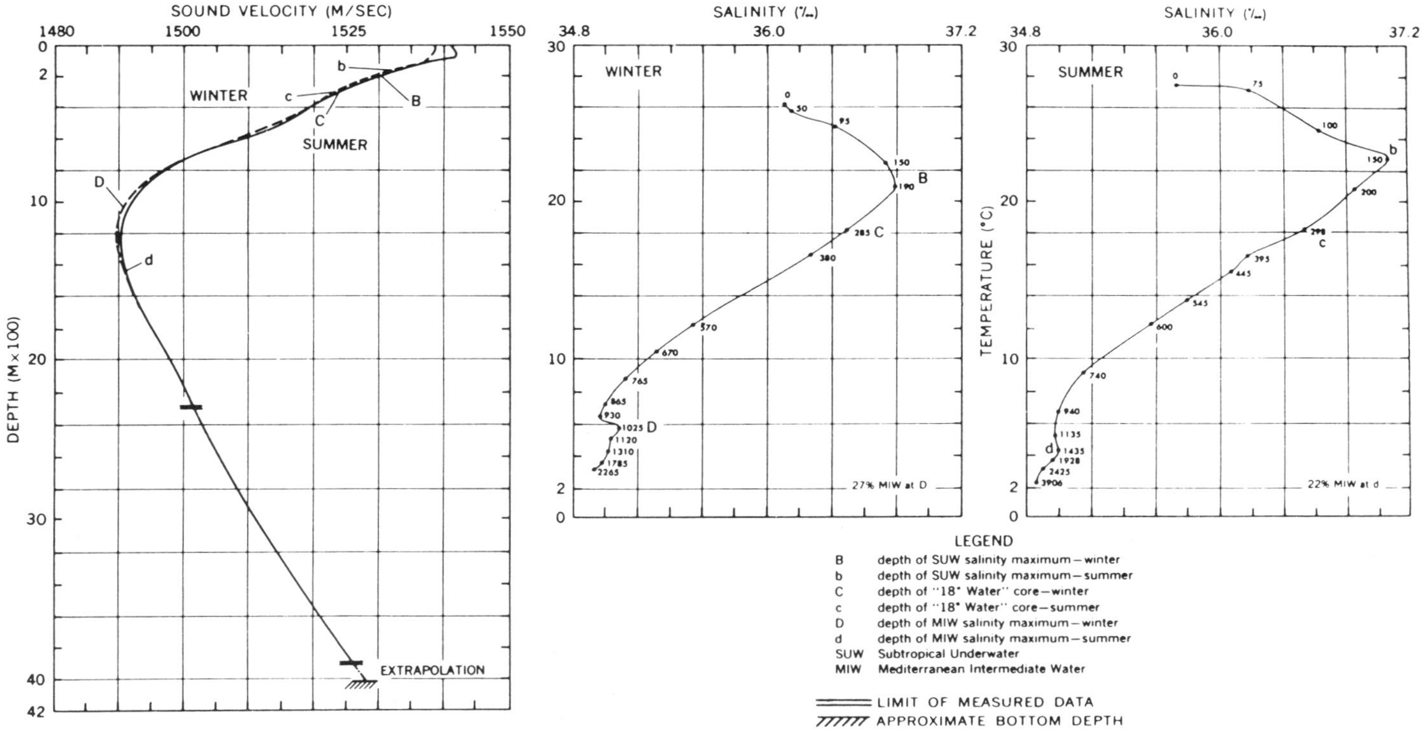

Fig. 2.6 Sound speed profiles (winter and summer) and T–S comparisons for the North Atlantic Ocean near 23°N, 70°W (Naval Oceanographic Office, 1972).

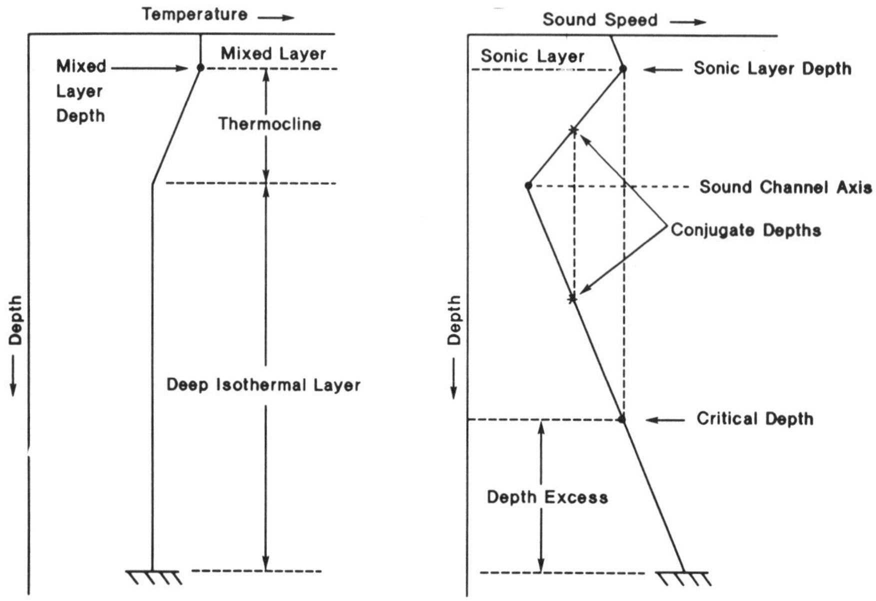

Fig. 2.7 Schematic relationship between temperature and sound speed profiles in the deep ocean.

located near the sea surface. The associated sound speeds on this axial surface generally decrease away from the equatorial regions.

In profiles containing a sound channel axis, a critical depth can be defined as that depth below the axis at which the sound speed equals the near-surface maximum value. (The near-surface maximum value of sound speed is usually located at the sonic layer depth). The vertical distance between the critical depth and the sea floor is referred to as the depth excess. Other pairs of points can be identified on the sound speed profile which have the same value of sound speed but lie on opposite sides of the sound channel axis; such pairs are referred to as conjugate depths. In Fig. 2.7, the critical depth is actually a conjugate of the sonic layer depth.

Additional sound speed profiles typical of the winter season in different ocean areas of the world are presented in Fig. 2.9 to demonstrate that the simple model for sound speed profiles described above is not applicable to high-latitude ocean areas or to some smaller water bodies.

The depth dependence of sound speed in the ocean poses a particular problem for echo sounders which use near-vertical acoustic paths to measure the depth of the sea floor based on the two-way travel time of the signal. Echo sounders are set to read the depth directly by assuming a constant speed of sound in the water column, usually 1463 m/s or 1500 m/s. When the depth-integrated (or mean) sound speed departs from the assumed value, a correction must be applied to the observed readings. Bialek (1966, p. 63), for example, has tabulated such corrections according to ocean area. Depending on the particular ocean area and water depth, these corrections can be on the order of several percent of the true water depth.

Underwater acoustic propagation problems involving long range may not be able to ignore the hotizontal variations in either sound speed or bathymetry. Modeling developments, therefore, generally distinguish between range-independent (the ocean varies only as a function of depth) and range-dependent (the ocean varies as a function of both depth and range) problems. Parameters other than sound speed and water depth may also be considered in range-dependent problems: surface losses, bottom losses and absorption, for example.

An east–west cross-section of the North Atlantic Ocean between 23°N and 24°N is presented in Fig. 2.10 to illustrate the variability of both sound speed and bathymetry over moderate range scales. Also noteworthy are changes in the depth of the sound channel axis and the absence of a critical (or limiting) depth in some basins. The sound speed profiles presented previously in Fig. 2.6 are consistent with the data of Fig. 2.10 at a longitude of about 70°W.

2.4 BOUNDARIES

The boundaries of the water column – the sea surface and the sea floor – can exert a profound influence on the propagation of acoustic energy through the action of reflection, scattering and absorption.

2.4.1 Sea surface

The surface of the sea is both a reflector and a scatterer of sound. If the sea surface were perfectly smooth, it would form an almost perfect reflector of sound due to the acoustic impedance mismatch at the air–water interface. As the sea surface becomes rough, as it does under the influence of wind (see Table 2.1), reflection losses are no longer near zero.

Sea-surface roughness is typically specified in terms of wave height. Observations of weather at sea generally record wind speed and not wave height as a description of sea state (Bowditch, 1977). A number of statistical relationships exist with which to quantitatively associate these two parameters (e.g. Earle and Bishop, 1984). These relationships are very precise as to the height above sea level at which wind speed is observed and the type of statistical wave height considered. Based on the Pierson–Moskowitz spectrum (Moskowitz, 1964; Pierson and Moskowitz, 1964; Pierson, 1964, 1991), a fully developed significant wave height can be calculated from the observed wind speed as

$$H_{1/3} = 0.55 \times 10^{-2} V^2 \tag{2.3}$$

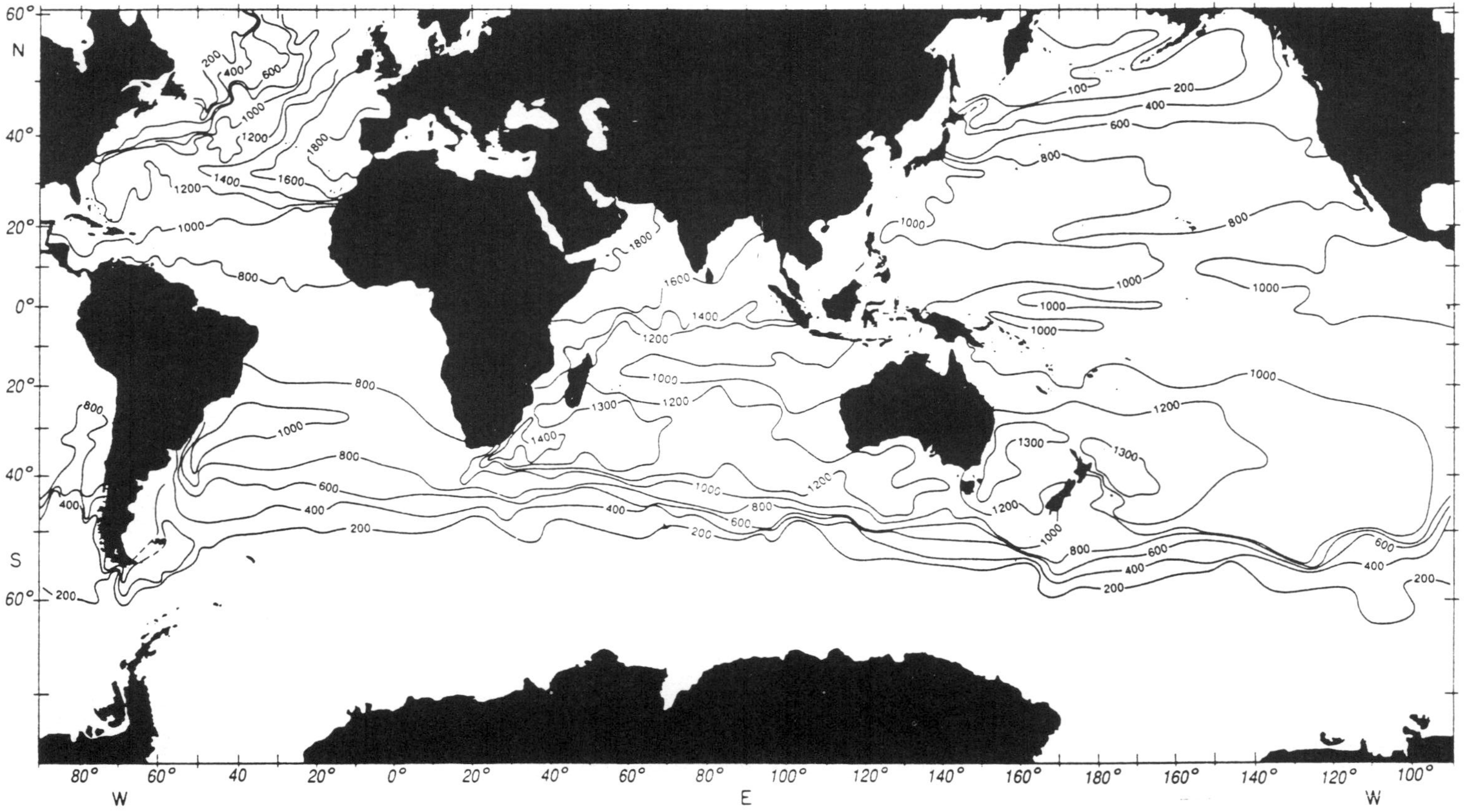

60°
N
40°
20°
0°
20°
40°
S
60°
80°
60°
40
20°
0°
20°
40°
60°
80°
100°
120°
140°
160°
180°
160°
140°
120°
100°
W
E
W

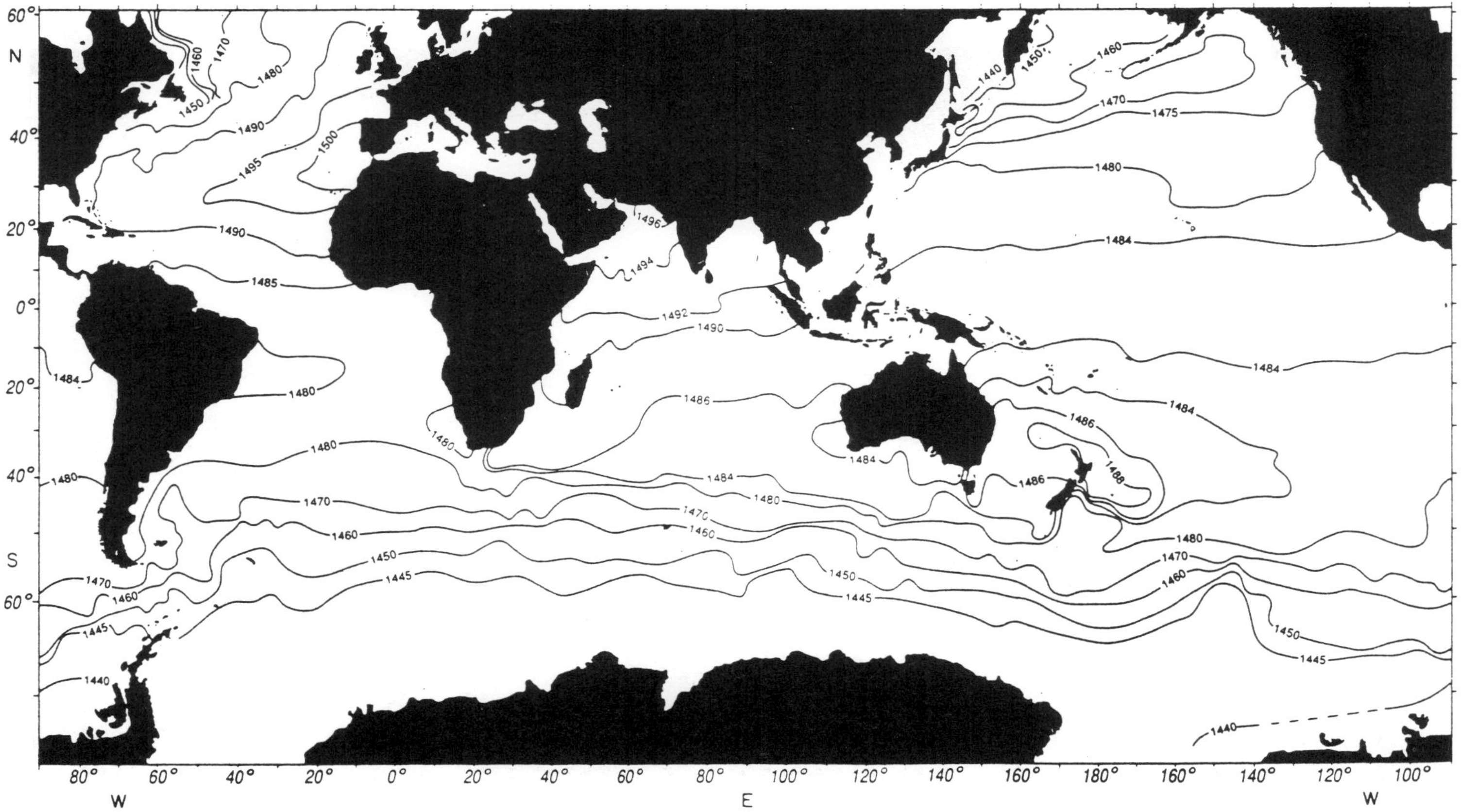

Fig. 2.8 Global maps of the sound channel axis. Upper panel: channel depth (m). Bottom panel: channel sound speed (m/s) (Munk and Forbes, 1989; *J. Phys. Oceanogr.*, **19**, 1765–78; copyright by the American Meteorological Society).

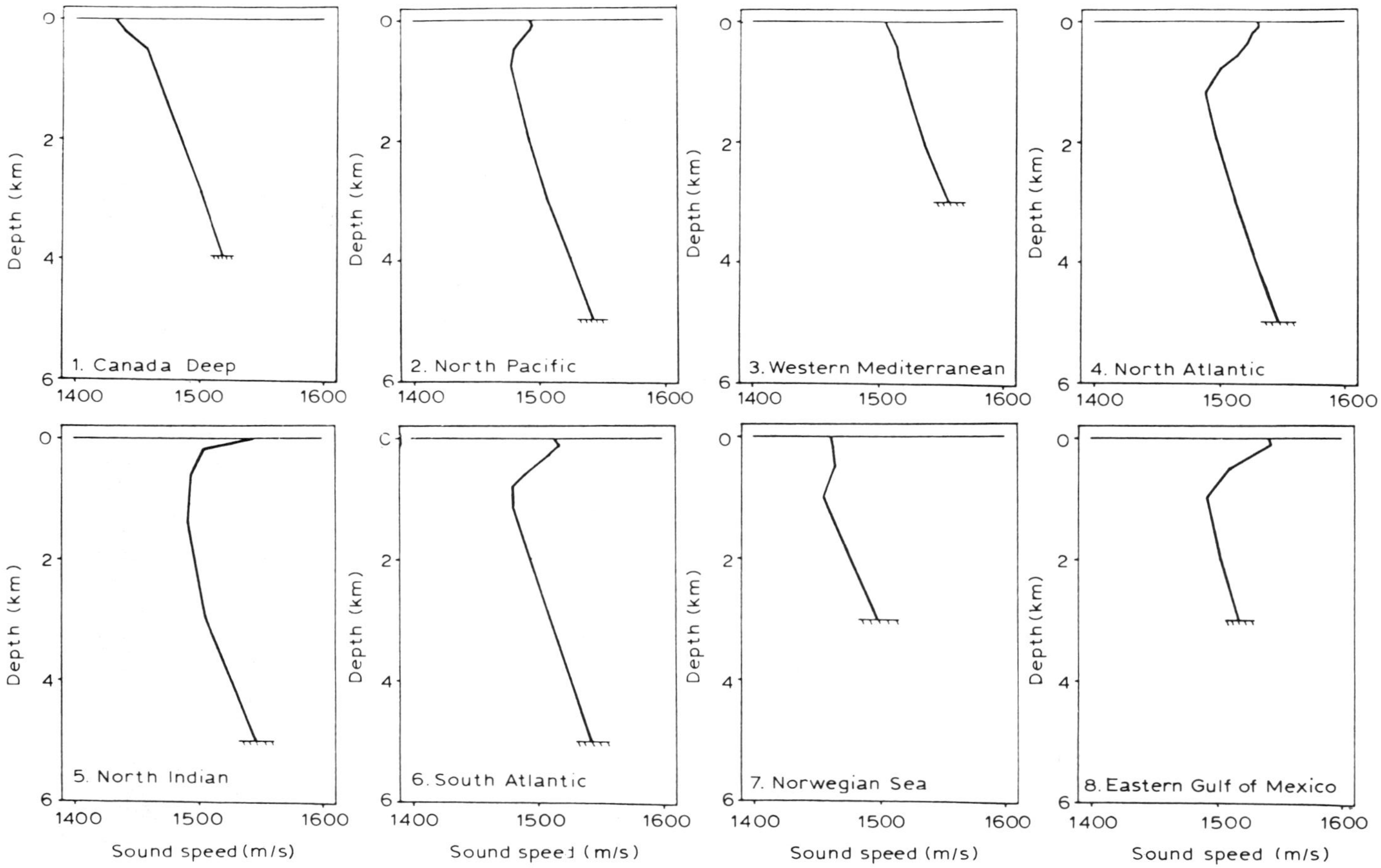

Fig. 2.9 Characteristic winter sound speed profiles for selected deep-ocean areas of the world.

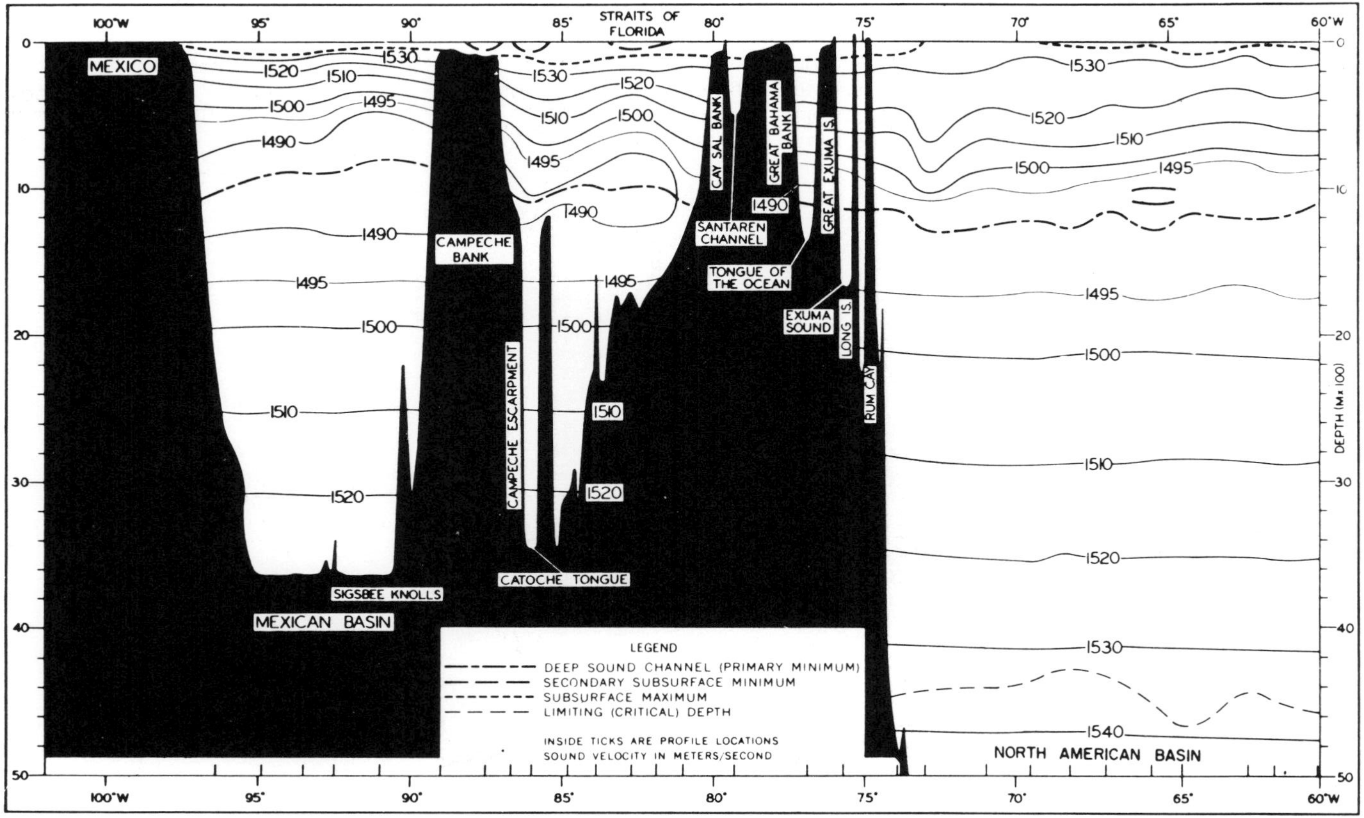

Fig. 2.10 East–west sound speed cross-section between 23°N and 24°N and between 60°W and 100°W for the period February–April (Naval Oceanographic Office, 1972).

Table 2.1 Beaufort Wind Scale with corresponding Sea State Codes (Bowditch, 1977)

Beaufort number or force	Wind speed				World Meteorological Organization (1964)	Estimating wind speed			Sea state	
	Knots	*mph*	*Meters per second*	*km per hour*		*Effects observed far from land*	*Effects observed near coast*	*Effects observed on land*	*Term and height of waves, in meters*	*Code*
0	Under 1	Under 1	0.0–0.2	Under 1	Calm	Sea like mirror	Calm	Calm; smoke rises vertically	Calm, glassy, 0	0
1	1–3	1–3	0.3–1.5	1–5	Light air	Ripples with appearance of scales; no foam crests	Fishing smack just has steerage way	Smoke drift indicates wind direction; vanes do not move		
2	4–6	4–7	1.6–3.3	6–11	Light breeze	Small wavelets; crests of glassy appearance, not breaking	Wind fills the sails of smacks which then travel at about 1–2 miles per hour	Wind felt on face; leaves rustle; vanes begin to move	Calm, rippled 0–0.1	1
3	7–10	8–12	3.4–5.4	12–19	Gentle breeze	Large wavelets; crests begin to break; scattered whitecaps	Smacks begin to careen and travel about 3–4 miles per hour	Leaves, small twigs in constant motion; light flags extended	Smooth, wavelets, 0.1–0.5	2
4	11–16	13–18	5.5–7.9	20–28	Moderate breeze	Small waves, becoming longer; numerous whitecaps	Good working breeze, smacks carry all canvas with good list	Dust, leaves and loose paper raised up; small branches move	Slight, 0.5–1.25	3
5	17–21	19–24	8.0–10.7	29–38	Fresh breeze	Moderate waves, taking longer form; many whitecaps; some spray	Smacks shorten sail	Small trees in leaf begin to sway	Moderate, 1.25–2.5	4
6	22–27	25–31	10.8–13.8	39–49	Strong breeze	Larger waves forming; whitecaps everywhere; more spray	Smacks have doubled reef in mainsail; care required when fishing	Larger branches of trees in motion; whistling heard in wires	Rough 2.5–4	5
7	28–33	32–38	13.9–17.1	50–61	Near gale	Sea heaps up; white foam from breaking waves begins to be blown in streaks	Smacks remain in harbor and those at sea lie-to	Whole trees in motion; resistance felt in walking against wind		

8	34–40	39–46	17.2–20.7	62–74	Gale	Moderately high waves of greater length; edges of crests begin to break into spindrift; foam is blown in well-marked streaks	All smacks make for harbor, if near	Twigs and small branches broken off trees; progress generally impeded	Very rough, 4–6	6
9	41–47	47–54	20.8–24.4	75–88	Strong gale	High waves; sea begins to roll; dense streaks of foam; spray may reduce visibility		Slight structural damage occurs; slate blown from roofs		
10	48–55	55–63	24.5–28.4	89–102	Storm	Very high waves with over-hanging crests; sea takes white appearance as foam is blown in very dense streaks; rolling is heavy and visibility reduced		Seldon experienced on land; trees broken or uprooted; considerable structural damage occurs	High, 6–9	7
11	56–63	64–72	28.5–32.6	103–117	Violent storm	Exceptionally high waves; sea covered with white foam patches; visibility still more reduced		Very rarely experienced on land; usually accompanied by widespread damage	Very high, 9–14	8
12	64 and over	73 and over	32.7 and over	118 and over	Hurricane	Air filled with foam; sea completely white with driving spray; visibility greately reduced			Phenomenal, over 14	9

where V is the wind speed (in knots), measured at a height of 19.5 m, and $H_{1/3}$ is the average height (in meters) of the one-third highest waves.

Two other commonly used wave height descriptors are the rms wave height (H_{rms}) and the one-tenth significant wave height ($H_{1/10}$). These are related to the one-third significant wave height as

$$H_{rms} = 0.70H_{1/3} \tag{2.4}$$

$$H_{1/10} = 1.80H_{rms} \tag{2.5}$$

These relationships can be compared with those presented in Table 2.1, which are based on the rms wave heights. It should be evident that care is needed when specifying the type of wave height being considered in the calculations.

Air bubbles are produced by the breaking of waves and are carried by turbulence beneath the surface. They are also generated in the wakes of ships where they persist for long periods of time. Free air bubbles in the sea are quite small since the larger bubbles tend to rise quickly to the surface. Bubbles only form a very small volumetric percentage of the sea; however, because air has a markedly different density and compressibility from that of sea water and because of the resonant characteristics of bubbles, the suspended air content of sea water has a profound effect upon underwater sound. Urick (1983, pp. 249–54) has summarized these effects which include resonance and changes in sound speed.

Aside from reflection losses, there are other acoustic effects associated with interactions with the sea surface. A moving sea surface produces frequency-smearing and shifting effects on constant-frequency signals. Large and rapid fluctuations in amplitude or intensity are also produced by reflection at the sea surface. Furthermore, Lloyd mirror, or image-interference, effects produce a pattern of constructive and destructive interference between direct and surface-reflected sound. When the sea surface is roughened by wind, this effect is lessened.

2.4.2 Ice cover

When the sea surface is covered by an ice canopy, as in the polar regions (see Figs 2.11 and 2.12), acoustic interaction with the surface is further complicated by an irregular under-ice surface.

The Arctic environment can be segregated into three distinct regions according to the type of ice cover: (1) pack ice, (2) Marginal Ice Zone (MIZ) and (3) open ocean. The differing geometrical configurations associated with the degree of ice cover and depth of water afford substantially different environmental impacts on acoustic system operation and performance.

Field measurements have shown that forward scatter from a rough anisotropic ice canopy is a function of acoustic frequency, geometry and the statistical (spatial correlation) properties of the under-ice surface.

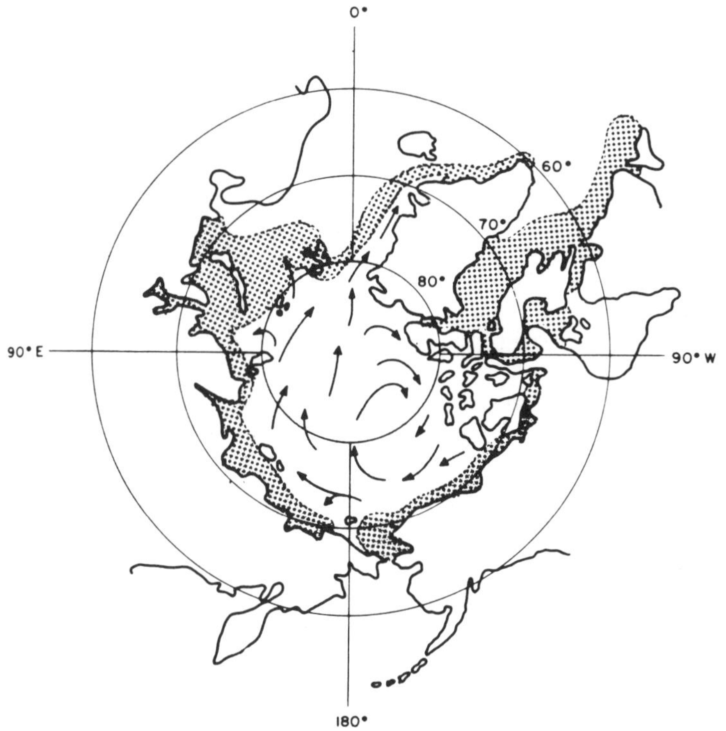

Fig. 2.11 Average boundaries of sea ice (coverage at least 5–8 tenths) in autumn and spring in the Arctic. Arrows indicate the general drift pattern. The width of the stippled area indicates the range of ice limits between autumn and spring (Untersteiner, 1966).

2.4.3 Sea floor

The sea floor is a reflecting and scattering boundary having a number of characteristics similar in nature to those of the sea surface. Its effects, however, are more complicated than those of the sea surface because of its diverse and multilayered composition. Specifically, the sea floor is often layered, with a density and sound speed that may change gradually or abruptly with depth or even over short ranges. Furthermore, the sea floor is more variable in its acoustic properties since its composition may vary from hard rock to soft mud. One feature that is distinct from the sea surface is that the bottom characteristics can be considered to be constant over time, whereas the configuration of the sea surface is statistically in a state of change as the wind velocity changes.

Because of the variable stratification of the bottom sediments in many areas, sound is often transmitted into the bottom and refracted or reflected

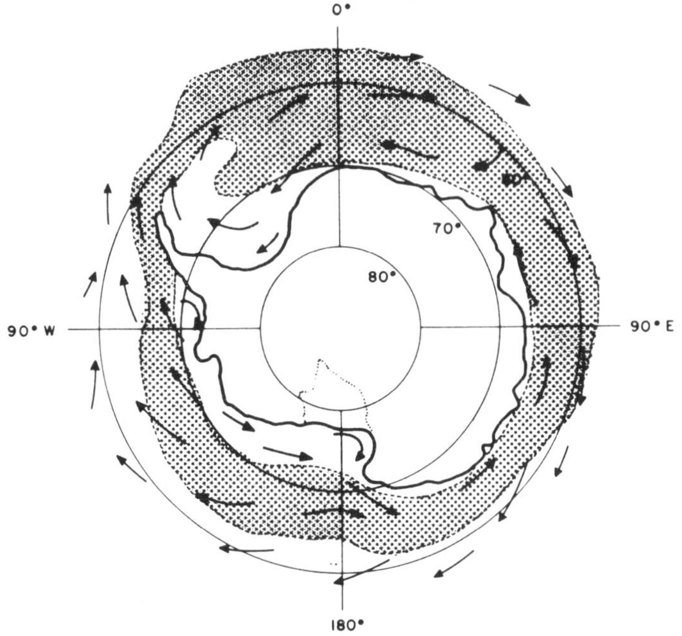

Fig. 2.12 Average boundaries of sea ice (coverage at least 5–8 tenths) in autumn and spring in the Antarctic. Arrows indicate the general drift pattern. The width of the stippled area indicates the range of ice limits between autumn and spring (Untersteiner, 1966).

internally. Thus, the bottom often becomes a propagating medium characterized by both shear and compressional sound speeds.

The topography of the sea floor exhibits a diversity of features not unlike those of the continental land masses. Figure 2.13 presents an artist's conception of common ocean basin features. The undersea features noted in Fig. 2.13 are defined in the Glossary of terms (Appendix B).

Underwater ridges and seamounts can effectively block the propagation of sound. When actively ensonified, seamounts can mask targets of interest by either providing false targets or by shadowing targets of interest.

Marine seismic studies have greatly improved our understanding of the crustal structure of that part of the earth covered by the oceans (Bryan, 1967); a typical bottom structure section is presented in Fig. 2.14. This typical section consists of 5–6 km of water, about 0.5 km of unconsolidated sediments, 1–2 km of basement rock and 4–6 km of crustal rock overlying the upper mantle.

Major portions of the sea floor are covered with unconsolidated sediments with an average thickness of approximately 500 m. Sediments can be classified according to their origin as either terrigenous or pelagic, although no single classification scheme has universal approval. The

Fig. 2.13 Common sea-floor features (Bowditch, 1977).

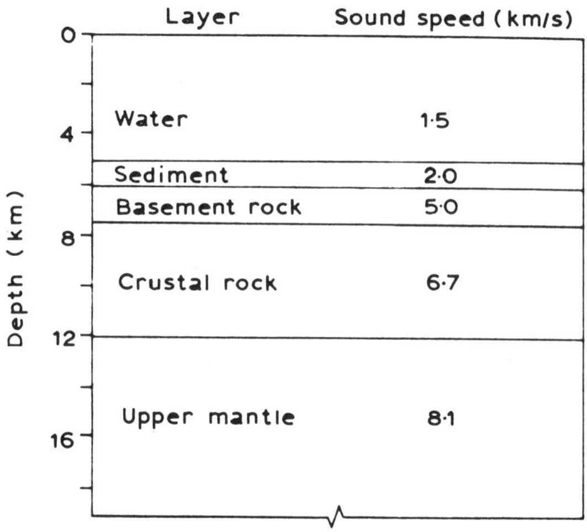

Fig. 2.14 Typical oceanic section with associated depths and sound speeds in the water column and subbottom layers (Bryan, 1967).

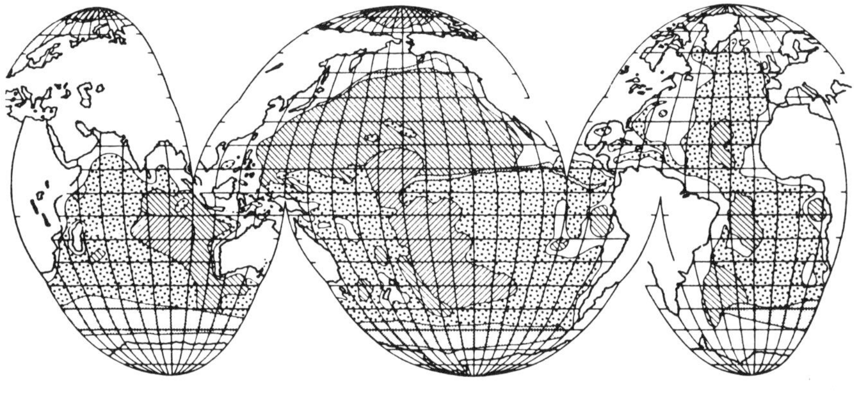

Fig. 2.15 Idealized distribution of marine sediments. Hatched areas indicate inorganic pelagic and terrigenous sediments; stippled areas denote organic pelagic sediments comprising calcareous and siliceous oozes (adapted from Arrhenius (1963); *The Sea*, Vol. 3, pp. 655–727; reprinted by permission of John Wiley and Sons, Inc., all rights reserved).

general distribution of sediments according to Arrhenius (1963) is presented in Fig. 2.15.

Terrigenous sediments are derived from land and are particularly prominent near the mouths of large rivers. These sediments are generally classified as silt, sand and mud.

Pelagic sediments are derived from two sources: organic and inorganic. Organic sediments comprise the remains of dead organisms and are classified as either calcareous or siliceous oozes. Inorganic pelagic sediments are derived from materials suspended in the atmosphere and are generally classified as clay.

2.5 DYNAMIC FEATURES

It is convenient to categorize dynamic features of the ocean according to characteristic time and space scales. While no precise terminology is universally accepted, it is common to recognize space scales as large (> 100 km), meso (100 m–100 km) and fine (< 100 m). Time scales are less precise, but generally distinguish among seasonal, monthly, inertial, tidal and other feature-specific time scales appropriate to currents, eddies, waves (both surface and internal) and turbulence, among others.

2.5.1 Large-scale features

The large-scale circulation of the ocean can be classified either as wind-driven or thermohaline. The former is due to wind stress acting on the sea

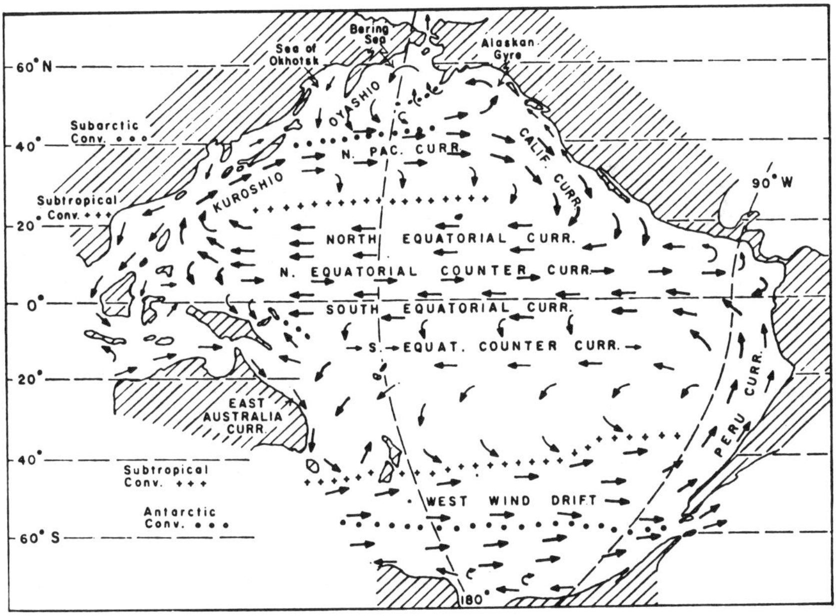

Fig. 2.16 Surface circulation of the Pacific Ocean (reprinted with permission from Pickard, 1963; *Descriptive Physical Oceanography*, copyright by Pergamon Press PLC).

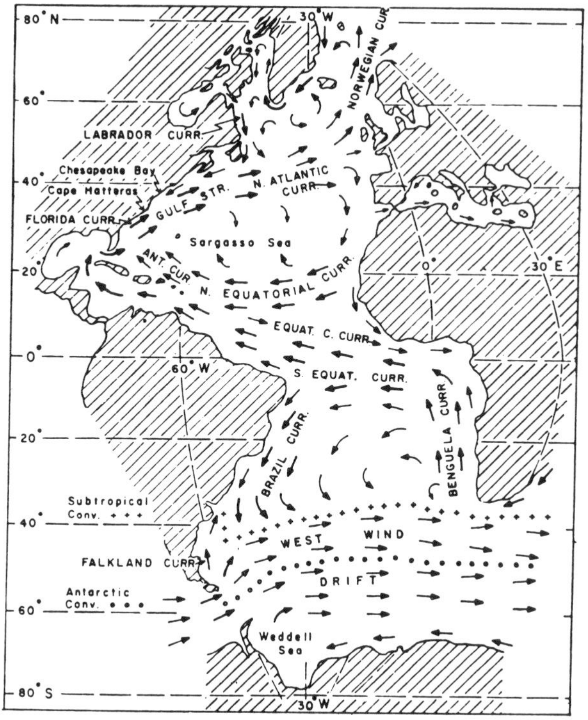

Fig. 2.17 Surface circulation of the Atlantic Ocean (reprinted with permission from Pickard, 1963; *Descriptive Physical Oceanography*, copyright by Pergamon Press PLC).

surface, while the latter is due, in part, to density changes arising from variations in temperature and salinity (e.g. Pickard, 1963).

The wind-driven component of circulation is usually horizontal in nature and is restricted primarily to the upper few hundred meters of the ocean. In the case of upwelling or downwelling near coasts, the original horizontal flow is forced to become vertical due to the basin geometry. In the open ocean, bands of upwelling and downwelling can be created by divergence and convergence, respectively, of wind-generated surface currents. Such currents are commonly referred to as Ekman drift currents.

The thermohaline component normally originates as a vertical flow, arising from imbalances in heat or freshwater (salt) fluxes near the surface, then sinks, eventually becoming a horizontal flow at great depth.

The circulation of the surface waters corresponds closely to the prevailing wind patterns. The water currents set in motion by the winds would encircle the globe in the absence of land masses. The land masses, however, obstruct the flow and force the water along the coasts, thus forming

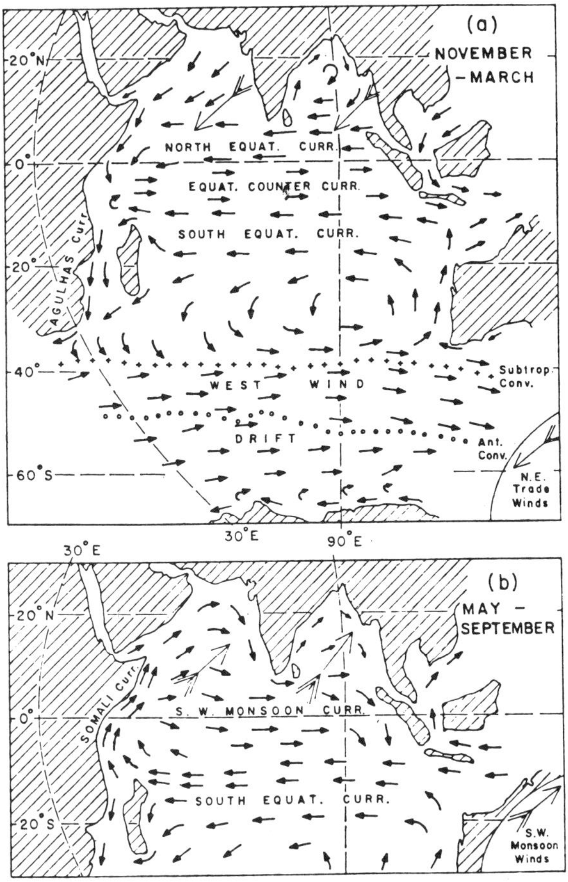

Fig. 2.18 Surface circulation of the Indian Ocean, showing effects of monsoon season: (a) NE trade winds; (b) SW monsoon winds (reprinted with permission from Pickard, 1963; *Descriptive Physical Oceanography*, copyright by Pergamon Press PLC).

completed loops (or gyres) in the Pacific (Fig. 2.16) and Atlantic Oceans (Fig. 2.17). The effect of changing wind patterns due to monsoons is evident in the Indian Ocean (Fig. 2.18). Only in the Antarctic Ocean does the absence of land masses permit true circumpolar flow at a latitude of about 60°S (Figs 2.16–2.18).

2.5.2 Meso-scale features

Meso-scale oceanic features of importance to underwater acoustics include fronts, eddies (or rings) and internal waves.

(a) Fronts and eddies

Ocean frontal features are frequently associated with major ocean currents, or with vertical circulation patterns in areas of upwelling or downwelling. An example of ocean fronts associated with major ocean currents is shown in Fig. 2.19 (Naval Oceanographic Office, 1967). This figure represents a range–depth section crossing the North Atlantic Ocean between Newfoundland (left) and Senegal (right). The hydrographic station numbers used in creating this section are shown at the top. Between stations 6 and 23, the isotherms are relatively horizontal. Two adjacent frontal features are evidenced by the vertically oriented isotherms between stations 3 and 6. These features correspond to the cold, southward-flowing Labrador Current (centered at station 4) and the warm, northward flowing Gulf Stream (centered at station 5). The effects of the Gulf Stream frontal system are evident at depths exceeding 1000 m.

Different water masses are commonly separated by a transition zone referred to as an ocean front. The degree of abruptness in the change of water mass characteristics (particularly temperature and salinity, and thus sound speed) determine whether the front is classified as a strong or weak front. Oceanic fronts are similar in concept to the more familar fronts encountered in meteorology which separate different air masses.

Cheney and Winfrey (1976) have conveniently summarized the classification and distribution of ocean fronts. For underwater acoustic applications, they recommend the following definition of an ocean front: A front is any discontinuity in the ocean which significantly alters the pattern of sound propagation and transmission loss. Thus, a rapid change in the depth of the sound channel, a difference in sonic layer depth, or a temperature inversion would denote the presence of a front. In terms of acoustics, Cheney and Winfrey (1976) identify the following significant effects:

1. Surface sound speed can change by as much as 30 m/s. Although this is due to the combined effect of changing temperature and salinity, temperature is usually the dominant factor.
2. Differences in sonic layer depth (SLD) on the order of 300 m can exist on opposite sides of a front during certain seasons.
3. A change in in-layer and below-layer gradient usually accompanies a change in surface sound speed and sonic layer depth.
4. Depth of the sound channel axis can change by 750 m when crossing from one water mass to the next.
5. Increased biological activity generally found along a front will increase ambient noise and reverberation levels.
6. Enhanced air–sea interaction along a frontal zone can cause a dramatic change in sea state and thus increase ambient noise levels and surface roughness.

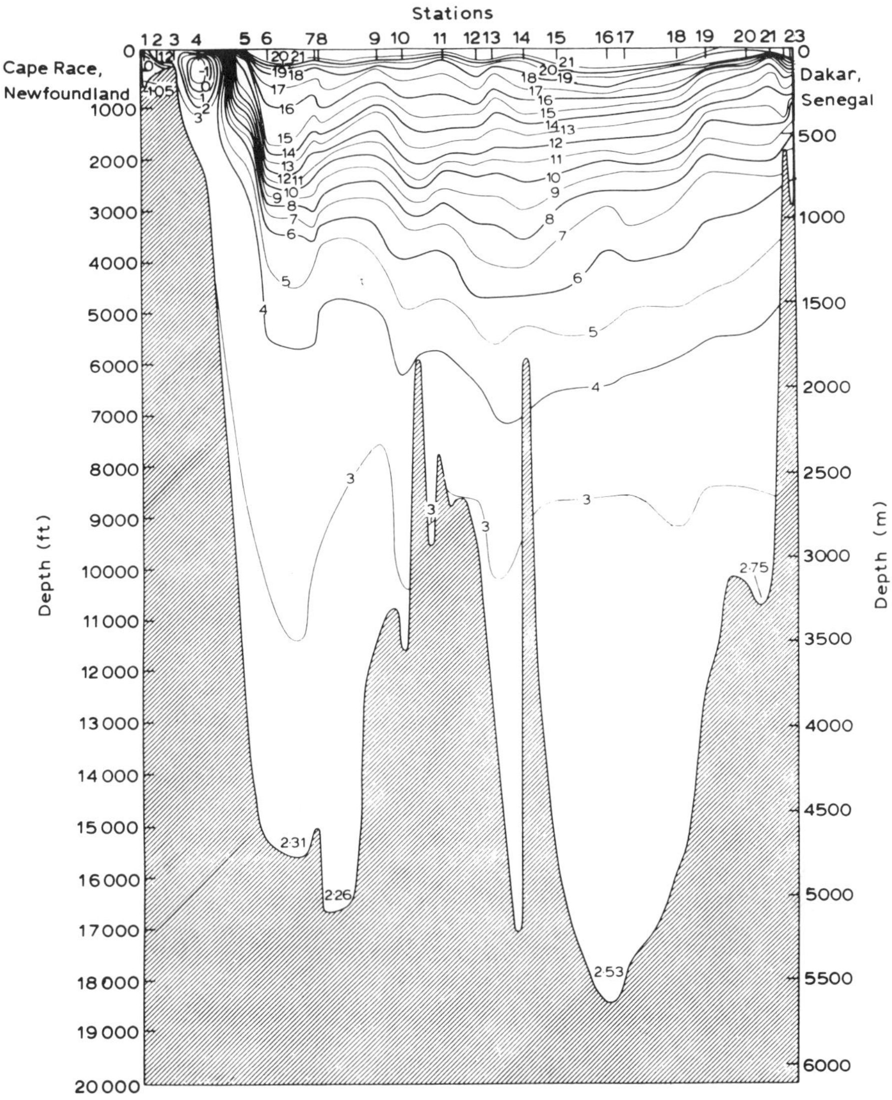

Fig. 2.19 Vertical section of water temperature (°C) for the transect between Cape Race, Newfoundland (left), and Dakar, Senegal (right) (Naval Oceanographic Office, 1967).

7. Refraction of sound rays as they pass through a front at an oblique angle can cause bearing errors in sonar systems.

A summary of the positions of prominent fronts is presented in Fig. 2.20 (accompanied by Table 2.2). Each front is characterized as strong, moder-

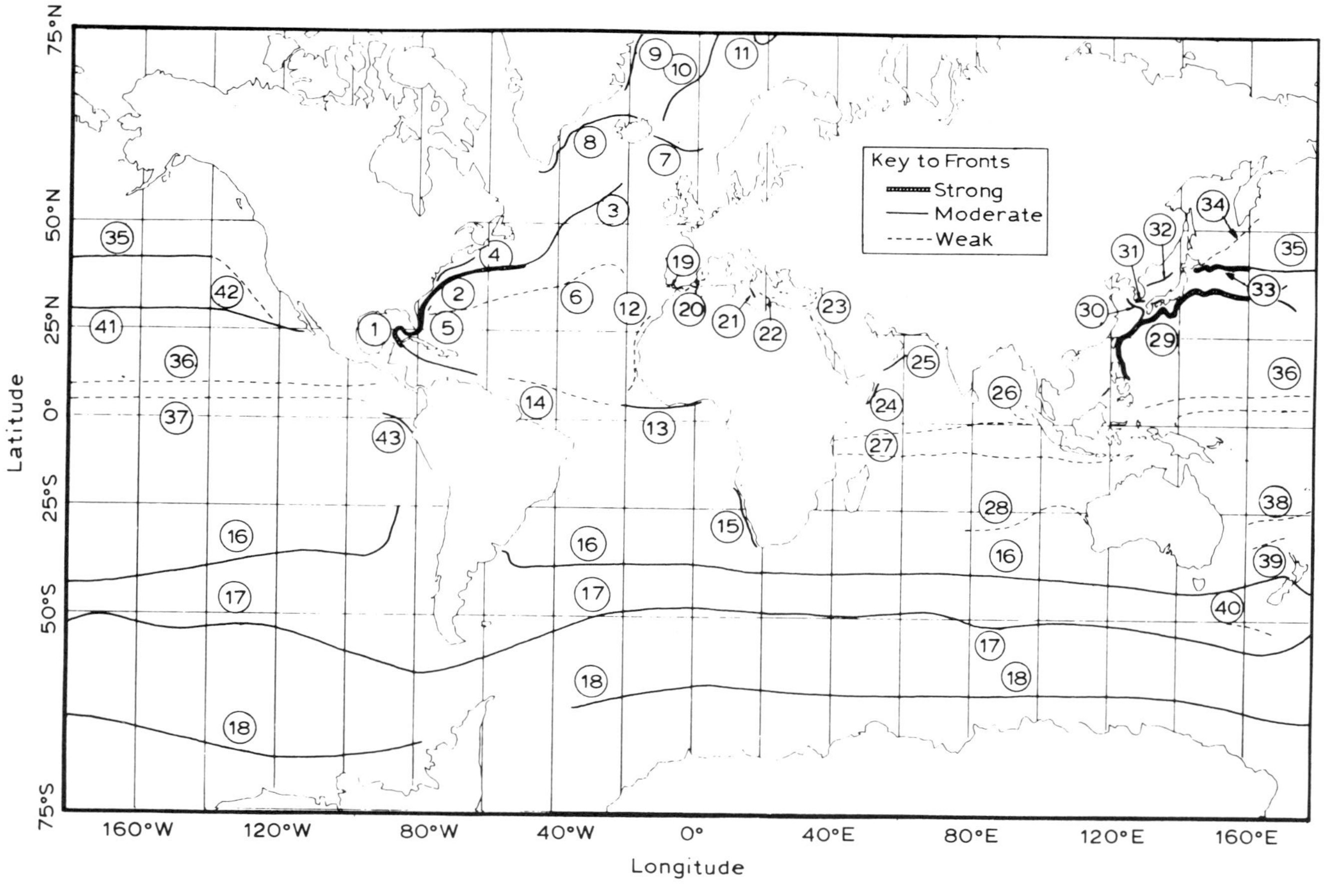

Fig. 2.20 The global distribution of ocean fronts. Numbers correspond to the generally accepted names, as listed in Table 2.2. Only the mean positions are presented. Due to seasonal variability, frontal locations can shift by as much as 200 km (Cheney and Winfrey, 1976).

Table 2.2 Names of Ocean Fronts Shown in Fig. 2.20

Atlantic Ocean fronts	
1	Loop current (Gulf of Mexico)
2	Gulf Stream
3	North Atlantic current (north polar front)
4	Slope front
5	Sargasso Sea front
6	Subtropical convergence
7	Iceland–Faeroe Island front
8	Denmark Strait front
9	East Greenland polar front
10	Greenland–Norwegian Sea front
11	Bear Island front
12	Northwest African upwelling
13	Gulf of Guinea front
14	Guiana current
15	Benguela upwelling
16	Subtropical convergence
17	Antarctic convergence (south polar front)
18	Antarctic divergence
Mediterranean Sea fronts	
19	Huelva front
20	Alboran Sea front
21	Maltese front
22	Ionian Sea front
23	Levantine Basin front
Indian Ocean fronts	
24	Somali upwelling
25	Arabian upwelling
26	Indian Ocean salinity front
27	Equatorial countercurrent fronts
28	West Australian front
Pacific Ocean fronts	
29	Kuroshio front
30	Yellow Sea warm current
31	Korean coastal front
32	Tsushima current
33	Oyashio front
34	Kuril front
35	Subarctic front
36	North Doldrum salinity front
37	South Doldrum salinity front
38	Tropical convergence
39	Mid Tasman convergence
40	Australian Subarctic front
41	Subtropical front
42	California front
43	East Pacific equatorial front

ate or weak depending upon representative values for: (1) the maximum change in sound speed across a front; (2) change of sonic layer depth; (3) depth to which the front extends; and (4) persistence.

Detached eddies, such as are found in the Atlantic Ocean near the Gulf Stream and in the Pacific Ocean near the Kuroshio current, represent another class of ocean fronts since they are separate water-mass entities contained within their own enclosed circulation and surrounded by water with different characteristics (Kerr, 1977).

Cold-core Gulf Stream rings form on the south side of the Gulf Stream. These rings are typically 100–300 km in diameter and rotate cyclonically (i.e. counterclockwise in the northern hemisphere) at speeds of up to 3 knots at the surface (The Ring Group, 1981). These rings generally drift in a southwestward direction in the Sargasso Sea and can persist for up to a year before losing their identity or being reabsorbed into the Gulf Stream (Fig. 2.21). It is has been estimated that five to eight cold-core rings form each year (Lai and Richardson, 1977). These rings have cold cores of slope

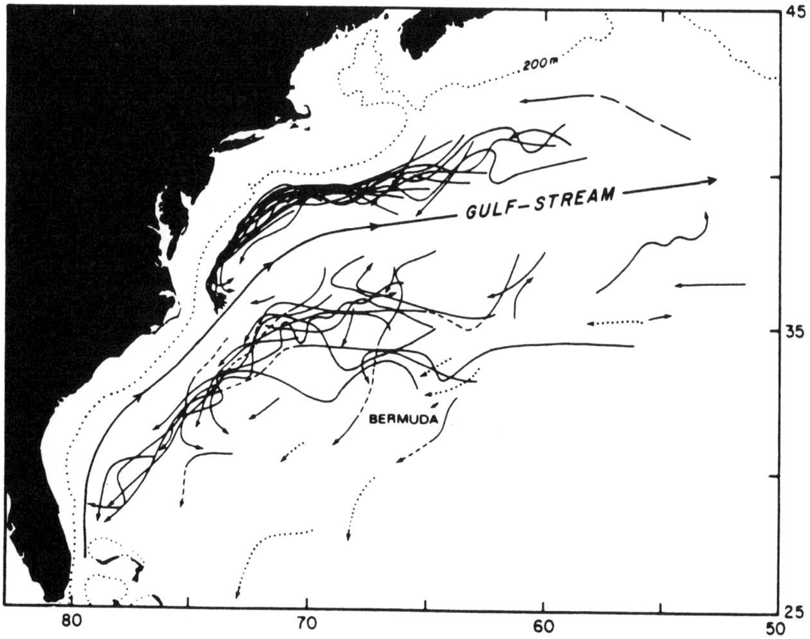

Fig. 2.21 Trajectories of Gulf Stream rings. Cold rings south of the stream: unbroken lines represent inferred ring time series; dashed and dotted lines indicate estimated trajectories of rings based on incomplete data. Warm rings north of the stream: trajectories are confined between the continental slope and the Gulf Stream (Lai and Richardson, 1977; *J. Phys. Oceanogr.*, **7**, 670–83; copyright by the American Meteorological Society).

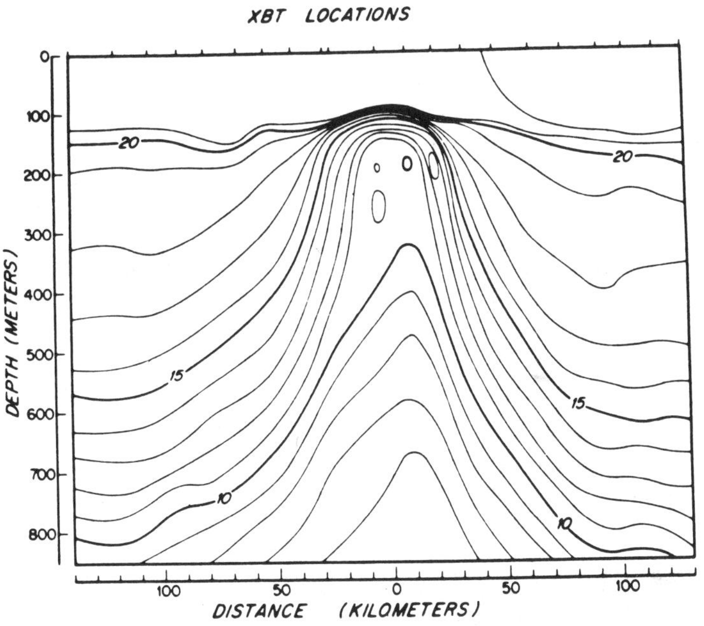

Fig. 2.22 Temperature (°C) section through a cyclonic (cold-core) Gulf Stream ring (oriented southwest to northeast) as observed during December 1975 near 36°N, 58°W. The central core of the ring contains cold, low-saline slope water with isotherms rising about 500 m above the normal depth in the Sargassso Sea. The overall size of the ring is approximately 200 km (Lai and Richardson, 1977, *J. Phys. Oceanogr.*, 7, 670–83; copyright by the American Meteorological Society).

water encircled by warm Sargasso Sea waters. They therefore represent strong thermal anomalies (Fig. 2.22).

Warm-core Gulf Stream rings form on the north side of the Gulf Stream, trapping warm Sargasso Sea water in the cold slope water (Fig. 2.21). The resulting circulation is anticyclonic (i.e. clockwise in the northern hemisphere). An average of 22 warm-core rings are formed annually between 75°W and 44°W (Auer, 1987).

(*b*) *Internal waves*

Internal waves are subsurface waves which propagate along the interface separating fluid layers of different densities. They can also exist within fluid layers where vertical density gradients are present. These waves can be generated by a number of mechanisms: surface waves, wind forcing, submarine earthquakes and submarine landslides, air pressure changes and current shears, among others.

In the open ocean, internal waves appear to take the form of progressive waves. In partially closed water bodies, standing internal waves are generally found. Internal waves are commonly observed over continental shelves. Such waves are probably generated by the scattering of the barotropic tide into baroclinic modes at the edge of the shelf, and then propagate shoreward on the shelf where they are absorbed and reflected as they break on the sloping bottom. Evidence suggests that such internal waves may have lifetimes of several days on the shelf.

The distribution of internal wave amplitudes as a function of depth is influenced by the vertical density distribution: the weaker the density boundary, the larger the amplitudes. Internal waves will normally have amplitudes several times greater than surface waves; crest-to-trough wave heights can be on the order of 10 m.

Wavelengths can range from a few hundred meters to many kilometers. At the long wavelength end of the internal wave spectrum are internal tides.

In theory, free internal waves can only exist in the frequency range bounded at the lower limit by the inertial frequency (which is a function of latitude) and at the upper limit by the buoyancy frequency (which is a function of depth). The inertial frequency (ω_i) is equivalent to the Coriolis parameter (f): $\omega_i = f = (2\pi \sin L)/12$ rad/h, where L is the latitude in degrees. The associated period (T) is 12 h/sin L. At a latitude of 30°N, the inertial period is 24 h. The buoyancy frequency (N), also referred to as the Brunt–Väisälä frequency, is related to water density (ρ), depth (z), gravitational acceleration (g) and sound speed (c) by

$$N = \sqrt{-g\left(\frac{1}{\rho}\frac{\mathrm{d}\rho}{\mathrm{d}z} + \frac{g}{c^2}\right)} \text{ rad/s} \tag{2.6}$$

where water depth (z) is measured in the negative (downward) direction (Apel, 1987, pp. 169–70). The buoyancy frequency decreases approximately exponentially with depth, with a maximum value ($N/2\pi$) of about 10 cycles/h near the surface, and an e-folding depth of about 1.5 km (Spindel, 1985)

When present, internal waves can be detected by making continuous temperature measurements at one location over a given period of time. Their presence is evidenced by oscillations in the temperature record with periods consistent with those of internal waves. Figure 2.23 shows the passage of internal waves as recorded in the changing depths of the isotherms.

Internal waves are considered to be a limiting factor in the propagation of acoustic energy, particularly in the frequency range 50 Hz-20 kHz. The effects are manifested as amplitude and phase variations. Internal waves may also limit both the temporal and spatial stability of acoustic paths

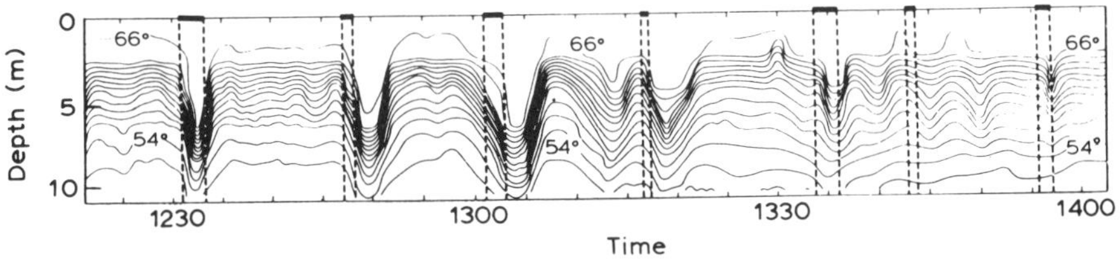

Fig. 2.23 Passage of internal waves past a stationary observation point as evidenced by temporal fluctuations in isotherm patterns. Water temperature is in °F. Vertical dashed lines indicate relationship between internal wave structure and the location of sea-surface slicks (LaFond, 1962; *The Sea*, Vol. 1, pp. 731–51; reprinted by permission of John Wiley and Sons, Inc., all rights reserved).

(Flatté, 1979). Below 50 Hz, the relatively long acoustic wavelengths (>30 m) are less likely to be affected by internal waves. Above 20 kHz (with acoustic wavelengths less than a few centimeters), the effects of fine-scale features are probably more important (Spindel, 1985).

Zhou *et al.* (1991) investigated the interaction of underwater sound with internal gravity waves in an attempt to explain the anomalous behavior of low-frequency (~300–1100 Hz) acoustic propagation conditions observed in some shallow-water areas. As a result of this investigation, it was noted that acoustic measurements could be employed in an inverse fashion for the remote sensing of internal wave activity in the coastal zone. Chin–Bing *et al.* (1993a) numerically simulated low-frequency backscatter from internal waves in shallow water.

One problem in testing theories of acoustic propagation in inhomogeneous media is the inability to determine accurately the spatial and temporal scales associated with fluctuations in the index of refraction. These fluctuations can be caused by tides, internal waves and fine-scale features. Ewart and Reynolds (1984) reported results from the Mid-Ocean Acoustic Transmission Experiment (MATE). This experiment was designed to measure phase and intensity fluctuations in sound pulses transmitted at 2, 4, 8 and 13 kHz over a wholly refracted path (i.e. no boundary interactions). Two receiver towers were placed on Cobb Seamount (in the northeastern portion of the North Pacific Ocean); a sister tower located 20 km to the southwest was used for placement of the acoustic transmitter tower. This geometry minimized transducer motion and also assured the presence of a wholly refracted path between the source and receiver. The environmental program conducted in support of MATE was specifically designed to oversample the internal-wave variability within the context of the Garrett–Munk model (Garrett and Munk, 1979). These environmental measurements can also be used to test future models of internal-wave variability.

2.5.3 Fine-scale features

One type of fine-scale oceanic feature is the 'thermohaline staircase'. These staircases are generally found in the main thermocline and are evidenced by layers of uniform temperature and salinity on the order of 10 m thick separated by thin, high-gradient interfaces on the order of a few meters in thickness (Fig. 2.24). The incidence of well-developed staircases appears to be limited to less than 10% of the available high-resolution profiles taken in the North Atlantic Ocean (Schmitt, 1987; Schmitt *et al.*, 1987). Staircase structures are most frequently associated with a strongly destabilizing vertical gradient of salinity. These features have been observed northeast of South America in the tropical Atlantic Ocean (just outside the Caribbean Sea), and also in the eastern Atlantic outside the Mediterranean Sea.

Before the dynamics of the thermohaline staircases were understood, these features were sometimes regarded as malfunctions in the oceanographic sensors which recorded them.

Research is not conclusive as to what effect these staircase features might have on underwater acoustic propagation. Chin–Bing *et al.* (1994) studied the effects of thermohaline staircases on low-frequency (50 Hz) sound propagation. Several propagation models were used to generate transmission loss as a function of range from source to receiver using a sound speed profile containing staircase features. A source was placed (in depth) at the center of the staircase features while receivers were placed above, below, and at the center of the features. These results were then compared

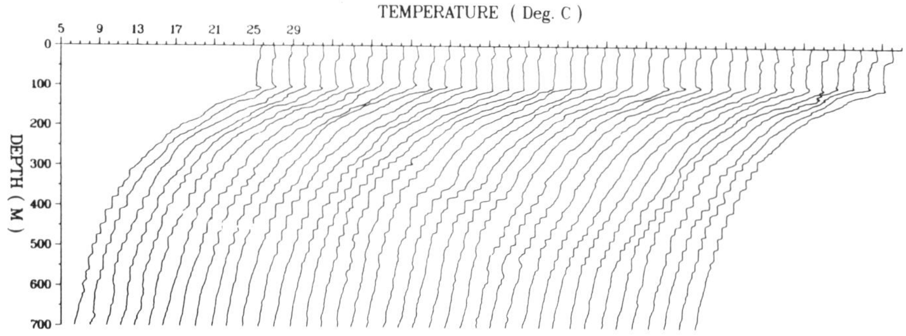

Fig. 2.24 Series of expendable bathythermograph (XBT) temperature profiles from an east–west section east of Barbados in the Atlantic Ocean. The profiles are separated by a distance of 5.5 km; the total distance covered is 220 km. The temperature scale is correct for the profile at the extreme left (west), and each subsequent profile is offset by 1.6°C. Undulations of the thermocline caused by internal waves and meso-scale eddies can also be seen (Schmitt, 1987; *EOS, Trans. Amer. Geophys. Union*, **68**, 57–60; copyright by the American Geophysical Union).

to baseline simulations using a profile in which the effects of the staircase features were effectively averaged out. The greatest effects were observed when both the source and receiver were placed at the center of the features. These effects were attributed to a redistribution of intensity caused by the staircase features. Chin–Bing *et al.* (1994) also noted that backscatter can occur when the step-structured discontinuities of the thermohaline staircases are on the order of an acoustic wavelength; thus, at frequencies greater than about 3 kHz, backscatter from the thermohaline steps could be significant.

2.6 BIOLOGICS

Marine organisms can simplistically be segregated into four major categories: plankton, nekton, benthos and algae.

Plankton (or floaters) include both plants (phytoplankton) and animals (zooplankton). The zooplankton have little or no swimming ability and thus drift with the currents. Phytoplankton are typically smaller than 0.5 mm while zooplankton are smaller than 1 cm.

Nekton (or free swimmers) are animals which are capable of swimming purposefully. Nekton include fish and mammals, and occur over the entire depth range of the ocean.

Benthos are dwellers on, in or near the bottom. Fouling organisms such as barnacles would also be included in this category. Algae include marine plant life, such as seaweed.

Biological organisms can affect underwater sound in several ways: through noise production; attenuation and scattering of signals; presentation of false targets; and fouling of sonar transducers.

Certain marine animals, many of which are found over the continental shelves, produce sounds that increase the background (or ambient) noise levels: snapping shrimp, whales, porpoises, and various fish such as croakers and drum fish.

Organisms that may cause attenuation are schools of fish, dense populations of plankton and floating kelp, for example. False targets are commonly presented to active sonars by whales or large schools of fish or porpoises.

While fouling organisms such as barnacles do not directly affect sound, indirectly they can degrade sonar performance by fouling sonar domes and transducer faces. Furthermore, such organisms can contribute to an increase in hull noise of ships and submarines through the generation of turbulence as the vessels move through the water. This effect is also referred to as self noise.

Perhaps the most notable impact of marine organisms on active sonars (particularly those operating at frequencies near 10 kHz) is known as the deep scattering layer (DSL). The DSL is a dense accumulation of marine organisms at depth below the surface. The strong scattering nature of the DSL is primarily attributable to fish and other marine animals with swim bladders and gas floats, although plankton and nekton are also present. The DSL is typically encountered in temperate regions and migrates vertically, being near the surface at night and at depths near 1000 m during the day.

3

Propagation I. Observations and physical models

3.1 BACKGROUND

The propagation of sound in the sea has been intensely studied since the beginning of World War II when it was recognized that an understanding of this phenomenon was essential to the successful conduct of Anti-Submarine Warfare (ASW) operations. These early measurements were quickly transformed into effective, albeit primitive, prediction tools. ASW has continued to be the primary motivation behind modern advances in propagation modeling.

The study of sound propagation in the sea is fundamental to the understanding of all other underwater acoustic phenomena. The essentiality of propagation models is thus inherent in the hierarchy of acoustic models, as illustrated previously in Fig. 1.1.

Advances in propagation modeling have been achieved by both marine seismologists and underwater acousticians, although the motivating factors have been quite different. Marine seismologists have traditionally used earthborne propagation of elastic waves to study the solid earth beneath the oceans. Underwater acousticians have concentrated on the study of waterborne, compressional-wave propagation phenomena in the ocean as well as in the shallow subbottom layers (Akal and Berkson, 1986). In recent years, research in underwater acoustics has been extended to frequencies below several hundred hertz. This overlaps with the spectral domain of marine seismologists. Moreover, marine seismologists have become more interested in exploring the velocity–depth structure of the uppermost layers of the sea floor using higher frequencies. This area of overlapping interests has been recognized as a subdiscipline of both communities and is referred to as 'ocean seismo-acoustics'.

The emphasis in this chapter is focused on underwater acoustic applications. Developments in marine seismology will be mentioned when the applications to sonar modeling are clearly evident. Much research has been performed in the marine seismology community that is theoretically and conceptually applicable to underwater acoustics. Such practical research includes the development of sophisticated yet robust mathematical methods.

Propagation models have continued to be used for the prediction of sonar performance. They have also found great utility in analyzing field measurements, in designing improved sonar systems, and in designing efficient field experiments.

As modeling has become prominent in many aspects of underwater acoustics, it was deemed timely to take stock of the state of the art in modeling techniques and their relationship to available measurements. Ideally, such an assessment should identify those areas requiring further measurement support as well as those that are firmly understood and hence properly-modeled.

This chapter addresses the observations that have been made in the field and the physical models that have been developed. Aspects of propagation phenomena – ducts and channels, boundary interactions, volumetric effects and coherence – are described. Chapter 4 will address the mathematical models that have been developed for underwater acoustic propagation. Specialized aspects of surface ducts, shallow water areas and Arctic regions are discussed in Chapter 5.

3.2 NATURE OF MEASUREMENTS

Field measurement programs are usually quite complex and typically involve multiple platforms (e.g. ships, buoys, towers, aircraft, submarines or satellites).

A wide variety of experimental field techniques have been used in underwater acoustic propagation studies. Some of the more typical types of measurement platforms and experimental geometries that have been utilized include (Urick, 1982, Chapter 1):

1. Two ships – one a source ship and the other a receiving ship; the range between them is changed as transmission runs are made in order to yield level versus range.
2. Single ship – using a suspended transmitter and either sonobuoys or a hydrophone array for reception.
3. Ship and aircraft – where the aircraft drops explosive sound sources while flying toward or away from the ship.
4. Single aircraft – using sonobuoys for reception and recording on board the aircraft.
5. Bottomed hydrophone array – with a cable connected to shore, receives signals transmitted from a ship or the explosive shots dropped by an aircraft.

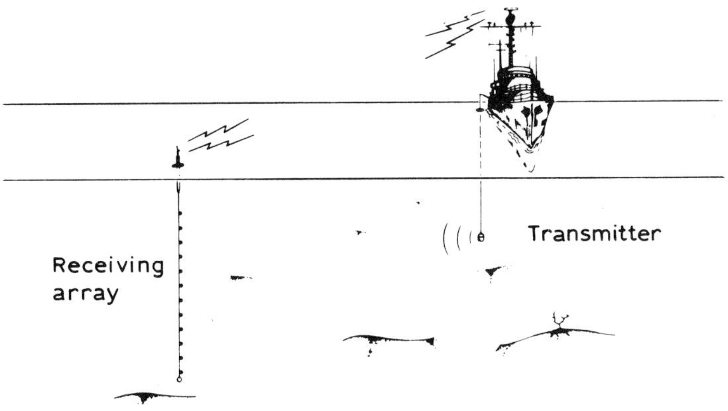

Fig. 3.1 Example of a simple experimental geometry (adapted from Ingenito *et al.*, 1978).

6. Two bottomed transducers – one acting as a source and the other as a receiver; this geometry is typically used in studies of the fluctuation of sound transmission between two fixed points in the sea.

A simple experimental geometry illustrating method (2) above is presented in Fig. 3.1. Here, the transmitting ship is receiving the signals via radio directly from the array.

Fully integrated oceanographic and acoustic field experiments are required in order to achieve a comprehensive portraiture of the temporal, spatial and spectral scales necessary for a full understanding of the governing acoustic phenomena.

3.3 BASIC CONCEPTS

The standard unit of measure of underwater acoustic propagation is acoustic intensity (I), which is sound pressure flow (power) per unit area:

$$I = \frac{p^2}{\rho c} \tag{3.1}$$

where p is the pressure amplitude, ρ is the density of sea water and c the sound speed in sea water. The product ρc is commonly referred to as the acoustic impedance.

Transmission loss (TL) can be defined as ten times the log (base 10) of the ratio of the reference intensity (I_{ref}) measured at a point 1 m from the source, to the intensity (I), measured at a distant point, and is expressed in units of decibels (dB):

$$\text{TL} = 10 \log \frac{I_{\text{ref}}}{I} \tag{3.2}$$

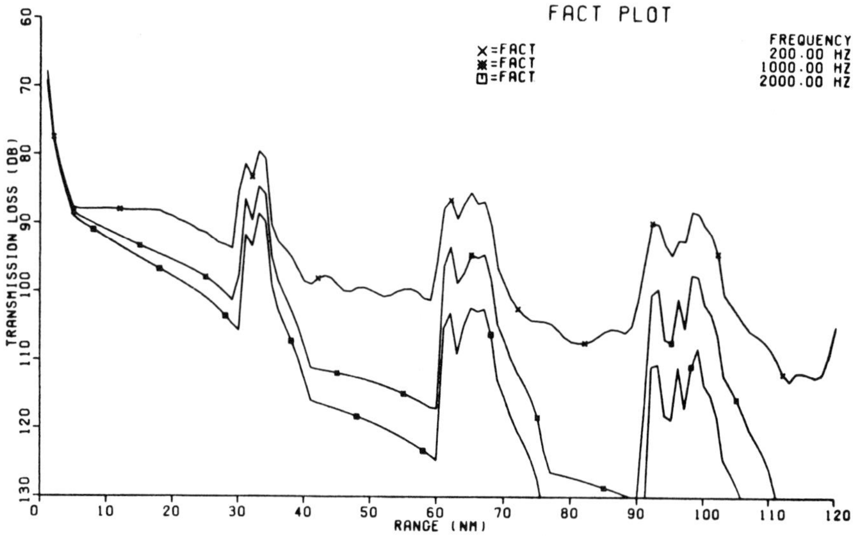

Fig. 3.2 Example of standard transmission loss curves generated by the FACT model for each combination of frequency and source–receiver geometry. Here, the source and receiver depths are fixed at 150 m and 90 m, respectively. The peaks (minimum TL values) correspond to convergence zones. Note the increase in transmission loss with increasing frequency due to absorption.

The standard metric unit for pressure (force per unit area) is 1 microPascal, which is equivalent to 10^{-6} newton/m^2, and is abbreviated μPa.

Transmission loss (TL) has conventionally been plotted for each frequency and source–receiver geometry as a function of range, as illustrated in Fig. 3.2. This type of display is easily generated by all propagation models. Certain types of propagation models can also generate and display acoustic transmission loss in the entire range–depth plane for all receiver positions, given a fixed source depth (Fig. 3.3).

Sonar performance is commonly described in terms of a figure of merit (FOM). The FOM is a quantitative measure of sonar performance: the larger the value, the greater the performance potential. Numerically, the FOM is equal to the allowable one-way transmission loss in passive sonars. The FOM is further described in Chapter 10 within the context of the sonar equations.

The display method illustrated in Fig. 3.2 is very useful in evaluating passive sonar performance. Specifically, once the FOM is calculated for a particular sonar, a horizontal line can be drawn on the plot equating the numerical value of the FOM to transmission loss. Then, any area below the transmission loss curve, but above the FOM line, represents a sonar detection area. Figure 3.4 shows a hypothetical relationship between the FOM

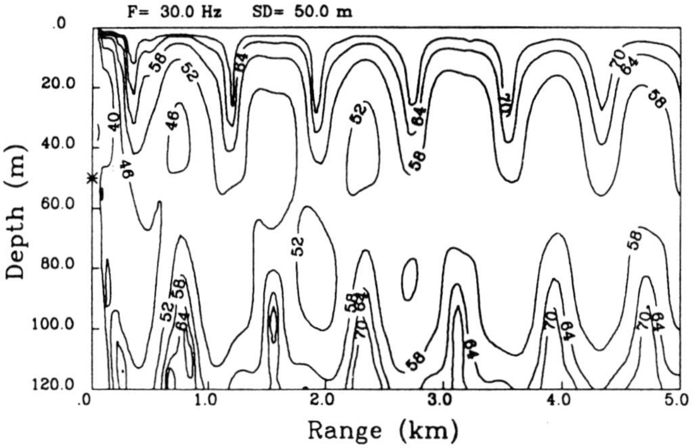

Fig. 3.3 Example showing contours of transmission loss plotted in the range–depth plane. This plot is valid for one frequency (30 Hz) and one source depth (50 m), but can be used to determine the transmission loss at any receiver location in the range–depth plane; the contour interval is 6 dB (Schmidt, 1988).

and the sonar detection areas, and the correspondence between the TL curve and the ray paths as propagated in the water column.

Sound propagates in the sea by way of a variety of paths. The particular paths traveled depend upon the sound speed structure in the water column and the source–receiver geometry. These paths are: direct path, surface duct, bottom bounce, convergence zone, deep sound channel and reliable acoustic path. These six basic paths are illustrated in Fig. 3.5. Various combinations of paths are possible and are referred to as multi-path propagation.

In basic ray tracing, Snell's law is used in one form or another. This law describes the refraction of sound rays in a medium in which sound speed varies as a function of depth, but is constant within discrete horizontal layers of the water column. Consider Fig. 3.6 where a ray (normal to the acoustic wavefronts) is traveling from medium 1 (with sound speed c_1) into medium 2 (with sound speed c_2), where $c_1 \neq c_2$. Let λ_1 be the distance between successive wavefronts (wavelength) in medium 1 and λ_2 the corresponding value in medium 2. Then, as defined in Fig. 3.6

$$\lambda_1 = \Delta x \sin\phi_1 = c_1 \Delta t \qquad \lambda_2 = \Delta x \sin\phi_2 = c_2 \Delta t$$

where Δt is an increment of time. Rearranging terms, we obtain the familiar relation

$$\frac{\sin\phi_1}{c_1} = \frac{\sin\phi_2}{c_2} \tag{3.3a}$$

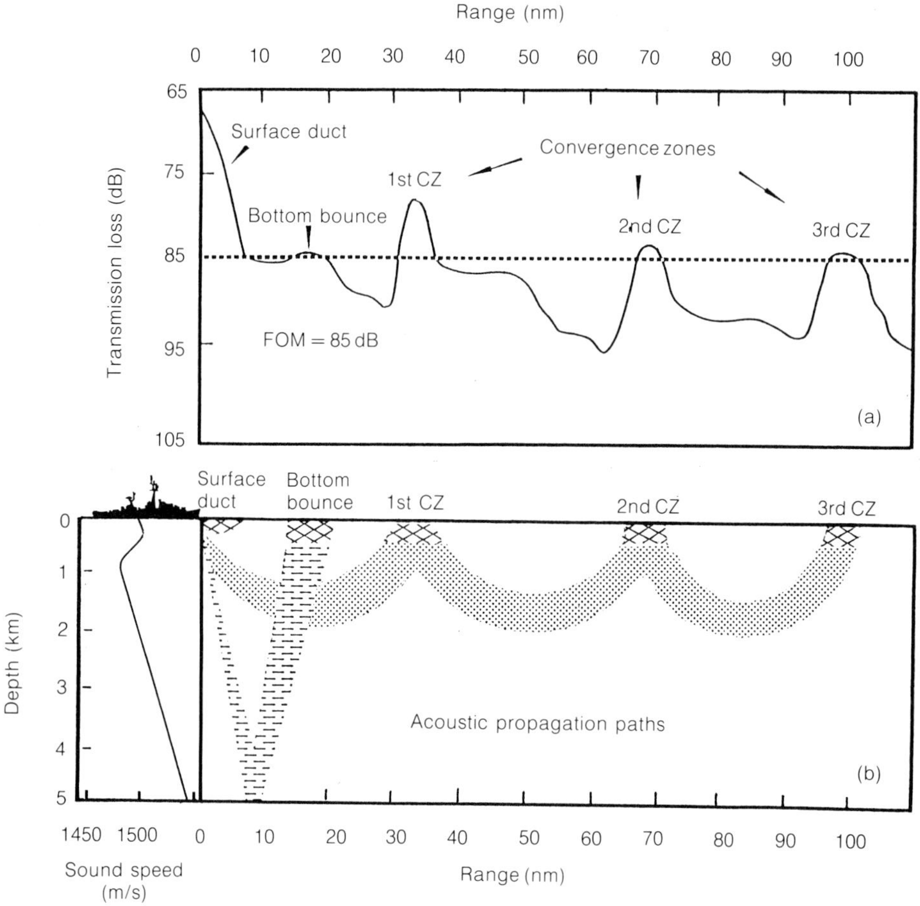

Fig. 3.4 Hypothetical relationship between (a) transmission loss (TL) curve and (b) the corresponding propagation paths and detection zones (cross-hatched areas near the surface) associated with a Figure of Merit (FOM) of 85 dB. A plausible sound speed profile is shown at the left of panel (b); both the source (target) and receiver (sonar) are positioned near the surface.

or equivalently, from Fig. 3.6,

$$\frac{\cos\theta_1}{c_1} = \frac{\cos\theta_2}{c_2} \tag{3.3b}$$

As a matter of convention, ϕ is referred to as the incidence angle while θ is referred to as the grazing angle.

In a homogeneous medium, acoustic transmission loss varies as the inverse of the range squared. This relationship is easily derived as follows:

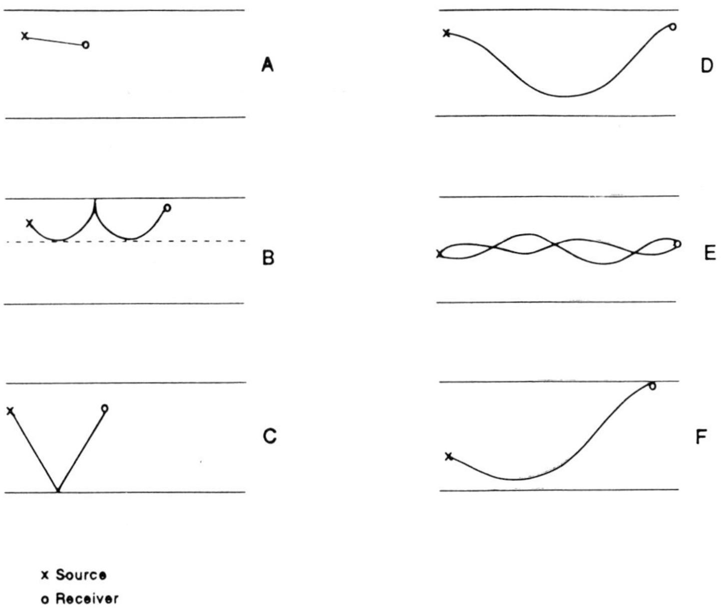

Fig. 3.5 Six basic propagation paths in the sea: A, direct path (DP); B, surface duct (SD); C, bottom bounce (BB); D, convergence zone (CZ); E, deep sound channel (DSC); and F, reliable acoustic path (RAP).

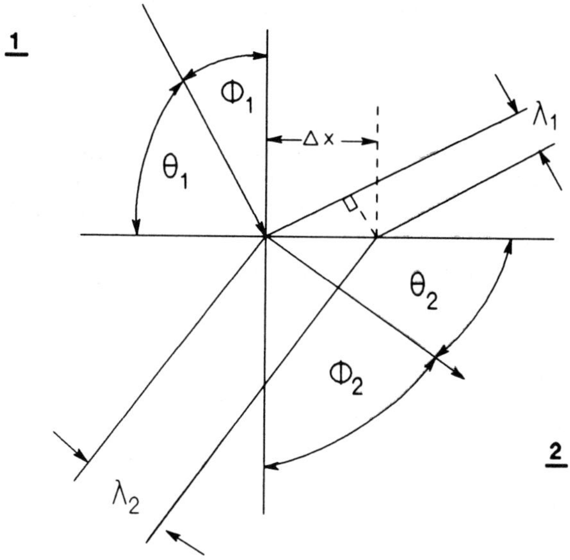

Fig. 3.6 Geometry for Snell's law.

Let I = intensity, P = power and A = area. Then

$$I = \frac{P}{A}$$

For spherical spreading (see Fig. 3.7(a))

$$I_1 A_1 = I_1(4\pi r_1^2) \qquad I_2 A_2 = I_2(4\pi r_2^2)$$

where r_1, r_2 are radii of concentric spherical sections.

Power (P) is conserved; therefore, $P = I_1 A_1 = I_2 A_2$ and

$$I_2 = I_1(r_1^2/r_2^2)$$

but r_1 = unit reference distance and thus

$$I_2 = I_1/r_2^2$$

Since intensity is power per unit area, and since the area of a sphere increases as the square of its radius, the intensity falls off as the inverse square of the radius (or range) in order that power remains constant.

The corresponding transmission loss (TL) is defined as

$$\text{TL} = 10 \log \frac{I_1}{I_2} = 10 \log r_2^2 = 20 \log r_2 \qquad (3.4)$$

This relationship would be valid for an isotropic deep ocean with no absorption effects.

An analogous expression can be derived for cylindrical spreading. This spreading law would be appropriate for a duct, or in shallow water, where the water is homogeneous and the boundaries are perfect reflectors. Then (referring to Fig. 3.7(b))

$$I_1 A_1 = I_1(2\pi r_1 H) \qquad I_2 A_2 = I_2(2\pi r_2 H)$$

where H is the depth of the duct or of the water column. Since power is conserved

$$P = I_1 A_1 = I_2 A_2 \qquad I_2 = I_1 \frac{r_1}{r_2}$$

For unit radius r_1

$$I_2 = \frac{I_1}{r_2}$$

Thus, the intensity falls off as the inverse of the radius (or range). The corresponding transmission loss (TL) is defined as

$$\text{TL} = 10 \log \frac{I_1}{I_2} = 10 \log r_2 \qquad (3.5)$$

Equations (3.4) and (3.5) are valid in isotropic ocean environments.

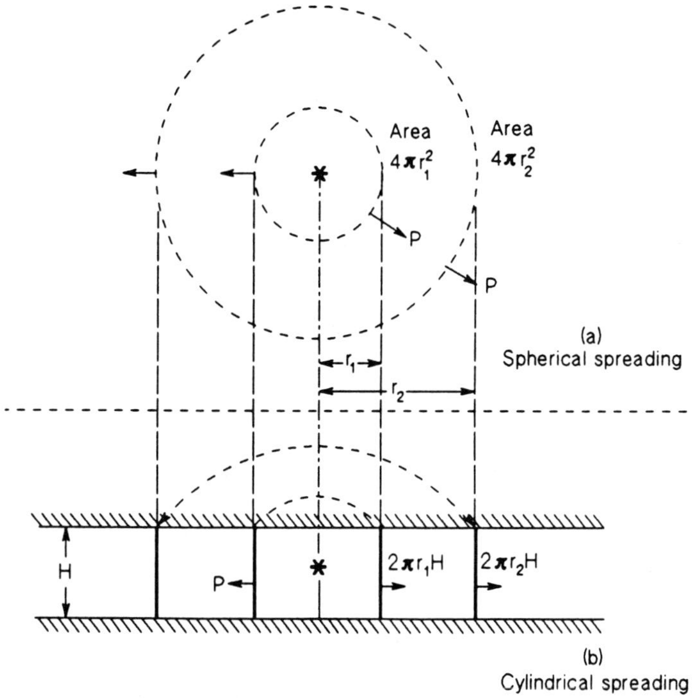

Fig. 3.7 Geometry for (a) spherical spreading and (b) cylindrical spreading (Urick, 1983; *Principles of Underwater Sound,* 3rd edn; reproduced with permission of McGraw-Hill Publishing Company).

3.4 SEA-SURFACE BOUNDARY

The sea surface affects underwater sound by providing a mechanism for:

1. forward scattering and reflection loss
2. image interference and frequency effects
3. attenuation by a layer of bubbles
4. noise generation at higher frequencies due to surface weather
5. backscattering and surface reverberation.

Urick (1982, Chapter 10) provides a comprehensive summary of sound reflection and scattering by the sea surface. Items (1) through (3) will be discussed below. Item (4) will be discussed in Chapter 6 and item (5) will be discussed in Chapter 8.

The mechanisms operating at the surface can be incorporated into mathematical models through the specification of 'boundary conditions'. These boundary conditions can range from simplistic to complex, depending upon the sophistication of the model and the availability of information concerning the state of the sea surface.

3.4.1 Forward scattering and reflection loss

When a plane sound wave in water strikes a perfectly smooth surface, nearly all of the energy is reflected at the boundary in the forward (or specular) direction as a coherent plane wave. As the sea surface roughens under the influence of wind, sound is also scattered in the backward and out-of-plane directions, and the intensity in the forward direction is accordingly reduced. The backward-directed (backscattered) energy gives rise to surface reverberation. Eckart (1953) developed a theoretical treatment of scattering by a sinusoidal boundary as a way to approximate reflection from a wind-roughened sea surface. Marsh *et al.* (1961) developed simple formulas to express scattering losses at the sea surface.

The sea surface is commonly modeled as a pressure release surface (see Kinsler *et al.*, 1982, pp. 126–7). This is a condition in which the acoustic pressure at the air–water interface is nearly zero, the amplitude of the reflected wave (in water) is almost equal to that of the incident wave, and there is a 180° phase shift.

It is also common practice to use the term 'reflection coefficient' to express the amount of acoustic energy reflected from a surface or boundary between two media. This coefficient depends upon the grazing angle and the difference in the acoustic impedance between the two media. A reflection loss is then defined as 10 log (reflection coefficient). This reflection loss is referred to as surface loss when describing the reflection of sound from the sea surface, or as bottom loss when describing the reflection of sound from the sea floor.

A measure of the acoustic roughness of the sea surface is provided by the Rayleigh parameter R through the relation

$$R = 2ka \sin \theta \tag{3.6}$$

where $k = 2\pi/\lambda$ is the acoustic wave number, λ is the acoustic wavelength, a is the root-mean-square (rms) amplitude of the surface waves ($2a$ = crest-to-trough wave height) and θ is the grazing angle relative to the horizontal plane. When $R \ll 1$, the sea surface is considered to be acoustically smooth; when $R \gg 1$, the surface is acoustically rough.

Sea-surface wave spectra can be numerically generated by executing available spectral ocean wave models in the hindcast mode. Hindcasting is usually the only means available for obtaining sufficiently long record lengths from which to generate reliable statistics. A statistical analysis of these hindcast data produces probability distributions of critical parameters for use in estimating future sea surface conditions.

Eller (1984a) has reviewed the availability of simple surface loss algorithms appropriate for incorporation into propagation models.

3.4.2 Image interference and frequency effects

When the surface is smooth, an interference pattern is produced between direct-path sound and sound reflected from the sea surface. The sound reflected from the sea surface may be considered to originate from an image source located on the opposite (mirror image) side of the surface (Fig. 3.8). This image signal will have an amplitude almost equal to that of the incident signal, but will be out of phase. This sound field may be divided into three parts (Fig. 3.9): (1) near field close to the source in which the image source is too far away, and the reflected sound is too weak to produce appreciable interference; (2) interference field in which there are strong peaks and nulls in the signal received as range increases; and (3) far field in which there is an increasingly out-of-phase condition between source and image, and the intensity falls off as the inverse fourth power of the range. This phenomenon, also known as the Lloyd Mirror effect, diminishes with increasing surface roughness.

Assuming an acoustically smooth sea surface ($R \ll 1$) and a shallow source depth (d) in deep water, the ranges r_1 and r_2 in Fig. 3.9 can be approximated as (Urick, 1982):

$$r_1 \approx \sqrt{2dH}, \qquad r_2 \approx \frac{4\pi dH}{\lambda}$$

where λ is the acoustic wavelength and H is the receiver depth.

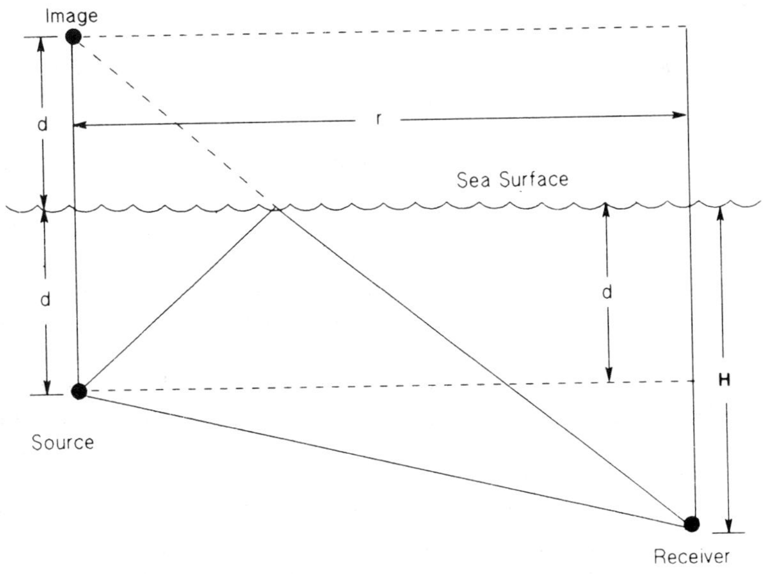

Fig. 3.8 Geometry for image interference effect.

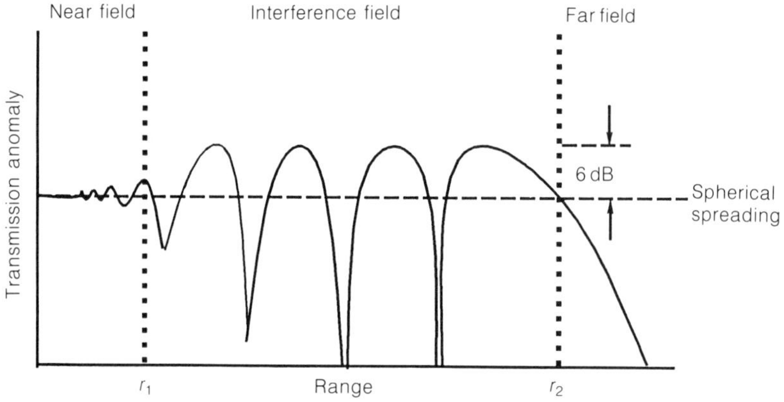

Fig. 3.9 Illustration of image interference effects (adapted from Urick, 1979). The transmission anomaly (TA) represents the difference between the observed transmission loss (TL) and losses due to the effects of spherical spreading [TA = $20 \log_{10}(r) - \text{TL}$], where the range ($r$) is measured in meters.

The image effect can be used to estimate the depth of a submerged target at short ranges. In Fig. 3.8, if the source is replaced by a submarine which has been ensonified by a single, short pulse, then the depth (d) of the submarine is approximated by

$$d = \frac{rc\Delta t}{2H} \tag{3.7}$$

where c is the speed of sound and Δt is the difference in time between receipt of the direct and the surface-reflected pulses (Albers, 1965, pp. 50–1).

The concept of surface interference can also be used to solve relatively simple propagation problems. The approach is called the 'method of images' and is valid for all frequencies. The solution is usually expressed as a sum of the contributions of all images within a multilayered space. This method is usually cumbersome and is commonly employed as a physical model against which to check the results of more elaborate mathematical models (to be discussed in Chapter 4). Kinsler *et al.* (1982, pp. 427–30) provide a more detailed discussion of this method together with several examples of its application. Tolstoy and Clay (1966, pp. 33–6) discuss solutions in waveguides.

When the sea surface is rough, the vertical motion of the surface modulates the amplitude of the incident wave and superposes its own spectrum as upper and lower sidebands on the spectrum of the incident sound.

When there is a surface current, the horizontal motion will appear in the scattered sound and cause a Doppler-shifted and Doppler-smeared spectrum.

3.4.3 Bubble layers

The presence of bubble layers near the sea surface further complicates the reflection and scattering of sound as a result of the change in sound speed, the resonant characteristics of bubbles and the scattering by the bubbly layers.

Hall (1989) developed a comperhensive model of wind-generated bubbles in the ocean. The effects on the transmission of short pulses in the frequency range 1.25–40 kHz were also examined. For long-range propagation, Hall concluded that the decrease in the near-surface sound speed due to bubbles does not significantly affect the intensity of the surface-reflected rays.

3.4.4 Ice interaction

Acoustic interaction with an ice canopy is governed by the shape of the under-ice surface and by the compressional wavespeeds (typically 1300–3900 m/s) and shear wavespeeds (typically 1400–1900 m/s) (see Untersteiner, 1966; Medwin *et al.*, 1988).

McCammon and McDaniel (1985) examined the reflectivity of ice due to the absorption of shear and compressional waves. They found that shear wave attenuation is the most important loss mechanism from 20° to 60° incidence for smooth ice at low frequencies (≤ 2 kHz).

In Arctic regions, the presence of a positive-gradient sound speed profile and a rough under-ice surface (with a distribution of large keels) may lead to significant out-of-plane scattering. The acoustic impacts of this scattering are twofold. First, significant beam widening may result from the multiple interactions with the randomly rough surface, similar to that observed in shallow water. Second, the presence of ice keels in the vicinity of the receiver leads to multiple source images or beam-steering errors from the interaction of the acoustic signal with the facets of the local under-ice surface.

Because of the overwhelming effect of ice on the propagation of sound in the Arctic, the magnitude of the excess attenuation observed under the ice should be determined by the statistics of the under-ice surface. Ice-ridge models can be categorized according to two classes: discrete models and continuous statistical models. These two classes of models are briefly described below.

Discrete ice-ridge models prescribe a representative ridge shape, or an ensemble of ridge shapes, to calculate the statistics of the surface from the discrete statistics of the known ice structure.

Continuous statistical ice-ridge models treat the under-ice surface as a stochastic process. This process is then analyzed using the techniques of time-series analysis in which the under-ice surface can be characterized by its autocorrelation function. These models can give a more complete description of the under-ice roughness than can the discrete models; however, they are limited in application to those surfaces that can be completely specified by a Gaussian depth distribution.

The model developed by Diachok (1976) will be discribed since it is considered to be representative of the class of discrete ice-ridge models known to exist and because of its intuitive appeal. The discrete models are also more robust (i.e. require less knowledge of the under-ice surface) than the continuous statistical models. Furthermore, Diachok's model has been incorporated into existing propagation models with some success.

According to Diachok's model, sea ice may be described as consisting of floating plates, or floes, about 3 m thick, occasionally interrupted by ridges, which are rubble piles formed by collisions and shear interactions between adjacent floes. Ridge dimensions vary widely, but are nominally about 1 m high, 4 m deep and 12 m wide, with the ridge length generally being much greater than the depth or width. A representative average spacing between ridges (which is random) is about 100 m. Ice-ridge orientation is commonly assumed to be directionally isotropic, although limited empirical data suggest that, at least locally, there may be a preferred orientation. The physical model of reflection developed by Twersky (1957) was used.

A comparison between measured contours and simple geometrical shapes suggests that ridge keel contours may reasonably be represented by a half-ellipse (as in Fig. 3.10), and that ridge sail contours may be described using a Gaussian distribution function. The relative dimensions of this geometrical model are indicated in Fig. 3.10. The exact solution of under-ice scattering off a flat surface with a single semi-elliptical cylindrical boss of infinite extent was developed by Rubenstein and Greene (1991).

LePage and Schmidt (1994) extended the applicability of perturbation theory to under-ice scattering at low frequencies (10–100 Hz) by including the scattering of incident acoustic energy into elastic modes, which then propagate through the ice.

3.4.5 Measurements

Three basic experimental techniques have been employed to measure forward reflection losses at the sea surface:

1. comparing the amplitude or energy of pulses returned from the surface with that of the direct arrival;
2. using the Lloyd Mirror effect and observing the depth of the minima as the frequency is varied;
3. measuring the attenuation in the surface duct.

Based on a compilation of results in the literature by Urick (1982, Chapter 10), it appears that surface losses are less than 1 dB (per bounce) at

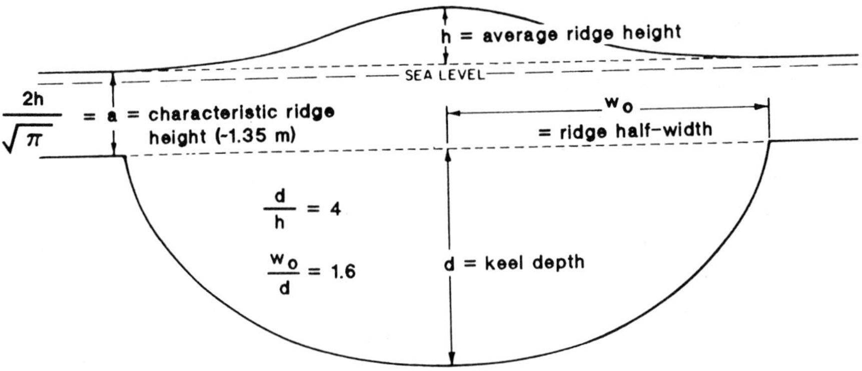

Fig. 3.10 Geometrical model of sea-ice ridges (Diachok, 1976).

frequencies below 1 kHz, and rise to about 3 dB (per bounce) at frequencies above 25 kHz.

3.5 SEA-FLOOR BOUNDARY

The sea floor affects underwater sound by providing a mechanism for:

1. forward scattering and reflection loss (but complicated by refraction in the bottom)
2. interference and frequency effects
3. attenuation by sediments
4. noise generation at lower frequencies due to seismic activity
5. backscattering and bottom reverberation.

Urick (1982, Chapter 11) provides a comprehensive summary of sound reflection and scattering by the sea floor. The single most important physical property that determines the acoustic characteristics of sediments is their porosity. Items (1) through (3) will be discussed below. Item (4) will be discussed in Chapter 6 and item (5) in Chapter 8.

Although the sea floor affects sound in ways similar to that of the sea surface, the return of sound from the bottom is more complex than from the surface for several reasons: (1) the bottom is more variable in composition; (2) the bottom is often stratified (layered) with density and sound speeds (both shear and compressional) varying gradually or abruptly with depth; (3) bottom characteristics (composition and roughness) can vary over relatively short horizontal distances; and (4) sound can propagate through a sedimentary layer and either be reflected back into the water by subbottom layers or be refracted back by the large sound speed gradients in the sediments.

These mechanisms can be incorporated into mathematical models through the specification of appropriate 'boundary conditions'. The complexity of these boundary conditions will depend upon the level of known detail concerning the composition and structure of the sea floor, and also to some degree on the sophistication of the mathematical model being used.

The specification of boundary conditions at the bottom has assumed greater importance due to increased interest in the modeling of propagation in shallow-water areas; such propagation, by definition, is characterized by repeated interactions with the bottom boundary. Acoustic interactions with highly variable sea-floor topographies and bottom compositions often necessitate the inclusion of both compressional and shear wave effects, particularly at lower frequencies. A fluid, by definition, cannot support shear stresses; therefore, in modeling acoustic propagation in an ideal (boundless) fluid layer, only compressional wave effects need be considered. As an approximation, saturated sediments are sometimes modeled as a fluid layer in which the sound speed is slightly higher than that of the overlying water column. The basement, however, can support both compressional and shear waves, and rigorous modeling of acoustic waves that interact with and propagate through such media must consider both types of wave effects. As an approximation, shear-wave effects are sometimes included in the form of modified attenuation coefficients.

3.5.1 Forward scattering and reflection loss

The ideal forward reflection loss of sound incident on a plane boundary between two fluids characterized only by sound speed and density was originally developed by Rayleigh (1945, Vol. II, p. 78). This model is commonly referred to as Rayleigh's law. In the simplest model incorporating absorption, the bottom can be taken to be a homogeneous absorptive fluid with a plane interface characterized by its density, sound speed and attenuation coefficient. In the case of sedimentary materials, all three of these parameters are determined by the porosity of the sediments.

In underwater acoustics, a common idealized model for the interaction of a point-source field with the sea floor is the so-called Sommerfeld model (after A.N. Sommerfeld). This model consists of an isospeed half-space water column overlying an isospeed half-space bottom. The bottom has a higher sound speed than the water. Thus, a critical angle exists in the plane wave reflection coefficient. For large grazing angles, energy is partially reflected and partially transmitted at the water–bottom interface; for small grazing angles, energy is totally reflected. Energy incident near the critical angle produces a complex phenomenon known as the lateral, or head, wave (Chin–Bing *et al.*, 1982, 1986; Westwood, 1989a). See also the discussion by Clay and Medwin (1977, pp. 262–3). This condition is termed an 'impedance boundary'.

Another commonly assumed boundary condition for the sea floor is the homogeneous Neumann bottom boundary condition. This condition requires that the normal component of the pressure gradient at the bottom be equal to zero. This condition is termed a 'rigid boundary'.

These simplistic models have largely been abandoned in favor of more rigorous geoacoustic models of the sea floor which more properly account for the propagation of sound in sediments (Anderson and Hampton, 1980a, 1980b; Hamilton, 1980). One early attempt at such modeling was provided by Hall and Watson (1967), who developed an empirical bottom reflection loss expression based largely on the results of AMOS (Marsh and Schulkin, 1955).

Westwood and Vidmar (1987) have summarized the current state of modeling of acoustic interaction with the sea floor. It is convenient to partition the discussion according to low-frequency and high-frequency bottom interaction. The transition between low and high frequencies is rather imprecise, but can be considered to occur at approximately 200 Hz.

At low frequencies and low grazing angles, the acoustic interaction with the sea floor in deep ocean basins is simple and well understood. The relatively long acoustic wavelengths are insensitive to the details of small-scale layering in the sediments. Moreover, for low grazing angles, there is little interaction with the potentially rough substrate interface. Accordingly, the sea floor can be accurately approximated as a horizontally stratified and depth-dependent fluid medium. The major acoustical processes affecting interaction with the sea floor are: (1) reflection and transmission of energy at the water–sediment interface, (2) refraction of energy by the positive sound speed gradient in the sediments, and (3) attenuation within the sediments. Modeling of this interaction is further enhanced by the availability of established methods for estimating the geoacoustic profile (i.e. sound speed, density and attenuation as functions of depth) of deep-sea sediments, given the sediment type and physiographic province.

In contrast, bottom interaction at high frequencies is not well understood. The relatively short wavelengths are now more sensitive to the small-scale sediment layering. These layers are reported to have an important effect on the magnitude and phase of the plane wave reflection coefficient. Stochastic techniques with which to analyze the effects of the near-surface sediment layering are being developed, but they do not yet incorporate potentially important acoustical processes such as refraction and shear wave generation. Modeling at high frequencies is further frustrated by the high spatial variability of sediment layering.

The concept of 'hidden depths' (Williams, 1976), which states that the deep ocean sediment structure well below the ray turning point has no acoustical effect, is important because it focuses attention on those low-frequency processes occurring in the upper regions of the sediments (see

Knobles and Vidmar, 1986). Westwood and Vidmar (1987) have developed a ray-theoretical approach, called CAPARAY, for simulating the propagation of broadband signals interacting with a layered ocean bottom. CAPARAY can simulate a time series at a receiver due to an arbitrary source waveform by constructing a frequency domain transfer function from the eigenray characteristics.

Geoacoustic models have been developed to account for bottom interaction processes in propagation models. As summarized by Holland and Brunson (1988), geoacoustic models of marine sediments can be formulated in one of three ways: (1) by empirically relating geoacoustic and geophysical properties of the sediments (e.g. Hamilton, 1980); (2) by using the Biot–Stoll model to relate sediment geoacoustic properties to geophysical properties on the basis of physical principles; and (3) by using an inversion technique to generate sediment geophysical parameters from bottom loss measurements [e.g. McCammon, 1991; Hovem *et al.*, 1991 (see especially Section 3, Modelling and inversion techniques); Rajan, 1992; Dosso *et al.*, 1993; Hovem, 1993; Frisk, 1994].

The Biot–Stoll model (Biot, 1956a, b; Stoll, 1974, 1980, 1989) provides a comprehensive description of the acoustic response of linear, porous materials containing a compressible pore fluid. The model predicts two types of compressional waves and one shear wave. Recent applications in underwater acoustics and references to the key historical literature are provided by Beebe *et al.* (1982) and Holland and Brunson (1988). Routine operational employment of this model is hampered by the input of more than a dozen geophysical parameters, some of which are difficult to obtain even in laboratory environments.

McCammon (1988) described the development of a geoacoustic approach to bottom interaction called the thin layer model. This model, which is based on an inversion technique, contains a thin surificial layer, a fluid sediment layer and a reflecting subbottom half-space. There are ten input parameters: sediment density, thickness, sound speed gradient and curvature, attenuation and attenuation gradient; thin layer density and thickness; basement reflectivity; and water–sediment velocity ratio (Fig. 3.11). The model generates bottom loss curves as a function of grazing angle over the frequency range 50–1500 Hz. The model makes several assumptions: it relies upon the 'hidden depths' concept of Williams (1976); the sediments are isotropic; the roughness of the sediment and basement interfaces, as well as multiple scattering within the layers, is neglected; and shear wave propagation is ignored.

Sample outputs from this thin layer model are presented in Fig. 3.12. A ratio $(c_s/c_w) > 1$ (where c_s is the sound speed in the upper sediment and c_w is the sound speed at the base of the water column) predicts a critical angle $[\theta = \cos^{-1}(c_w/c_s)]$ below which most of the incident energy is reflected; that is, the bottom loss is nearly zero. By comparison, a ratio $(c_s/c_w) < 1$

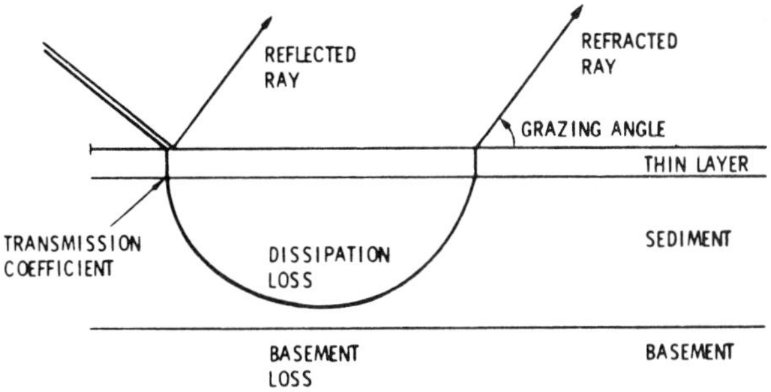

Fig. 3.11 Thin layer model for sediment reflected and refracted paths (McCammon, 1988; *J. Geophys. Res.*, **93**, 2363–9; published by the American Geophysical Union).

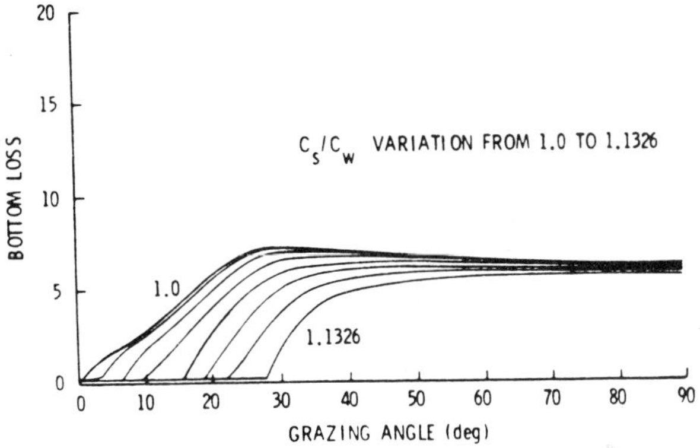

Fig. 3.12 Variation of bottom loss (dB) as a function of grazing angle for a frequency of 1000 Hz. Curves are presented for various values of the ratio of the upper sediment sound speed (c_s) to the sound speed at the base of the water column (c_w) (McCammon, 1988; *J. Geophys. Res.*, **93**, 2363–9; published by the American Geophysical Union).

would refract the energy into the sediments and result in greater losses at small angles.

A qualitative comparison of bottom loss versus grazing angle for $c_s/c_w \geq 1$ and for $c_s/c_w < 1$ is presented in Fig. 3.13. It has been demonstrated that low-porosity sediments (e.g. hard sands) have compressional wave speeds greater than that of the overlying water while high-porosity sediments (e.g. muds and silts) have sound speeds less than that of the overlying water (e.g. Urick, 1983, pp. 138–9; Apel, 1987, p. 386). Qualitatively, then, the comparison presented in Fig. 3.13 contrasts the effects of high-porosity and low-porosity sediments.

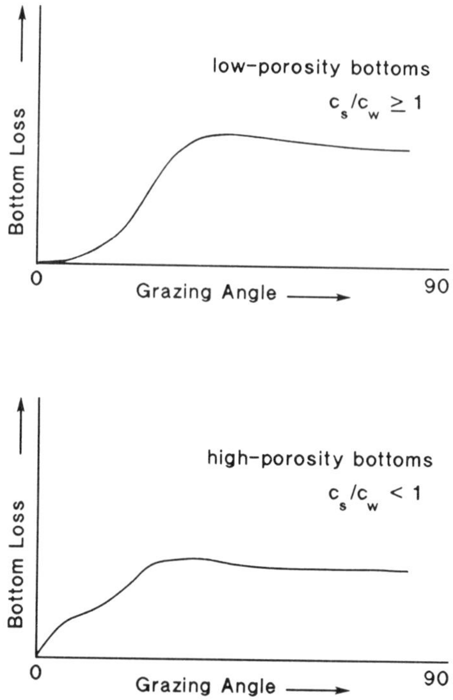

Fig. 3.13 Qualitative illustration of bottom loss versus grazing angle for low-porosity and high-porosity bottoms.

The Bottom Loss Upgrade (BLUG) model, which is a modular upgrade designed for incorporation into existing and future propagation models to treat bottom loss, is also based on a geoacoustic (or inverse) approach.

3.5.2 Interference and frequency effects

Stratification and attendant scattering within the bottom produce pulse distortion, as does reflection at grazing angles less than the critical angle.

Zabal *et al.* (1986) have developed a simple geometric–acoustic model to predict frequency and angle spreads as well as coherence losses to sonar systems. The sea floor is modeled by homogeneous and isotropic slope statistics. The facets are planar and reflect specularly, thus giving rise to the name 'broken mirror' model.

3.5.3 Attenuation by sediments

A variety of sediment attenuation units are commonly used in the underwater acoustic and marine seismology communities. The relationships among these units can become very confusing when attempting to enter values into propagation models.

Mathematically, acoustic attenuation (α) is expressed in the exponential form $e^{-\alpha x}$ using the units of nepers per unit distance for α. The acoustic attenuation can be converted to the units of dB per meter using the relationship

$$\alpha(\mathrm{dB/m}) = 20(\log \mathrm{e})\alpha(\mathrm{neper/m})$$
$$= 8.6859\alpha(\mathrm{neper/m})$$

Some propagation models require that the attenuation be specified in units of dB per wavelength (λ):

$$\alpha(\mathrm{dB}/\lambda) = \lambda(\mathrm{m})\alpha(\mathrm{dB/m})$$

Another term commonly encountered in acoustic modeling is the attenuation coefficient (k) which is based on the concept that attenuation and frequency are related by a power law:

$$\alpha(\mathrm{dB/m}) = kf(\mathrm{kHz})^n$$

Typically, n is assumed to be unity and $k = \alpha(\mathrm{dB/m*kHz})$. Over the frequency range of interest to underwater acousticians, attenuation is approximately linearly proportional to frequency.

A compilation of sediment attenuation measurements made by Hamilton (1980) over a wide frequency range showed that the attenuation in natural, saturated sediments is approximately equal to $0.25f$ dB per meter, when f is in kHz. There was a tendency for the more dense sediments (such as compact sands) to have a higher attenuation than the less dense, high-porosity sediments (such as muds). The attenuation in sediments is several orders of magnitude higher than in pure water.

In the Arctic, acoustical parameters of the sea floor and subbottom are poorly known. Difficulties in obtaining direct core samples to great depths limit the database from which to extract the parameters needed to determine many of the major acoustical processes in bottom interaction. Estimating these geoacoustical parameters based on data from contiguous areas may not be meaningful because the basic processes of sedimentation at work under the pack ice are unique to that environment. Sedimentation rates are very low, being dominated by material carried by the ice rather than by material of biologic origin, as is the case in more temperate areas. The ice pack may also carry large boulders of glacial origin and deposit them in the Arctic Ocean. The low sedimentation rate leaves them exposed as potential scatterers for acoustic energy over a wide range of frequencies.

3.5.4 Measurements

The standard method for measuring bottom loss is to use pings or explosive pulses and to compare the amplitude, intensity or energy density (integrated intensity) of the bottom pulse with that of the observed or computed pulse traveling via a direct path.

Bottom loss data typically show a loss increasing with angle at low angles, followed by a nearly constant loss extending over a wide range of higher angles (refer back to Fig. 3.13). High-porosity bottoms (having a sound speed less than that of water) tend to have a maximum loss at an angle of 10–20° where an angle of intromission (i.e. no reflection, but complete transmission into the bottom) would be expected to occur in the absence of attenuation in the bottom. When narrow-band pulses are used, measured losses often are irregular and variable, showing peaks and troughs due to the interference effects of layering in the bottom. Measured data rarely show a sharp critical angle (as would be inferred from the Rayleigh reflection model) because of the existence of attenuation in the bottom (Urick, 1983, Chapter 5).

3.6 ATTENUATION AND ABSORPTION

Sound losses in the ocean can be categorized according to spreading loss and attenuation loss. Spreading loss includes spherical and cylindrical spreading losses in addition to focusing effects. Attenuation loss includes losses due to absorption, leakage out of ducts, scattering and diffraction. Urick (1982, Chapter 5) has summarized the relevant literature on this subject.

Absorption describes those effects in the ocean in which a portion of the sound intensity is lost through conversion to heat. Measurements of the absorption loss, typically expressed in dB per kilometer using the symbol α, span the frequency range 20 Hz–60 kHz. The dependence of α on frequency is complicated, reflecting the effects of different processes or mechanisms operating over different frequency ranges. The equation developed by Thorp (1967) is probably the best known, and is valid at frequencies below 50 kHz:

$$\alpha = 1.094\left[\frac{0.1f^2}{1+f^2} + \frac{40f^2}{4100+f^2}\right] \tag{3.8}$$

where α is the attenuation coefficient (dB/km) and f the frequency (kHz). The factor 1.094 converts the original formula from units of dB/kyd to dB/km.

In practice, the effects of attenuation and absorption are considered jointly. Then, the frequency dependence can conveniently be segregated into four distinct frequency regions over which the controlling mechanisms can be readily identified. These regions are (in order of ascending frequency): (1) large-scale scattering or leakage; (2) boric acid relaxaton; (3) magnesium sulfate relaxation; and (4) viscosity. Fisher and Simmons

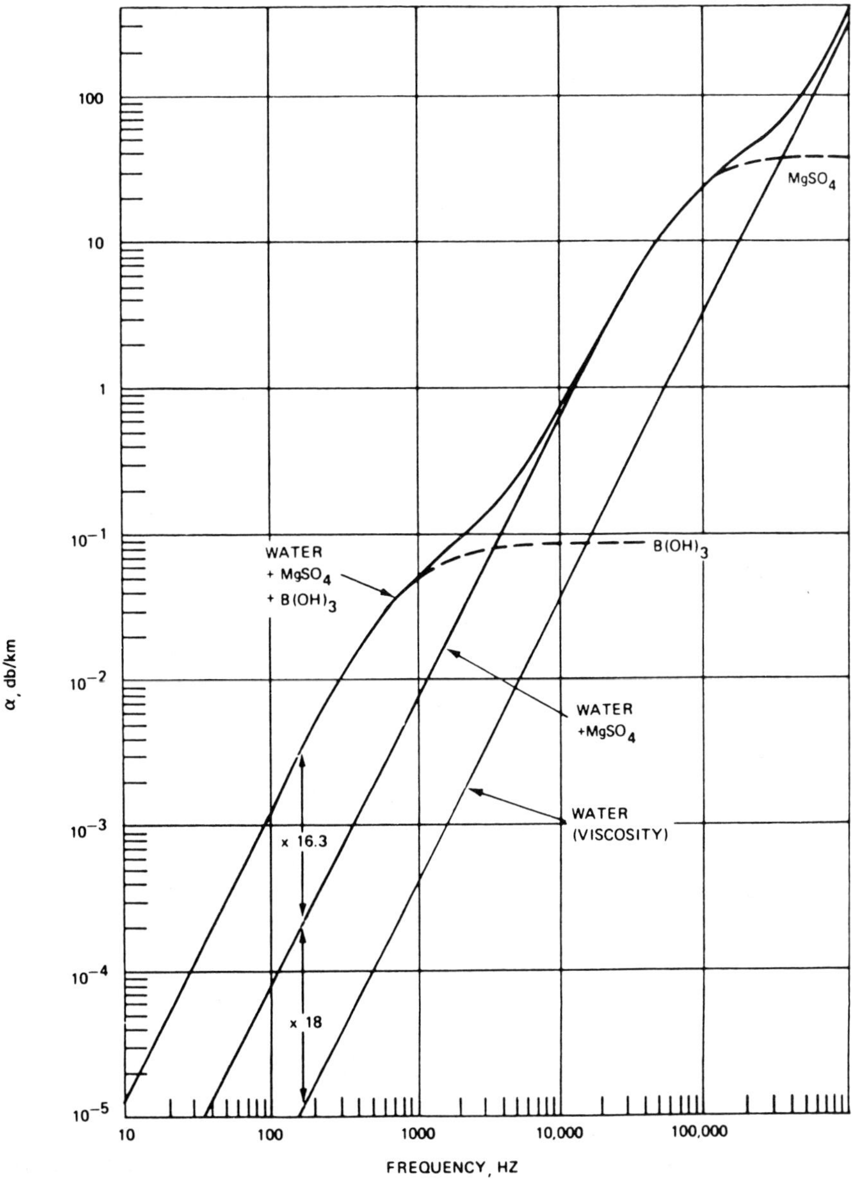

Fig. 3.14 Absorption coefficients for sea water at a temperature of 4°C at the surface (Fisher and Simmons, 1977). Dashed lines indicate contributing absorption rates due to relaxation processes.

(1977) have summarized these effects graphically (Fig. 3.14). Additional research has been conducted by other investigators which provides regional formulas for absorption (e.g. Mellen *et al.*, 1987a, 1987b, 1987c). Absorption is regionally dependent due mainly to the pH-dependence of the boric acid relaxation.

3.7 SURFACE DUCTS

Sound travels to long distances in the ocean by various forms of ducted propagation. When sound travels in a duct, or sound channel, it is prevented from spreading in depth and remains confined between the boundaries of the duct.

The surface duct is a zone bounded above by the sea surface and below by the Sonic Layer Depth (SLD) in which sound rays are alternatively refracted and reflected. A surface duct exists when the negative temperature gradient within it does not exceed a value determined by the effect of pressure on sound speed (refer to Chapter 2). In isothermal water (and ignoring the effects of salinity), the pressure effect will produce a positive sound speed gradient of 0.017/s.

The surface duct is the acoustical equivalent of the oceanographic mixed layer, although they are defined differently. Specifically, the Sonic Layer Depth (SLD) is normally defined only in terms of the sound speed gradient, whereas the Mixed Layer Depth (MLD) is defined in terms of temperature or, more precisely, in terms of density (which is a function of temperature, salinity and pressure). The mixed layer is a quasi-isothermal layer of water created by wind–wave action and thermohaline convection. The depth of the mixed layer is an important variable used in the prediction of surface duct propagaton.

3.7.1 Mixed layer distribution

Oceanographers have extensively studied the dynamics of the mixed layer. Variations in the temperature and depth of the mixed layer are closely related to the exchange of heat and mass across the air–sea interface and are thus of interest to scientists engaged in studies of the global climate.

Lamb (1984) presented bimonthly charts of the mean mixed layer depths for the North and tropical Atlantic Oceans; Bathen (1972) presented monthly charts of mixed layer depths for the North Pacific Ocean.

Levitus (1982) presented charts of mixed layer depths on a global basis. Distributions of mixed layed depths were calculated using both a temperature criterion and a density criterion. The temperature criterion was based on a temperature difference of 0.5 °C between the surface and the depth referred to as the mixed layer depth. The density criterion was based on a difference of 0.125×10^{-3} g/cm^3 between the surface and the depth of the mixed layer. The use of the density criterion recognizes the importance of salinity in determining the stability of the mixed layer and hence, from an acoustics viewpoint, the true depth of the sonic layer. For example, in subarctic regions, isothermal conditions, or even temperature profiles with inversions, combine with a salinity profile that stabilizes the water column to control the depth of mixing. Mixed layer depths for the months of March and September are presented in Figs 3.15 and 3.16, respectively, comparing

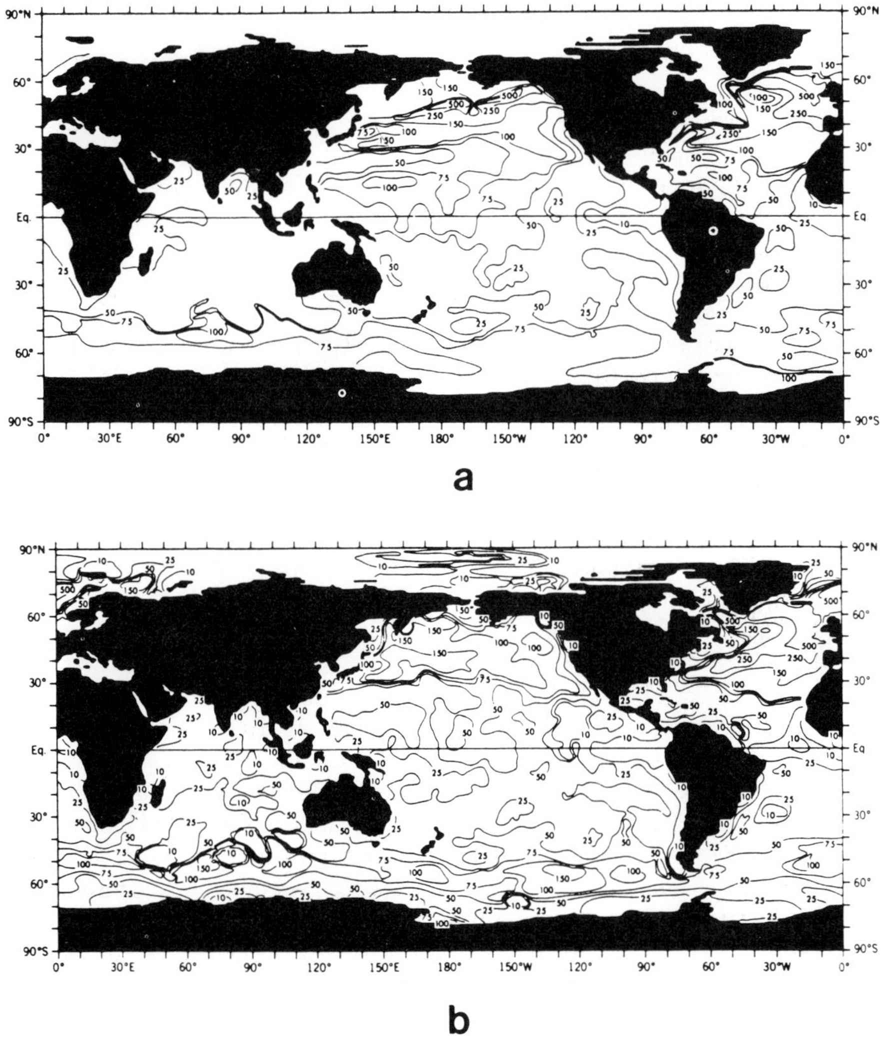

Fig. 3.15 Mixed layer depths (in meters) for March based on (a) a temperature criterion of 0.5°C and (b) a density criterion of 0.125×10^{-3} g/cm^3 (Levitus, 1982).

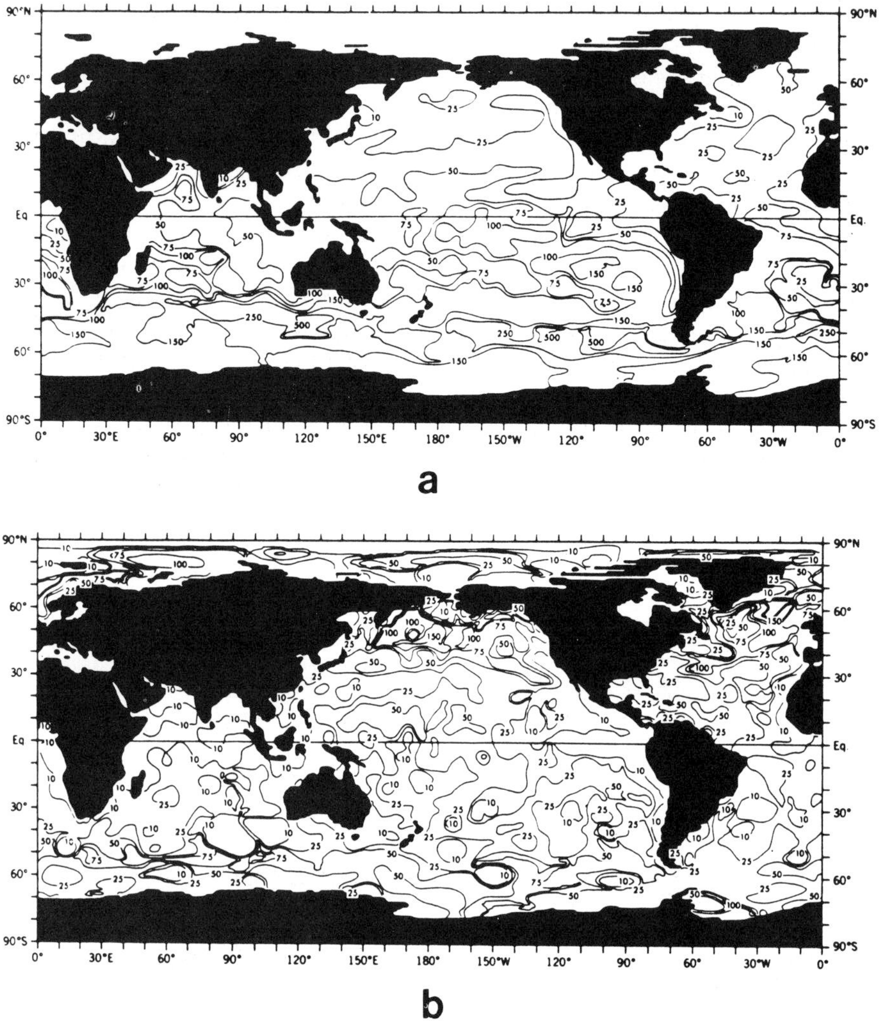

Fig. 3.16 Mixed layer depths (in meters) for September based on (a) a temperature criterion of 0.5°C and (b) a density criterion of 0.125×10^{-3} g/cm^3 (Levitus, 1982).

the global distributions resulting from both the temperature and density criteria.

Traditionally, it has been assumed that the mixed layer is the vertical extent of the turbulent boundary layer. Peters and Gregg (1987) have pointed out that the turbulence initiated by the exchanges of energy, buoyancy or momentum across the air–sea interface may actually penetrate the lower boundary of the mixed layer as defined by the previous criteria. Thus, precise discussions of the mixed layer are often frustrated by imprecise terminology.

Some regional examples will serve to clarify features of the mixed layer as evidenced by the thermal structure of the water column. For these examples, the continental shelf region off the Texas–Louisiana coast in the Gulf of Mexico (Etter and Cochrane, 1975) will be explored. Figure 3.17 presents the annual variation of water temperature down to 225 m. In this graphical representation, the seasonal variation of the depth of the mixed layer is portrayed as a uniform layer of temperature varsus depth. In January, for example, the mixed layer is approximately 100 m deep, while in August the mixed layer depth is seen to shallow to about 25 m. These variations are further illustrated in Figs 3.18 and 3.19, which are vertical sections across the shelf. In winter (February), for example, the isotherms tend to be vertically oriented, evidencing strong mixing and vertical (versus horizontal) stratification. In summer (July), however, the isotherms tend to be more horizontal, a result of weak mixing with attendant horizontal stratification.

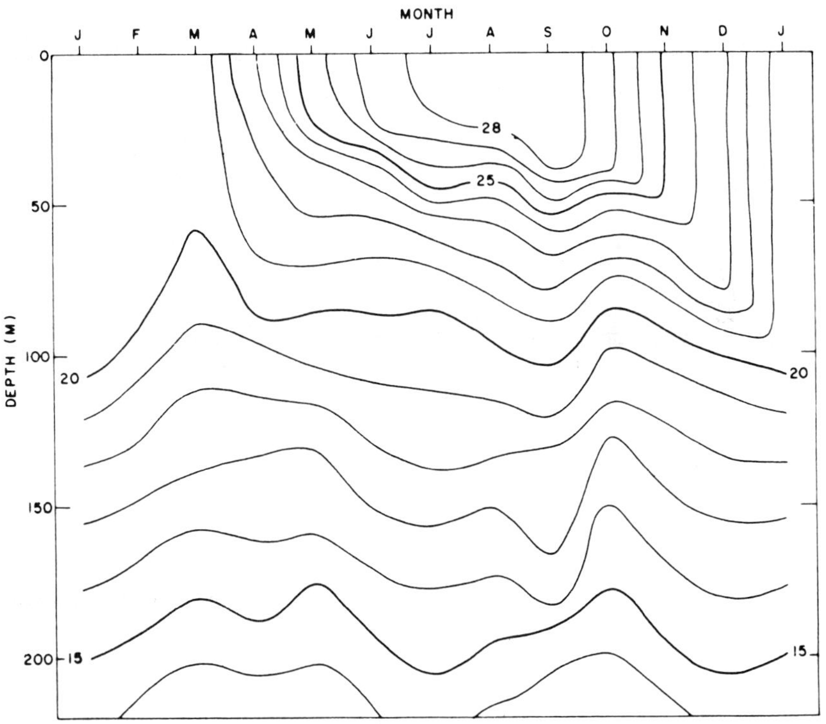

Fig. 3.17 Temperature contours (°C) for a region of the Texas–Louisiana continental shelf in the Gulf of Mexico (Etter and Cochrane, 1975).

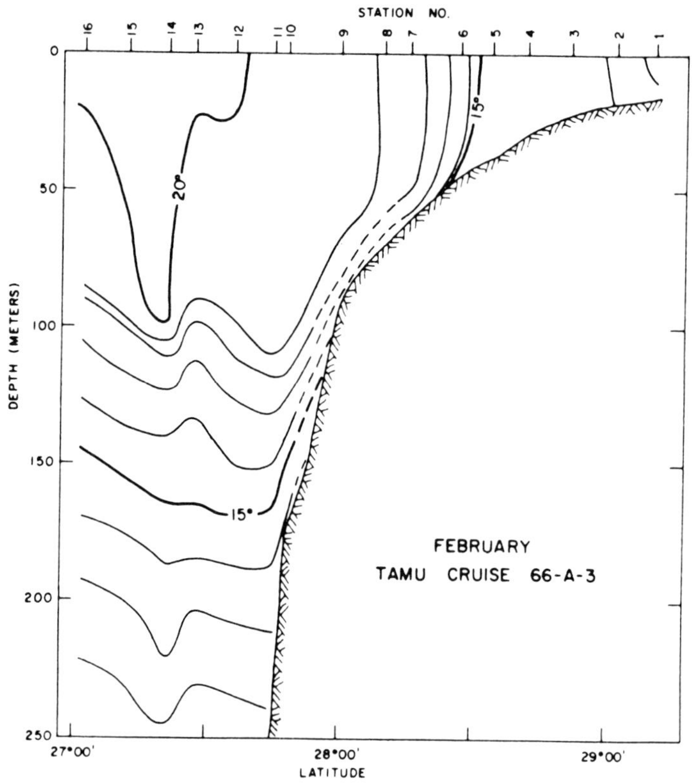

Fig. 3.18 Vertical section of water temperatures (°C) over the Texas–Louisiana continental shelf in the Gulf of Mexico in February (Etter and Cochrane, 1975).

Algorithms for the prediction of surface duct propagation will be discussed in Chapter 5. These algorithms use the depth of the mixed layer as an input variable.

3.7.2 General propagation features

A computer-generated ray diagram for a source in a typical mixed layer is shown in Fig. 3.20. Under the conditions for which the diagram was drawn, the ray leaving the source at an angle of 1.76° becomes horizontal at the base of the layer. Rays leaving the source at smaller angles remain in the layer, and rays leaving the source at greater angles are refracted downward to greater depths. A shadow zone is produced beneath the layer at ranges beyond the immediate sound field. The shadow zone is not completely devoid of acoustic energy but is insonified by diffraction and by sound scattered from the sea surface. The rate of sound leakage out of

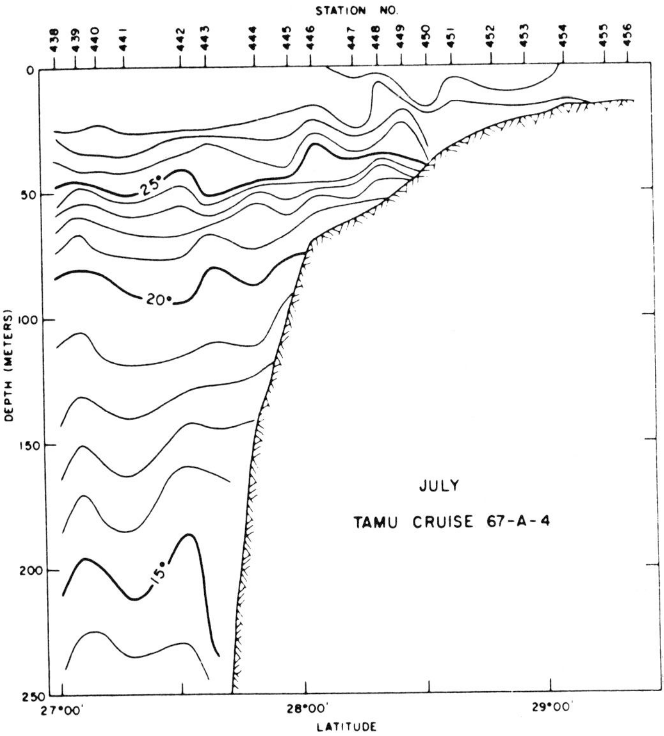

Fig. 3.19 Vertical section of water temperatures (°C) over the Texas–Louisiana continental shelf in the Gulf of Mexico in July (Etter and Cochrane, 1975).

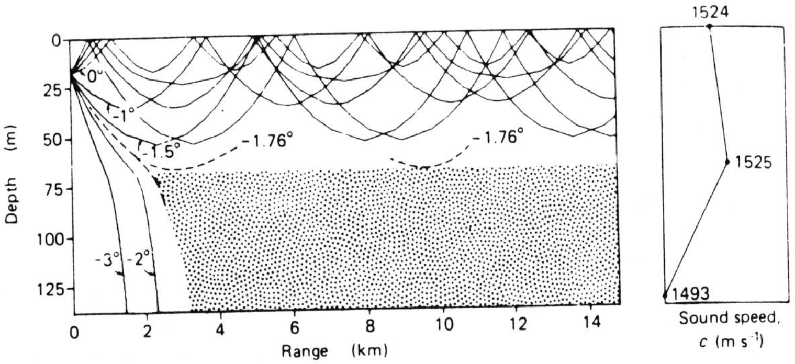

Fig. 3.20 Ray diagram for a sound source located at a depth of 16 m, showing propagation and trapping of energy in the surface duct, with a shadow zone below 60 m, for the sound speed profile shown at the right (Urick, 1983; *Principles of Underwater Sound*, 3ed edn; reproduced with permission of McGraw-Hill Publishing Company).

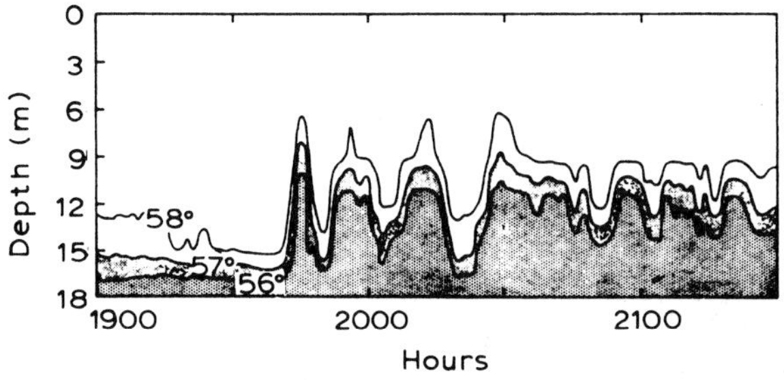

Fig. 3.21 Fluctuations in the depth of the mixed layer caused by the passage of internal waves. Water temperature is in °F; the mixed layer depth is defined by the 58°F isotherm (LaFond, 1962; *The Sea*, Vol. 1, pp. 731–51; reprinted by permission of John Wiley and Sons, Inc., all rights reserved).

the channel is given by an empirical 'leakage coefficient' (α_L) that expresses the attenuation (in dB/km) of sound trapped in the channel. The leakage coefficient varies with surface roughness (wave height), duct thickness, sound speed gradient below the layer and acoustic frequency (Urick, 1983, pp. 153–4).

Over short distances, or even at a fixed location over time, the thickness of the mixed layer may vary because of internal waves. These waves propagate along the density discontinuity at the base of the mixed layer (Fig. 3.21). Internal waves complicate the propagation of underwater sound by causing variations in transmission loss over short distances. Fluctuations of echoes between a source and receiver separated by the mixed (or sonic) layer depth are attributable in part to the existence of internal waves.

Urick (1982, Chapter 6) has reviewed the available open-literature algorithms for estimating transmission loss in the surface duct.

An analysis of the extensive data set collected during the AMOS program (Marsh and Schulkin, 1955) has been used to characterize transmission loss in the surface duct in the frequency range 2–8 kHz. Condron *et al.* (1955) have graphically summarized these results (Figs 3.22 and 3.23). These figures demonstrate the importance of source–receiver geometry in surface duct propagation, especially in those cases where the source and receiver are situated on opposite sides of the layer depth (referred to as cross-layer geometries).

3.7.3 Low-frequency cutoff

At very low frequencies, sound ceases to be trapped in the surface duct. The maximum wavelength for duct transmission may be found from the theory of radio propagation in ground-based radio ducts to

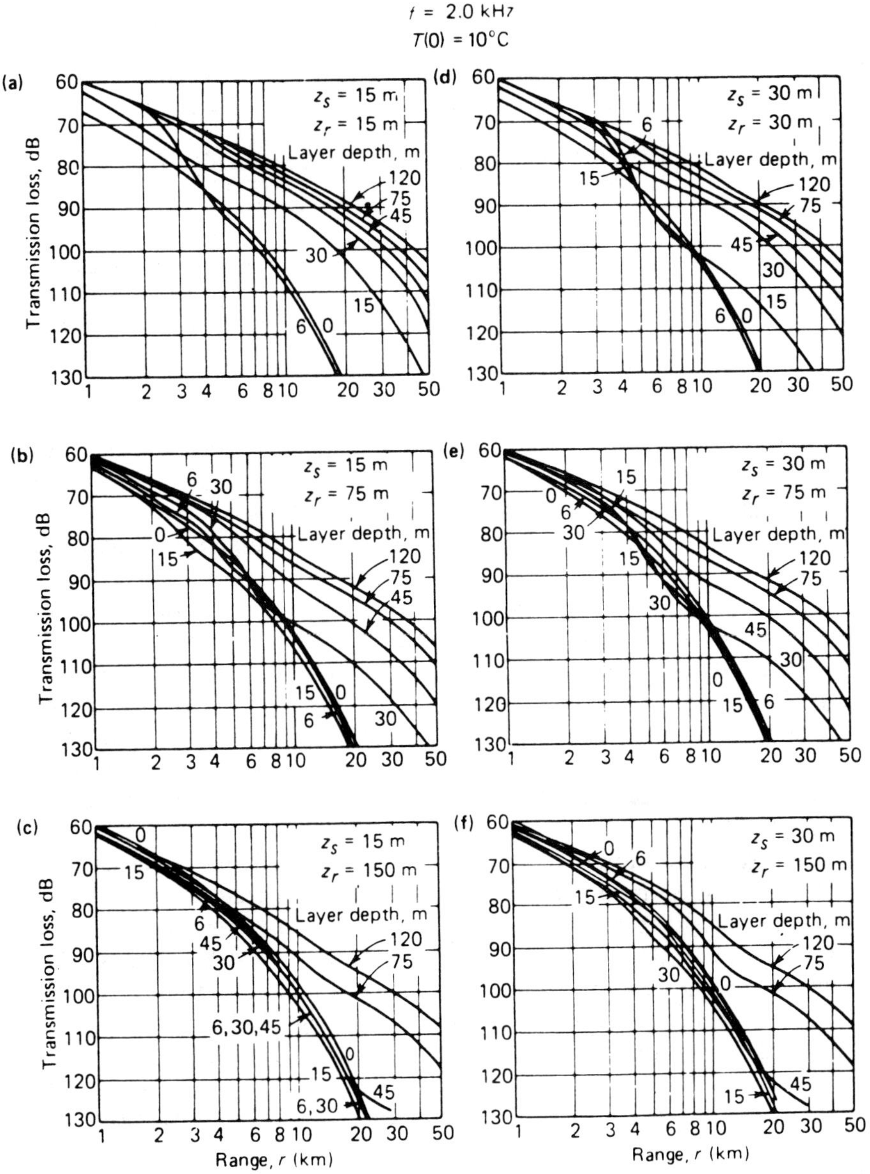

Fig. 3.22 Surface duct transmission loss estimates generated from the AMOS empirical relationships at a frequency of 2 kHz for various combinations of mixed layer depths, source depths (z_s) and received depths (z_r). (Adapted from Condron *et al.*, 1955, by Apel, 1987.)

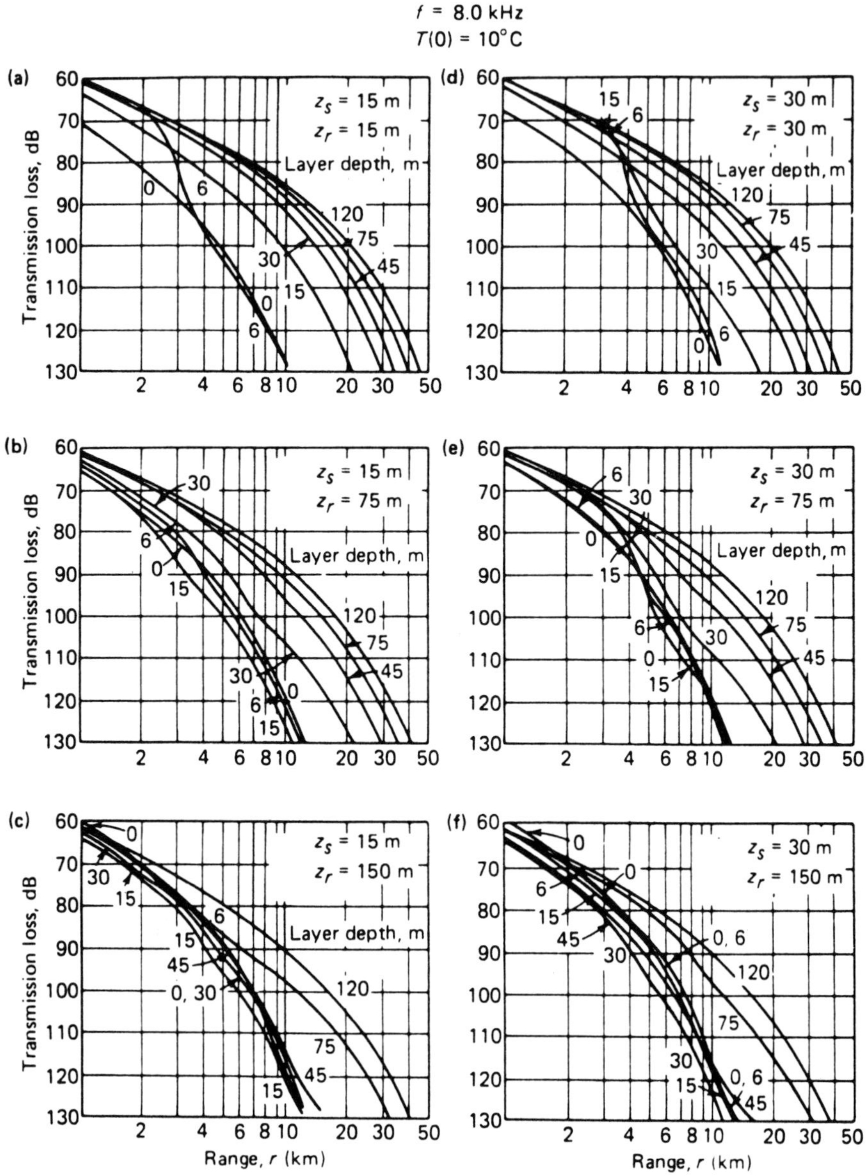

Fig. 3.23 As for Fig. 3.22, except at a frequency of 8 kHz. (Adapted from Condron *et al.*, 1955, by Apel, 1987).

be (Kerr, 1951, p. 20)

$$\lambda_{max} = (8/3)\sqrt{2}\int_0^H \sqrt{n(z) - n(H)}\,dz \tag{3.9}$$

where $n(z)$ is the index of refraction at any depth z in the duct and $n(H)$ is the index of refraction at the base of the duct. Using values of sound speed and sound speed gradient appropriate for sound propagation in the mixed layer, we obtain

$$\lambda_{max} = 8.51 \times 10^{-3} H^{3/2} \tag{3.10}$$

for the maximum wavelength, in meters, trapped in a mixed layer duct H m thick. For example, for a mixed layer 30 m thick, the maximum trapped wavelength λ_{max} is 1.4 m, corresponding to a frequency of approximately 1070 Hz. Although this does not represent a sharp cutoff, wavelengths much longer than this are strongly attenuated. Wavelengths much shorter are attenuated by absorption and leakage. Thus, for a mixed layer of a given thickness, it follows that there is an optimum frequency for propagation at which the leakage of sound out of the layer is a minimum (Fig. 3.24).

3.8 DEEP SOUND CHANNEL

The deep sound channel, sometimes referred to as the SOFAR (Sound Fixing and Ranging) channel, is a consequence of the sound speed profile characteristic of the deep ocean (see Chapter 2). This profile has a minimum at a depth which varies from about 1000 m at mid-latitudes to near the surface in polar regions. This sound speed minimum causes the ocean to act like a lens: above and below the minimum, the sound speed gradient continually refracts the sound rays back toward the depth of minimum sound speed. This depth is termed the axis of the sound channel. A portion of the acoustic energy originating in the deep sound channel thus remains within the channel and encounters no losses by reflection from the sea surface or sea floor.

The properties of the deep sound channel were first investigated by Ewing and Worzel (1948). The exceptional ducting characteristics of this channel have been used to advantage by oceanographers in the design and conduct of acoustic tomography experiments (see section 5.6.9).

In terms of the sound speed profile, the upper and lower limits of the channel are defined by the two depths of equal maximum sound speed in the profile between which a minimum exists. In Fig. 3.25, these limits of the deep sound channel are the depths A and A'; the depth A' is referred to as the critical (or conjugate) depth. Different ray paths from a source in the channel exist depending on whether or not the channel extends to the sea surface or sea floor. In Fig. 3.25(a), the sound speeds at the surface and sea

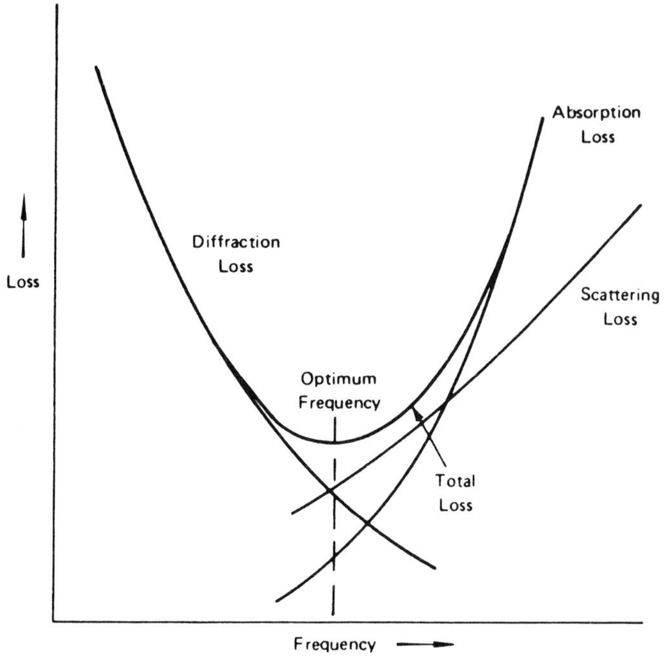

Fig. 3.24 Loss to a fixed range in a surface duct. The optimum frequency is that at which the total loss is a minimum (Urick, 1979).

floor are the same. All depths in the water column then lie within the channel, and sound is propagated via paths that are either refracted (path 1) or reflected (with consequent losses) at the surface and floor of the sea (path 2). In Fig. 3.25(b), the upper limit of the deep sound channel lies at the sea surface (which may happen at high latitudes, for example). Here, in addition to paths 1 and 2, refracted-surface-reflected (RSR) paths occur (path 3) involving losses at levels intermediate to those suffered by paths 1 and 2. In Fig. 3.25(c), the channel is cut off by the sea floor, and refracted-bottom-reflected (RBR) paths exist (path 4). The entirely refracted paths and the low transmission losses associated with these paths do not exist when the source or the receiver is outside the depth limits A and A' of the channel.

3.9 CONVERGENCE ZONES

Urick (1983, pp. 163–8) described the formation of convergence zones in the ocean (Fig. 3.25). Specifically, there must exist a refracted ray which leaves the source horizontally. If this ray is reflected at either the sea surface or the sea floor (ray types 2 or 4 in Fig. 3.25), then the required caustic

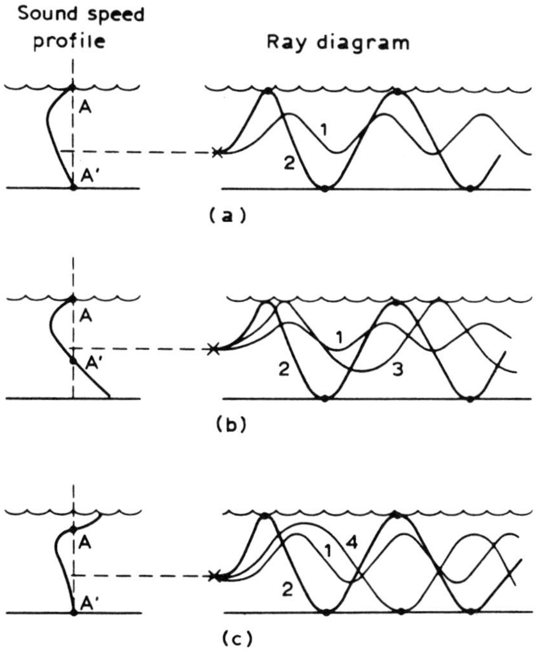

Fig. 3.25 Ray paths for a source in the deep sound channel. In (a) the channel extends between the sea surface and the sea floor; it is cut off by the sea surface in (b) and by the sea floor in (c). The depth A' is referred to as the critical depth (Urick, 1983; *Principles of Underwater Sound*, 3rd edn; reproduced with permission of McGraw-Hill Publishing Company).

pattern is destroyed and no convergence is possible. Another requirement for convergence is that the water depth must be greater than the critical depth (depth A' in Fig. 3.25) in order to allow the rays traveling downward to refract without striking the bottom and to converge (ray type 3). There must also be a depth excess (i.e. a vertical distance between the critical depth and the bottom) on the order of a few hundred meters. For water depths less than critical (Figs 3.25(a) and 3.25(c)), the rays that would converge if the water were deeper are cut off by the bottom and become bottom-reflected without convergence.

Because the deep waters of the ocean are of a fairly uniform low temperature (near 1 °C), the speed of sound at great depths at any particular region of the ocean is largely a function of depth (or pressure) only. Near the surface, however, the speed of sound is determined largely by the water temperature. Thus, water temperature near the sea surface and the water depth in a particular area will largely determine whether sufficient depth excess exists and therefore whether or not a convergence zone will occur. Charts of surface temperature and water depth can then be used as basic prediction tools for ascertaining the existence of convergence zones. A convergence zone range slide rule (TACAID 6-10) was developed in 1973 by the Naval Underwater Systems Center based on an analysis of oceanographic data performed by E.M. Podeszwa. This slide rule could be used in the North Atlantic and North Pacific Oceans, and the Mediterranean, Norwegian and Caribbean Seas.

In the North Atlantic Ocean convergence zones are seen to appear at intervals of approximately 35 nm (65 km), with widths of about 2 nm (4 km). Successive convergence zones get wider with increasing range until, at a range beyond a few hundred nautical miles, they coalesce. The results is a smooth variation of loss with range, with cylindrical spreading plus attenuation determining the transmission loss at long ranges.

3.10 RELIABLE ACOUSTIC PATH

When a source is located at the critical depth (depth A' in Fig. 3.25), and provided sufficient depth excess exists, propagation to moderate ranges can take place via the so-called reliable acoustic path (or RAP), as illustrated in Fig. 3.5(F). Such paths are termed 'reliable' because they are insensitive to near-surface effects or bottom interaction.

3.11 SHALLOW-WATER DUCTS

Two definitions of shallow water may be made: hypsometric and acoustic. The hypsometric definition is based on the fact that most continents have continental shelves bordered by the 200 m contour line, beyond which the bottom generally falls off rapidly into deep water. Therefore, shallow water is taken to mean continental shelf waters shallower than 200 m; shallow water represents about 7.5% of the total ocean area.

Acoustically, shallow-water conditions exist whenever the propagation is characterized by numerous encounters with both the sea surface and the sea floor. By this definition, some hypsometrically shallow-water areas are acoustically deep; alternatively, the deep ocean may be considered shallow when low-frequency, long-range propagation conditions are achieved through repeated interactions with both the surface and bottom. Shallow-water regions are distinguished from deep-water regions by the relatively greater role played in shallow water by the reflecting and scattering boundaries. Also, differences from one shallow-water region to another are primarily driven by differences in the structure and composition of the sea floor. The sea floor is perhaps the most important part of the marine environment that distinguishes shallow-water propagation from deep-water propagation.

The most common shallow-water bottom sediments are sand, silt and mud (see Chapter 2), with compressional sound speeds greater than that of the overlying water. Sediments are also characterized by shear waves that are not present in the water column. Acoustic energy that strikes the sea floor at sufficiently small grazing angles is nearly totally reflected back into the water. This results in a slightly lossy duct with transmission loss approximately equal to that associated with cylindrical spreading in the frequency range 100–1500 Hz. At low frequencies, the acoustic field can extend into the bottom with sound being returned to the water by subbottom reflection or subbottom upward refracted paths (Eller, 1984b).

The tendency toward cylindrical spreading is illustrated in Fig. 3.26, which shows one-third-octave-band transmission loss data at a center frequency of 200 Hz as a function of range. An omnidirectional hydrophone was located at a depth of 91 m in water approximately 210 m deep. The sources were set at a depth of 91 m on a track along which the water depth gradually increased from about 220 m to 300 m. The sound speed profile was nearly constant, and the bottom sediments were reported to be silty-sand near the beginning range and sand-silt-clay at greater ranges. Figure 3.26 also presents reference curves depicting spherical and cylindri-

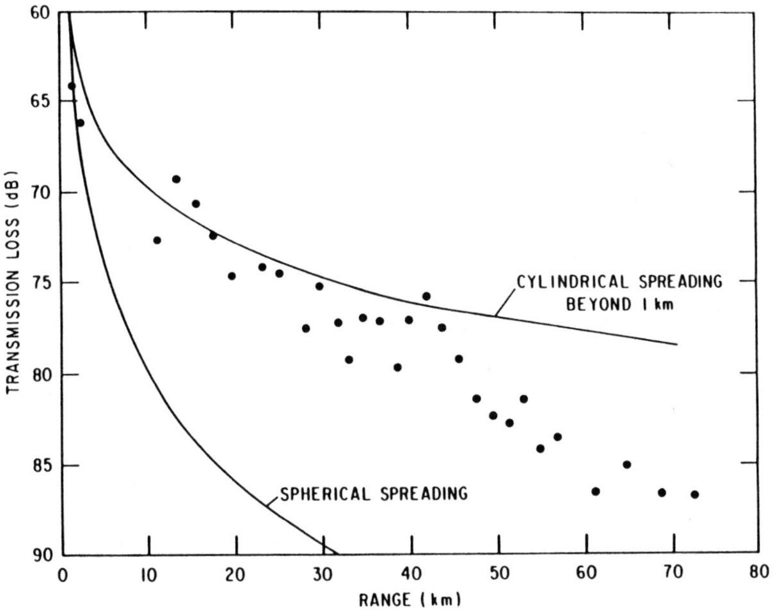

Fig. 3.26 Example of transmission loss data in shallow water illustrating the cylindrical spreading associated with energy trapping by the waveguide. Data are for the one-third octave band centered at 200 Hz (Eller, 1984b).

cal spreading (beyond 1 km). The acoustic energy is seen to be effectively trapped in the shallow-water duct at ranges less than about 40 km, beyond which the data drop below the reference curve for cylindrical spreading.

3.12 ARCTIC HALF-CHANNEL

The acoustic waveguide in the Arctic is determined by the geometry of the ocean (type of ice cover and water depth) and by a positive gradient sound speed profile. This waveguide forms a half-channel, that is, the lower half of the deep sound channel, with the axis of the sound channel now located at the surface.

The continuously upward-refracting propagation conditions which generally prevail in Arctic waters (as demonstrated by the ray trace in Fig. 3.27) cause repeated interactions with the ice canopy and tend to create a low-pass filter, favoring the propagation of low-frequency (<30 Hz) signals. The situation is further complicated when the geometry of the sea floor and the acoustic wavelength combine to yield shallow water conditions with the attendant increase in bottom interaction opportunities.

Measurements of transmission loss in the Arctic are limited since access to this region is generally restricted to the spring season when conditions are favorable for manned ice camps. Consequently, there is a relatively poor understanding of the seasonal variability.

One of the principal characteristics of acoustic propagation measurements under an ice cover is a rapid increase in transmission loss with range at frequencies above about 30 Hz. The loss mechanism has been attributed

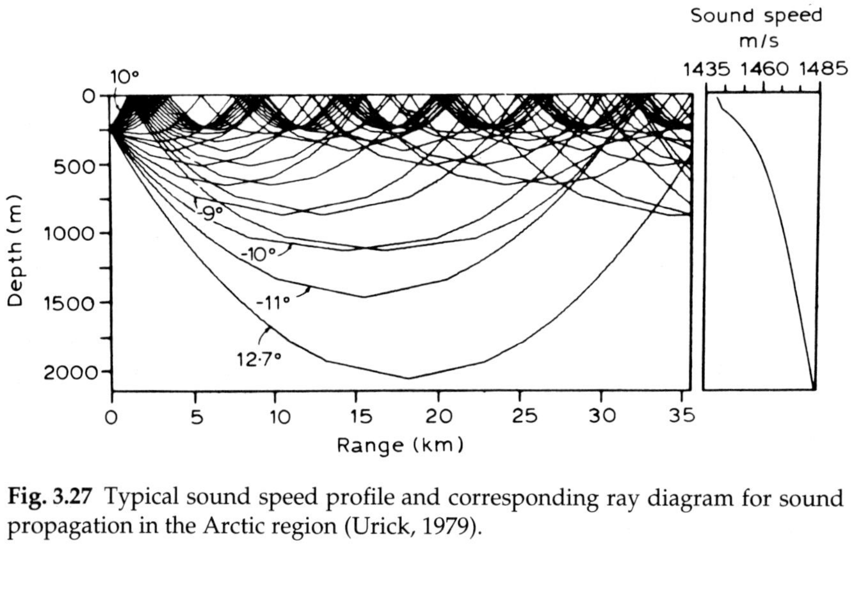

Fig. 3.27 Typical sound speed profile and corresponding ray diagram for sound propagation in the Arctic region (Urick, 1979).

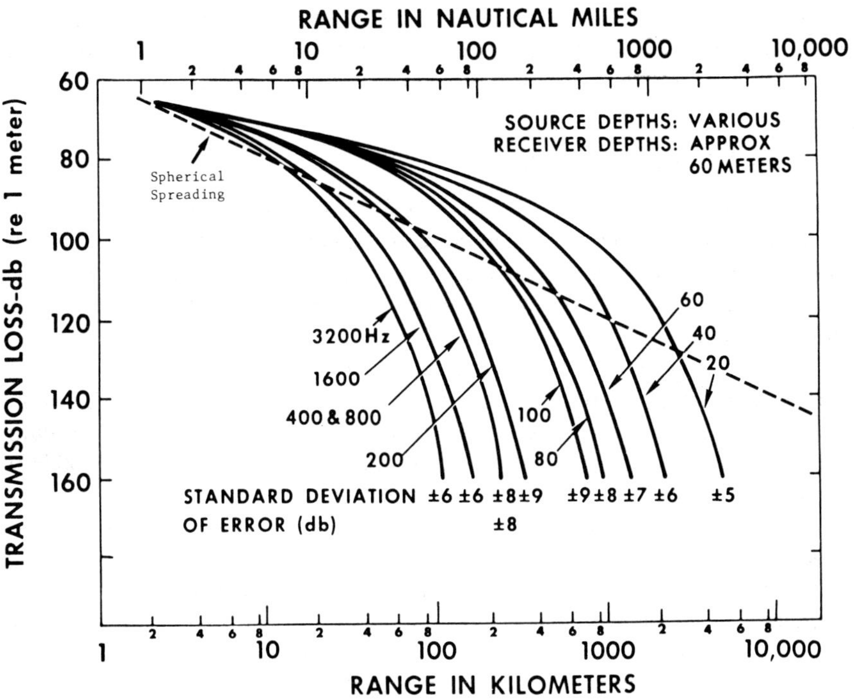

Fig. 3.28 Average curves of Arctic transmission loss versus range based on measured data (Buck, 1968).

primarily to the scattering at the ice–water interface. Other possible mechanisms include dissipative processes in the ice canopy, conversion of waterborne energy into energy traveling in and confined to the ice canopy, and increased absorption in the water column.

Buck (1968) has summarized available Arctic transmission loss data (Fig. 3.28). These curves give the average measured transmission loss in the Arctic at a number of frequencies; the standard deviation of the various data points from the smooth curve is stated for each frequency. The dashed line shows spherical spreading ($TL = 20 \log r$). It is evident that sound propagation in the Arctic degrades rapidly with increasing frequency, particularly above about 30 Hz. At short-to-moderate ranges, ducting improves propagation relative to spherical spreading; at long ranges, the repeated encounters with the under-ice surface degrade propagation.

3.13 COHERENCE

Coherence can be defined as a measure of the phase and amplitude relationships between sets of acoustic waves. In the ocean, which is characterized by both temporal and spatial variations, the effects of the medium on small-amplitude wave propagation can be described in terms of coherence time, coherence bandwidth, spatial coherence and angular coherence. As described by Ziomek (1985, Chapter 7), for example, this information can be obtained from the generalized coherence (or autocorrelation) function.

Temporal coherence, or fluctuations, refers to changes in a received signal (relative to a steady signal) over of time. Spatial coherence refers to the changes in the signals received at different locations in the ocean at a given time.

Signal fluctuations can be caused by a number of physical processes, including:

1. source or receiver motion
2. oceanic fine-scale features
3. sea surface motion (waves and tides)
4. scatterers
5. internal waves and tides
6. ocean currents and eddies

Urick (1982, Chapters 12 and 13) discusses additional aspects of temporal and spatial coherence.

4

Propagation II. Mathematical models (Part One)

4.1 BACKGROUND

Chapter 1 described a framework within which all underwater acoustic models could be categorized. It was shown that propagation models formed the foundation for the category of models classified as basic acoustic models which, in turn, supported the more spacialized category of sonar performance models. Propagation models are the most common (and numerous) type of underwater acoustic models in use. Their application is fundamental to the solution of all types of sonar performance problems.

Chapter 3 described the observations and physical models that are available to support the mathematical modeling of sound in the sea. Conceptually, it was convenient to separate propagation phenomena into the categories of boundary interactions, volumetric effects and propagation paths. A similar approach will be adopted here in the description of the mathematical models.

The mathematical models will first be distinguished on the basis of their theoretical treatment of volumetric propagation. Then, as appropriate, further distinctions will be made according to specification of boundary conditions and treatment of secondary volumetric effects such as absorption and diffraction. These latter effects are generally accommodated by using many of the physical models described in Chapter 3. Special propagation paths such as surface ducts, shallow water and Arctic half-channels will be discussed in Chapter 5.

The various physical and mathematical models all have inherent limitations in their applicability. These limitations are usually manifested as restrictions in the frequency band or in specification of the problem geometry. Such limitations are collectively referred to as 'domains of applicability', and vary from model to model. Most problems encountered in model usage involve some violation of these domains. In other words, the models are misapplied in practice. Therefore, considerable emphasis is placed on these restrictions and on the assumptions which ultimately give rise to them. Finally, model selection criteria are provided to guide potential users to those models most appropriate to their needs; comprehensive summaries identify the available models and associated documentation.

The emphasis in this chapter, as throughout the book, is placed on sonar (versus seismic) applications. Reviews of mathematical models of seismo-acoustic propagation in the ocean have been provided by Tango (1988), who placed particular emphasis on the very low frequency (VLF) band (5–35 Hz), and by Schmidt (1991).

4.2 THEORETICAL BASIS FOR PROPAGATION MODELING

The theoretical basis underlying all mathematical models of acoustic propagation is the wave equation.

The earliest attempts at modeling sound propagation in the sea were motivated by practical problems in predicting sonar performance in support of anti-submarine warfare (ASW) operations during World War II. These early models used ray-tracing techniques derived from the wave equation to map those rays defining the major propagation paths supported by the prevailing environment. These paths could then be used to predict the corresponding sonar detection zones. This approach was a forerunner to the family of techniques now referred to as ray-theoretical solutions.

An alternative approach, referred to as wave-theoretical solutions, was first reported by Pekeris (1948), who used the normal mode solution of the wave equation to explain the propagation of explosively generated sound in shallow water.

As modeling technology has matured over the intervening decades, the attendant sophistication has complicated the simple categorization of ray versus wave models. The terminology is still useful in distinguishing those models based principally on ray-tracing techniques from those using some form of numerical integration of the wave equation. Occasionally, a mixture of these two approaches is used to capitalize on the strengths and merits of each and to minimize weaknesses. Such combined techniques are referred to as hybrid approaches. Related developments in propagation modeling have been reviewed by Harrison (1989), McCammon (1991), Buckingham (1992), Porter (1993), and Dozier and Cavanagh (1993).

4.2.1 Wave equation

The wave equation is derived from the more fundamental equations of state, continuity and motion. Rigorous derivations have been carried out in numerous basic texts in physics. A particularly lucid derivation has been presented by Kinsler *et al.* (1982, Chapter 5). Accordingly, the derivation will not be repeated here. Rather, the mathematical developments described in this book will build directly upon the wave equation.

Formulations of acoustic propagation models generally begin with the three-dimensional, time-dependent wave equation. Depending upon the governing assumptions and intended applications, the exact form of the wave equation can vary considerably (DeSanto, 1979; Goodman and Farwell, 1979). For most applications, a simplified linear, hyperbolic, second-order, time-dependent partial differential equation is used:

$$\nabla^2\Phi = \frac{1}{c^2}\frac{\partial^2\Phi}{\partial t^2} \tag{4.1}$$

where ∇^2 is the Laplacian operator $[=(\partial^2/\partial x^2)+(\partial^2/\partial y^2)]+(\partial^2/\partial z^2)]$, Φ is the potential function, c is the speed of sound, and t the time.

A somewhat more general form of the wave equation which includes gravitational and rotational effects has been derived by DeSanto (1979).

Subsequent simplifications incorporate a harmonic (single-frequency, continuous wave) solution in order to obtain the time-independent Helmholtz equation. Specifically, a harmonic solution is assumed for the potential function Φ:

$$\Phi = \phi\,\mathrm{e}^{-i\omega t} \tag{4.2}$$

where ϕ is the time-independent potential function and ω the source frequency ($2\pi f$). Then the wave equation (4.1) reduces to

$$\nabla^2\phi + k^2\phi = 0 \tag{4.3a}$$

where $k=\omega/c=2\pi/\lambda$ is the wave number and λ is the wavelength. In cylindrical coordinates, we obtain

$$\frac{\partial^2\phi}{\partial r^2}+\frac{1}{r}\frac{\partial\phi}{\partial r}+\frac{\partial^2\phi}{\partial z^2}+k^2(z)\phi=0 \tag{4.3b}$$

The Helmholtz equation (4.3a) is referred to as the time-independent (or frequency-domain) wave equation. In cylindrical coordinates, equation (4.3b) is commonly referred to as the elliptic-reduced wave equation.

Various theoretical approaches are applicable to the Helmholtz equation. The approach used depends upon the specific geometrical assumptions made for the propagation and the type of solution chosen for ϕ, as will be discussed in the following sections. To describe the different approaches more effectively, it is first useful to develop a classification scheme, with associated terminology, based on five canonical solutions to the wave equation: ray theory, normal mode, multipath expansion, fast field and parabolic equation techniques.

Throughout the theoretical development of these five techniques, the potential function ϕ normally represents the acoustic field pressure. When this is the case, the transmission loss (TL) can easily be calculated as

$$TL = 10 \log [\phi^2]^{-1} = -20 \log |\phi|$$

This relationship necessarily follows from equations (3.1) and (3.2). If phases are considered, the resulting transmission loss is referred to as coherent; otherwise, phase differences are ignored and the transmission loss is termed incoherent.

4.2.2 Classification of modeling techniques

Although acoustic propagation models can be classified according to the theoretical approach employed, the cross-connections that exist among the various approaches complicate a strict classification, or taxonomic, scheme. Consequently, as the schemes become more detailed, more cross-connections will appear. A generalized classification scheme has been constructed using five categories corresponding to the five canonical solutions of the wave equation (also see Jensen and Krol, 1975; DiNapoli and Deavenport, 1979; Weston and Rowlands, 1979).

Within these five categories, a further subdivision can be made according to range-independent and range-dependent types. Range independence means that the model assumes a cylindrical symmetry for the environment (i.e. a horizontally stratified ocean in which properties vary only as a function of depth). Range dependence indicates that some properties of the ocean medium are allowed to vary as a function of range (r) and azimuth (θ) from the receiver, in addition to a depth (z) dependence. Such range-varying properties commonly include sound speed and bathymetry. Range dependence can further be regarded as two-dimensional (2D) for range and depth variations, or three-dimensional (3D) for range, depth and azimuthal variations.

In order to illustrate the relationships among the five approaches used to solve the wave equation, the rather elegant scheme developed by Jensen and Krol (1975) will be adopted with slight modifications (Fig. 4.1). According to this classification scheme, there are three avenues which connect the five basic approaches applicable to underwater acoustic propagation modeling. These five categories of propagation models will be described in detail in the following sections, and Fig. 4.1 will serve as a useful road map. For convenience, the general functions and equations represented in Fig. 4.1 have been identified with the letters F and G. In the discussions which follow, different symbols will be substituted to facilitate identification with relevant physical properties or with other well-known mathematical functions.

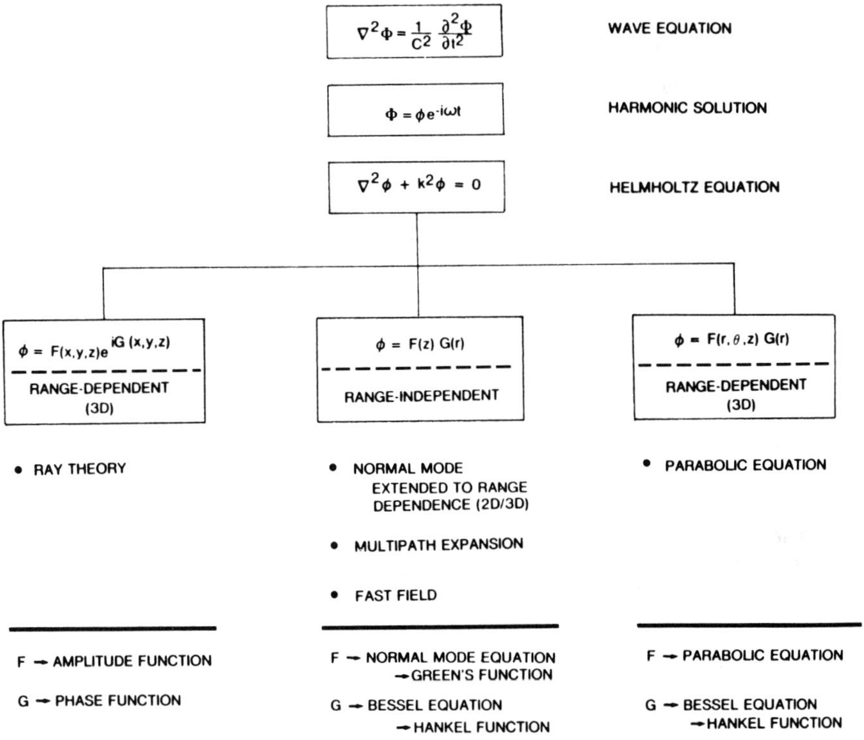

Fig. 4.1 Summary of relationships among theoretical approaches for propagation modeling (adapted from Jensen and Krol, 1975).

4.3 RAY THEORY MODELS

4.3.1 Basic theory

Ray-theoretical models calculate transmission loss on the basis of ray tracing (National Defense Research Committee, 1946). Ray theory starts with the Helmholtz equation. The solution for ϕ is assumed to be the product of a pressure amplitude function $A = A(x, y, z)$ and a phase function $P = P(x, y, z)$: $\phi = Ae^{iP}$. P is commonly referred to as the eikonal, a Greek word meaning 'image'. Substituting this solution into the Helmholtz equation (4.3a) and separating real and imaginary terms yields

$$\frac{1}{A}\nabla^2 A - [\nabla P]^2 + k^2 = 0 \tag{4.4}$$

and

$$2[\nabla A.\nabla P] + A\nabla^2 P = 0 \tag{4.5}$$

Equation (4.4) contains the real terms and defines the geometry of the rays. Equation (4.5), also known as the transport equation, contains the imaginary terms and determines the wave amplitudes. The separation of functions is performed under the assumption that the amplitude varies more slowly with position than does the phase (geometrical acoustics approximation). The geometrical acoustics approximation is a condition in which the fractional change in the sound speed gradient over a wavelength is small compared to the gradient c/λ, where c is the speed of sound and λ is the acoustic wavelength. Specifically

$$\frac{1}{A}\nabla^2 A \ll k^2 \tag{4.6}$$

In other words, the sound speed must not change much over one wavelength. Under this approximation, equation (4.4) reduces to

$$[\nabla P]^2 = k^2 \tag{4.7}$$

Equation (4.7) is referred to as the eikonal equation. Surfaces of constant phase ($P =$ constant) are the wavefronts, and the normals to these are the rays. Eikonal refers to the acoustic path length as a function of the path endpoints. Such rays are referred to as eigenrays when the endpoints are the source and receiver positions. Differential ray equations can then be derived from the eikonal equation. Typically, four sets of eigenrays are considered (Fig. 4.2): direct path (DP), refracted-surface-reflected (RSR), refracted-bottom-reflected (RBR) and refracted-surface-reflected-bottom-reflected (RSRBR).

The physical models described in Chapter 3 are generally incorporated into ray models to account for boundary interaction and volumetric effects.

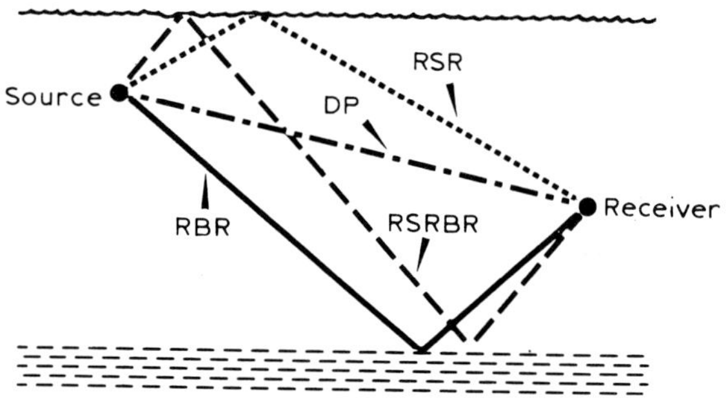

Fig. 4.2 Four basic types of eigenrays: DP, direct path; RSR, refracted-surface-reflected; RBR, refracted-bottom-reflected; RSRBR, refracted-surface-reflected-bottom-reflected.

The geometrical acoustics approximation effectively limits the ray-theoretical approach to the high-frequency domain. An approximate guideline for defining high frequency is provided by the relation

$$f > 10c/H \tag{4.8}$$

where f is the frequency, H is the duct depth, and c is the speed of sound.

The computation of pressure amplitude can be accomplished using the transport equation (4.5) and by invoking the principle of conservation of energy flux for a ray bundle (Fig. 4.3). Assuming constant density (Tolstoy and Clay, 1966, p. 57)

$$A_2 = \left[\frac{c_2\,d\sigma_1}{c_1\,d\sigma_2}\right]^{1/2} A_1 \tag{4.9}$$

where

$$A_2 = A(x_2, y_2, z_2) \qquad A_1 = A(x_1, y_1, z_1)$$

are the signal amplitudes;

$$c_2 = c(x_2, y_2, z_2) \qquad c_1 = c(x_1, y_1, z_1)$$

are the sound speeds; and

$$d\sigma_1 = \text{ray bundle cross-section at } x_1, y_1, z_1$$
$$d\sigma_2 = \text{ray bundle cross-section at } x_2, y_2, z_2$$

As $d\sigma_2$ approaches zero, A_2 approaches infinity; thus, ray theory does not hold in the vicinity of focal surfaces (caustics) and focal points.

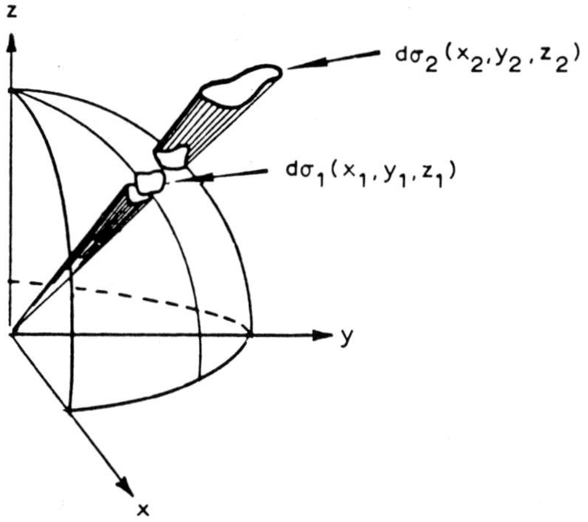

Fig. 4.3 Geometry of ray bundle.

4.3.2 Caustics

Focal surfaces, or caustics, are formed when the refractive properties of the ocean environment focus a number of adjacent rays into very close proximity. There are two types of caustics: smooth and cusped. A cusp is actually the intersection of two smooth caustics. Examples of smooth and cusped caustics are presented in Fig. 4.4.

In the vicinity of caustics, a higher-order approximation can be used to yield predictions on the caustic itself, and also in the nearby shadow zone. One theory, developed by Sachs and Silbiger (1971), is essentially an approximate asymptotic method which predicts a spatially oscillating field amplitude on the illuminated side of the caustic. In the shadow zone, the field is damped with increasing distance from the caustic boundary. Boyles (1984, Chapter 5) provides a lucid description of caustic formation together with appropriate corrections to ray theory. With appropriate frequency-dependent (diffraction) corrections and proper evaluation of caustics, ray theory can be extended to lower frequencies. Under these conditions, the approach is commonly termed ray theory with corrections.

4.3.3 Gaussian beam tracing

Another method useful in dealing with caustics is Gaussian beam tracing (e.g. Porter and Bucker, 1987; Bucker, 1994), which has been adapted from seismic applications. This method associates with each ray a beam with a Gaussian intensity profile normal to the ray. A pair of differential equations which govern the beamwidth and curvature are integrated along

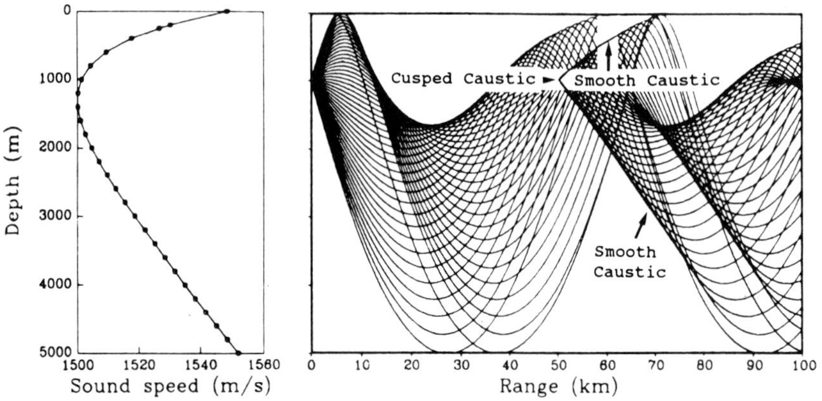

Fig. 4.4 Sound speed profile and associated ray trace showing the formation of smooth and cusped caustics (adapted from Jensen, 1988; *IEEE J. Oceanic Engr.*, **13**, 186–97; copyright by IEEE).

with the standard ray equations to compute the beam field in the vicinity of the central ray of the beam. This method avoids certain ray tracing artifacts such as perfect shadows and infinite energy levels at caustics. Furthermore, this technique is attractive for high-frequency, range-dependent applications in which wave-theoretical approaches might not be practical alternatives.

4.3.4 Range dependence

Although ray-tracing techniques are theoretically applicable to fully range-dependent (3D) problems (Fig. 4.1), they are rarely implemented as such. The mathematical complexity discourages 3D versions in favor of 1D or 2D versions. The 2D versions can be implemented by one of three methods: (1) by mapping rays over discrete range intervals in which the environment remains constant (Weinberg and Dunderdale, 1972; Weinberg and Zabalgogeazcoa, 1977); (2) by dividing the range–depth plane into triangular regions (Roberts, 1974; Watson and McGirr, 1975); or (3) by allowing the environment to vary smoothly as a function of range, as through the use of cubic splines (Foreman, 1983). These methods are explained in the following paragraphs.

A significant problem which confronts range-dependent ray-tracing programs is the proper representation of the transition of sound speed profiles between adjacent measurement points in the range dimension. Two aspects of this problem are important from a practical standpoint: the interpolated intermediate sound speed profiles should be physically plausible, and it is desirable that the resulting ray trajectories be analytically computable (versus numerically integrated) in order to maximize computational efficiency.

In method (1) (Fig. 4.5(a)), rays are traced in the first range interval. Selected rays are then mapped into the second interval, and the process continues throughout the remaining intervals. Drawbacks to this approach stem from discontinuities in ray tracing at the boundaries between adjacent intervals, and also from the potential omission of important rays in the selection process at each boundary. In particular, as the water (or duct) depth decreases, rays can properly be eliminated. As the depth increases, however, there is no valid and consistent process for adding new rays into consideration.

In method (2) (Fig. 4.5(b)), the input sound speed profiles are represented as piecewise-linear functions of depth. Then, the range–depth plane between specified profiles can be divided into triangular sectors within which the sound speed varies linearly in both range and depth. In each sector, ray trajectories then correspond to arcs of circles for which analytical ray-tracing programs are available (e.g. Urick, 1983, pp. 124–8). Each triangular sector is selected to provide a smooth and physically realizable

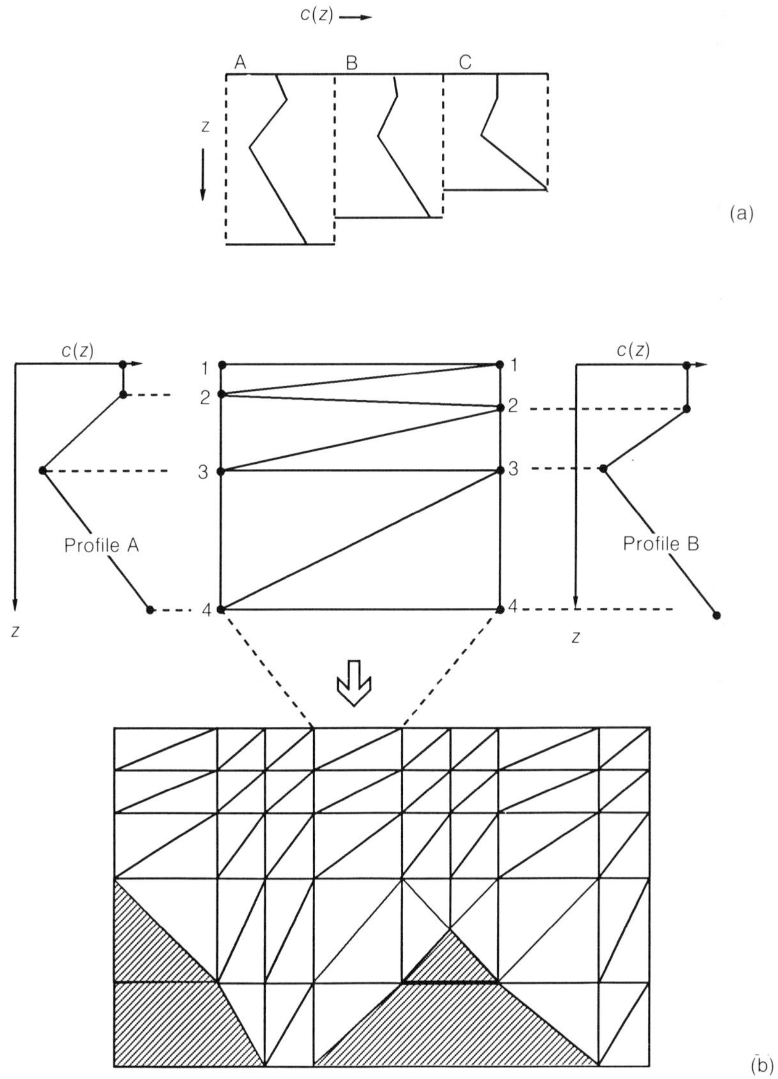

Fig. 4.5 Range-partitioning techniques: (a) discrete range intervals; (b) typical network of triangular regions formed by connecting corresponding features between two adjacent sound speed profiles (Roberts, 1974; Watson and McGirr, 1975).

transition between adjacent profiles. The vertexes are commonly selected at those depths representing sound speed minima; otherwise, the triangle sides are selected to subtend the gradient which departs least from the gradient of the preceding triangle. These procedures generally require the expertise of an oceanographer to ensure the generation of realistic intermediate sound speed profiles. One tries to maintain continuity in features such as the sonic layer depth and the sound channel axis.

In method (3), the sound speed profiles and bathymetry are fitted with cubic splines (Solomon *et al.*, 1968) or quadratic functions. The domains thus formed are rectangular, much like that in method (1) above. However, linear interpolations in range are performed between adjacent profiles to obtain continuous range derivatives. The differential ray geometry and amplitude equations (derived from equations (4.4) and (4.5), respectively) are then solved numerically (Foreman 1983).

The proper use and application of complex range-dependent models requires a great deal of planning, especially with regard to the selection and spacing of the range-dependent environmental data inputs. Improper spacing of inputs can overlook important ocean frontal systems or bathymetric features (Henrick, 1983).

Limited three-dimensional ray-tracing techniques have been developed to account more properly for the effects of horizontal gradients of oceanic parameters which are manifested as cross-range variations in the transmission loss patterns. Examples include the use of a hybrid approach involving horizontal rays and vertical modes (Weinberg and Burridge, 1974; Burridge and Weinberg 1977), and the use of three-dimensional Hamiltonian ray tracing (Jones, 1982; Jones *et al.*, 1986). A compact three-dimensional ray-tracing algorithm suitable for implementation on small computers has been developed by Einstein (1975).

4.3.5 Arrival structure

One useful property of ray models is their ability to calculate arrival structure. Similar information can be obtained from some wave-theoretical models, but only with much additional computation. Arrival structure curves indicate the horizontal range from source to receiver that will be traversed by a ray leaving the source (or arriving at the receiver) at the specified angle for each propagation path of interest. These curves are not frequency dependent, but they do depend on the source and receiver depths.

These curves are also referred to as $\theta - r$ diagrams. An alternative way of viewing these curves is that r represents the range at which a ray leaving the source (receiver) at angle θ crosses a horizontal line representing the receiver (source) depth.

To be properly utilized, these curves must indicate a reference to either the source or the receiver angle, and the sign convention must be explicitly stated. For example, relative to the receiver, a negative angle (θ) usually (but not always) signifies rays arriving from above the horizontal axis (i.e. rays traveling downward), and a positive angle signifies rays arriving from below (i.e. rays traveling upward).

Arrival structure information is not just an academic curiosity. This information is essential for the proper computation of the vertical directionality of ambient noise (see Chapter 7) and for the proper evaluation of the performance of vertically oriented hydrophone arrays. Arrival structure is also important in the calculation of volumetric and boundary reverberation (see Chapter 9).

In arrival structure diagrams, particular ray families are evidenced by distinctive sets of patterns. For example, in Fig. 4.6, the inverted L-shaped trace at very short range is indicative of direct path arrivals, while the more graceful arching curves on either side of the horizontal axis are associated with bottom-bounce paths. The sets of curves (for both positive and negative arrival angles) that represent bottom-bounce paths are referred to as order contours. Rays with the same number of bottom reflections belong to the same order. Thus, the first set of curves represents one bottom bounce and is of order 1. The curve associated with the positive angle within this first order has traveled from the source to receiver over a refracted-bottom-reflected (RBR) path and arrives at the receiver from below the horizontal axis. The corresponding curve with the negative angle has traveled from source to receiver over a refracted-bottom-reflected-surface-reflected (RBRSR) path and arrives at the receiver from above the horizontal axis.

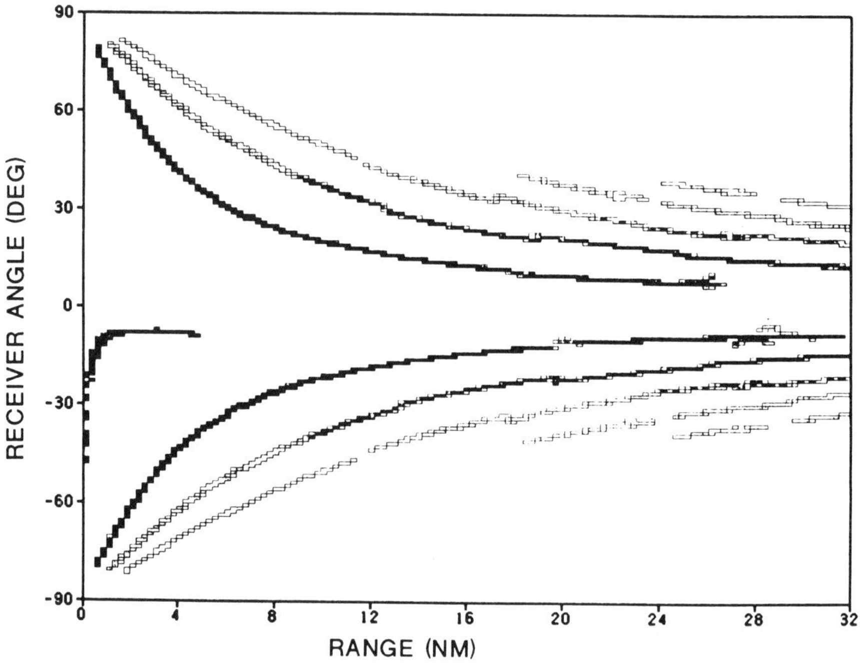

Fig. 4.6 Example of a vertical arrival structure diagram.

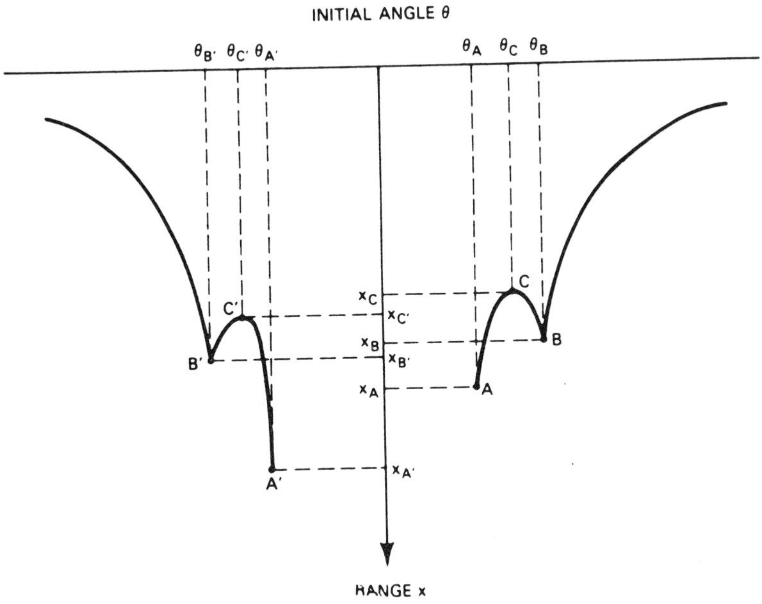

Fig. 4.7 Smoothed order contours showing caustic behavior (Franchi *et al.*, 1984).

These order contours can be used to identify all ray paths connecting the source and receiver that encounter the bottom at any range r. This information is very useful in the calculation of bottom reverberation; for example, only those rays encountering the bottom will contribute to bottom reverberation. The concept of order contours also applies to convergence zone (CZ) paths. Practical examples of $\theta - r$ diagrams will be provided later in this chapter during discussions of the RAYMODE propagation model.

The $\theta - r$ curves can also be used to identify caustic formations. An instructive example has been provided by Franchi *et al.* (1984). Figure 4.7 illustrates order contours obtained for a typical deep ocean sound speed profile. The two rays that just graze the bottom at ranges X_B and $X_{B'}$ determine points B and B′ on the contour. The two rays that just graze the surface at ranges X_A and $X_{A'}$ occur at shallower angles and determine points A and A′ on the countour. Thus, rays with angles in the intervals ($\theta_{B'}$, $\theta_{A'}$) and (θ_A, θ_B) hit the surface but not the bottom. Caustics are associated with points C and C′ on these contours.

4.3.6 Beam displacement

Propagation models based on ray-tracing techniques generally treat bottom reflection as specular and reduce the intensity through application of

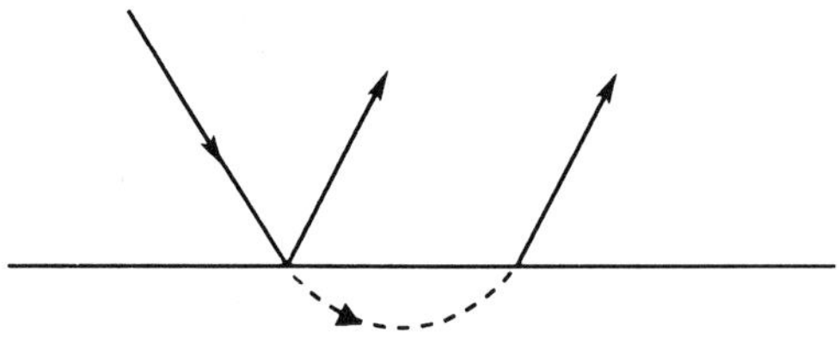

Fig. 4.8 Geometry for beam displacement.

a bottom reflection loss. We have seen, however, that energy can actually be transmitted into the bottom where it is subsequently refracted, attenuated and even transmitted back into the water column at some distance down range (Fig. 4.8). This spatial offset is referred to as beam displacement. Time displacements would also be associated with this process while the ray is absent from the water column.

In their initial study of the effects of beam displacement in ray calculations, Tindle and Bold (1981) considered a simple two-fluid Pekeris model as a good first approximation to many shallow-water environments. The Pekeris model consists of a fluid (water) layer of depth H, density ρ_1, and sound speed c_1 overlying a semi-infinite fluid (sediment) layer of density ρ_2 and sound speed c_2, where $c_2 > c_1$. Attenuation was neglected and only totally reflected rays were considered to propagate to ranges (r) greater than the water depth (H). The vertical wavenumbers (γ) in the two layers were defined as

$$\gamma_1 = \left(\frac{\omega}{c_1}\right)\sin\theta, \qquad \gamma_2 = \left(\frac{\omega}{c_1}\right)\sqrt{\cos^2\theta - \frac{c_1^2}{c_2^2}}$$

where θ is the grazing angle and ω is the angular frequency of the wave field. The lateral displacement (Δ) of a beam of finite width undergoing reflection at the water–sediment interface is

$$\Delta = \frac{2k_h\rho_1\rho_2(\gamma_1^2 + \gamma_2^2)}{\gamma_1\gamma_2(\rho_1^2\gamma_2^2 + \rho_2^2\gamma_1^2)}$$

where k_h is the horizontal wavenumber in the water layer:

$$k_h = \left(\frac{\omega}{c_1}\right)\cos\theta$$

A typical ray path for this simple model is presented in Fig. 4.9. The source and receiver depths are z_0 and z, respectively. A ray leaving the source at an angle θ relative to the horizontal travels in a straight (unrefracted) path in the water layer. There is no beam displacement at the sea surface and the ray is reflected in the conventional manner. At the bottom, the beam is displaced horizontally by an amount Δ before traveling upward again at angle θ.

Recent efforts to improve ray theory treatments of bottom attenuation and beam displacement have been conducted by Siegmann *et al.* (1987) and Westwood and Tindle (1987), among others. Modified ray theory with beam and time displacements has advantages over wave theory in that the interaction of the acoustic energy with the bottom can be intuitively visualized. Moreover, Jensen and Schmidt (1987) have numerically obtained complete wave theory solutions for a narrow Gaussian beam incident on a water–sediment interface near the critical grazing angle. They observed that the fundamental reflectivity characteristics of narrow beams can be entirely explained within the framework of linear acoustics.

4.4 NORMAL MODE MODELS

4.4.1 Basic theory

Normal mode solutions are derived from an integral representation of the wave equation. In order to obtain practical solutions, however, cylindrical symmetry is assumed in a stratified medium (i.e. the environment changes as a function of depth only). Then, the solution for the potential function ϕ in equation (4.3b) can be written in cylindrical coordinates as the product of a depth function $F(z)$ and a range function $S(r)$:

$$\phi = F(z).S(r) \tag{4.10}$$

Next, a separation of variables is performed using ξ^2 as the separation constant. The two resulting equations are

$$\frac{d^2F}{dz^2} + (k^2 - \xi^2)F = 0 \tag{4.11}$$

and

$$\frac{d^2S}{dr^2} + \frac{1}{r}\frac{dS}{dr} + \xi^2 S = 0 \tag{4.12}$$

Equation (4.11) is the depth equation, better known as the normal mode equation, which describes the standing wave portion of the solution; equation (4.12) is the range equation which describes the traveling wave portion of the solution. Thus, each normal mode can be viewed as a traveling wave in the horizontal (r) direction and as a standing wave in the depth (z) direction.

The normal mode equation (4.11) poses an eigenvalue problem. Its solution is known as the Green's function. The range equation (4.12) is the zero-order Bessel equation. Its solution can be written in terms of a zero-order Hankel function ($H_0^{(1)}$). The full solution for ϕ can then be expressed by an infinite integral, assuming a monochromatic (single-frequency) point source:

$$\phi = \int_{-\infty}^{\infty} G(z, z_0; \xi) . H_0^{(1)}(\xi r) . \xi \, d\xi \tag{4.13}$$

where G ( ) is the Green's function, $H_0^{(1)}$ is a zero-order Hankel function of the first kind, and z_0 is the source depth. Note that ϕ is a function of the source depth (z_0) and the receiver depth (z).

The properties of these various functions are described in standard mathematical handbooks such as Abramowitz and Stegun (1964).

4.4.2 Normal mode solution

To obtain what is known as the normal mode solution to the wave equation, the Green's function is expanded in terms of normalized mode functions (u_n). The eigenvalues, which are the resulting values of the separation constants, are represented by ξ_n. These eigenvalues (or characteristic values) represent the discrete set of values for which solutions of the mode functions u_n exist. The infinite integral in equation (4.13) is then evaluated by contour integration (denoted by the symbol $\oint$):

$$\phi = \oint \sum \frac{u_n(z) . u_n(z_0)}{\xi^2 - \xi_n^2} H_0^{(1)}(\xi r) . \xi \, d\xi + \text{branch-cut integral} \tag{4.14}$$

The contour integral represents the trapped (or discrete) modes which propagate through the water column. The branch-cut integral is associated with the continuous mode spectrum which represents those modes propagating through the ocean floor and which are strongly attenuated. The branch-cut integral describes the near-field conditions and corresponds in ray theory to those rays striking the bottom at angles greater than the critical angle. Thus, the contribution from the branch-cut integral is often neglected, particularly when the source and receiver are separated by horizontal distances greater than several water depths. For acoustic propagation problems using an impedance boundary condition at the bottom, the solution actually comprises three spectral intervals: the continuous, discrete, and evanescent. The evanescent spectrum is associated with interface waves that decay exponentially away from the boundary (Jensen *et al.*, 1994). Interface waves that propagate along a fluid–solid boundary (i.e. water–sediment interface) are called Scholte waves. In seismology, interface waves propagating along the boundaries between solid layers are called Stoneley waves.

By neglecting the branch-cut integral, evaluating the contour integral, and replacing the Hankel function expression by its asymptotic expansion

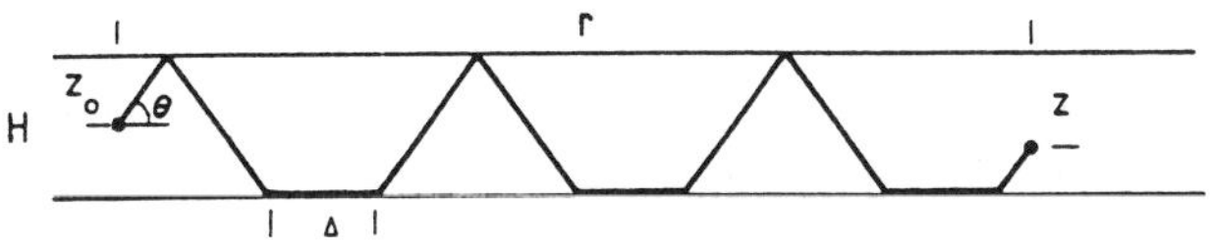

Fig. 4.9 A typical ray path with beam displacement included (Tindle and Bold, 1981).

for large arguments

$$H_0^{(1)}(\xi r) \approx \sqrt{\frac{2}{\pi \xi r}}\, e^{i(\xi r - \pi/4)} \qquad \text{for } \xi r \gg 1$$

(where $\xi r \gg 1$ is the far-field approximation), a simple solution for the potential function ϕ can be obtained:

$$\phi = g(r, \rho) \sum \frac{u_n(z) . u_n(z_0)}{\sqrt{\xi_n}} \exp\left(i[\xi_n r - \pi/4]\right) \exp\left(-\delta_n r\right) \tag{4.15}$$

where $g(r, \rho)$ is a general function of range (r) and water density (ρ). Each of the n-terms in equation (4.15) corresponds to the contribution of a single normal mode of propagation; each of these modal contributions is propagated independently of the others. Under idealized conditions, there is usually an upper limit on the number of modes to be calculated, and this number increases with increasing frequency.

The attenuation coefficient δ_n can be of the form (Miller and Wolf, 1980)

$$\delta_n = \varepsilon \gamma_n + \varepsilon_c \gamma_n^{(c)} + \varepsilon_s \gamma_n^{(s)} + S_n^{(0)} + S_n^{(1)} + \alpha_n \tag{4.16}$$

where ε = plane-wave attenuation coefficient in the sediment layer
$\varepsilon_c, \varepsilon_s$ = compressional and shear plane-wave attenuation coefficients of the basement
$\gamma_n, \gamma_n^{(c)}, \gamma_n^{(s)}$ = measures of nth mode interactions with sediment, and with basement compressional and shear wave mechanisms
$S_n^{(0)}, S_n^{(1)}$ = attenuation of the modal field due to the interaction of the nth mode with statistically rough boundaries at the pressure-release boundary (sea surface) and the water–sediment boundary, respectively
α_n = attenuation due to absorption by the water

One advantage of normal mode solutions over the ray-theoretical methods is that the transmission loss can easily be calculated at a given frequency and source depth (z_0) combination for all potential receiver depths (z) and ranges (r) (refer to equation (4.14)). Ray models, on the other hand, must be executed for each change in source or receiver depth (compare Figs 3.2 and 3.3).

A disadvantage associated with normal mode solutions is the degree of information required concerning the structure of the sea floor. In order to

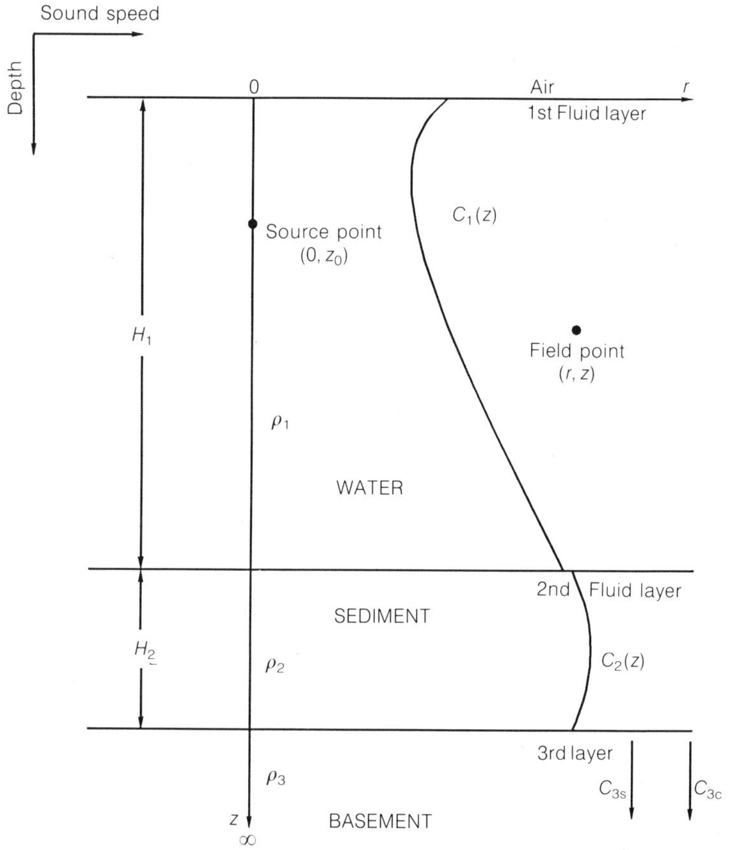

Fig. 4.10 Simple physical model depicting an infinite half-space consisting of two fluid layers with depths H_1, H_2; densities ρ_1, ρ_2; and sound speeds c_1, c_2. A semi-infinite third layer (representing the bottom) has density ρ_3, and compressional (c_{3c}) and shear (c_{3s}) sound speeds (adapted from Miller and Ingenito, 1975).

execute effectively, this type of model generally requires knowledge of the density as well as the shear and compressional sound speeds within the various sediment layers. Figure 4.10 shows an ideal three-layer physical model usable in normal mode solutions. The fluid layer boundaries are defined by duct depths or by other notable features in the sound speed profile. Accordingly, more layers are added to the physical model as the complexity of the profile increases. The computational intensity increases in proportion to this complexity.

Some normal mode models (Stickler, 1975; Bartberger, 1978b) partition the sound speed profile into N layers such that the square of the index of refraction in each layer can be approximated by a straight line and the density by a constant. Under these conditions, it is then possible to represent the depth-dependent portion of the pressure field in terms of Airy functions and thus improve computational efficiency.

Fast finite difference methods have been used to accurately determine the real (versus imaginary) eigenvalues (Porter and Reiss, 1984). Errors in these eigenvalues would otherwise appear as phase shifts in the range dependence of the acoustic field.

4.4.3 Dispersion effects

Unlike ray-theoretical solutions, the wave-theoretical solutions inherently treat dispersion effects. The phenomenon of dispersion is the condition in which the phase velocity is frequency dependent. If present, dispersion effects are most noticeable at low frequencies. In oceanic waveguides, dispersion depends on the characteristics and geometry of the waveguide and is referred to as geometrical dispersion. This is distinguished from intrinsic dispersion as might result from sound propagation through bubbly water layers near the sea surface (Clay and Medwin, 1977, pp. 311–12).

4.4.4 Experimental measurements

Ferris (1972) and Ingenito *et al.* (1978) summarized results from field measurements at a site near Panama City, Florida. The experiments were conducted over tracks having a range-independent water depth and sound speed profile. The sea floor was composed of hard-packed sand. The modal field distribution was calculated on the basis of equation (4.14). If all acoustic and environmental parameters except for receiver depth are held constant, then the dependence of the pressure amplitude (ϕ_n) of the nth-order mode on receiver depth (z) is directly proportional to the corresponding mode function (u_n):

$$\phi_n(z) \propto u_n(z)$$

If range and receiver depth are held constant, then the dependence of the pressure amplitude (ϕ_n) of the nth-order mode on source depth (z_0) is

$$\phi_n(z_0) \propto u_n(z_0)$$

Comparisons of measured and calculated pressure amplitude distributions at the receiver (using an arbitrary amplitude scale) are presented in Fig. 4.11 for the first and second modes. Two frequencies (400 Hz and 750 Hz) were considered, as were three different sound speed profiles (positive gradient, negative gradient and isospeed). The influence of the sound speed gradient on the low-order modes is evident. Specifically, the first mode is nearly symmetrically distributed about mid-depth in the isospeed case (Fig. 4.11(c)). The sound speed gradients alter this symmetry and

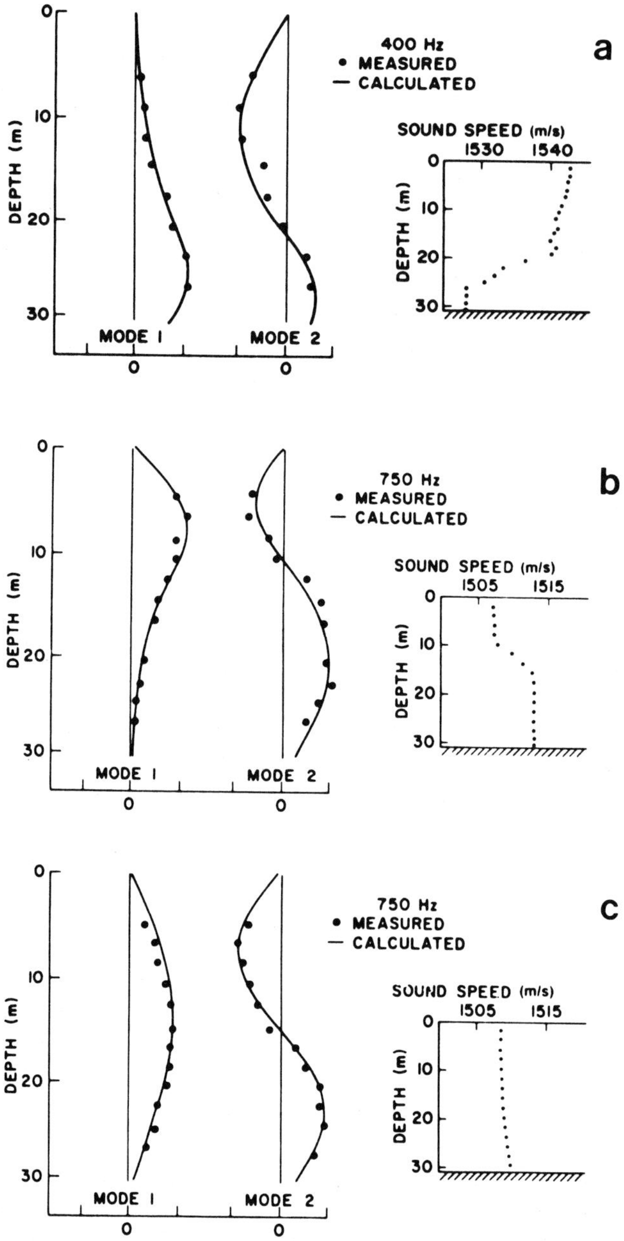

Fig. 4.11 Comparison of measured and calculated amplitude functions (using arbitrary scales) for the first and second modes: (a) at 400 Hz with a downward-refracting profile; (b) at 750 Hz with an upward-refracting profile; and (c) at 750 Hz with a nearly constant sound speed profile (Ferris, 1972).

concentrate the energy in the low-speed portion of the water column (Figs 4.11(a), (b)).

Boyles (1984, Chapter 6) has provided a comprehensive discussion which demonstrated that certain groupings of normal modes can be associated with sound propagation in particular oceanic waveguides. This was accomplished by explicitly calculating the transmission loss associated with selected subsets of modes using equation (4.14). These calculations were then compared to the full transmission loss curve. One example involved a typical North Atlantic Ocean sound speed profile during the winter half of the year (refer back to Fig. 2.7). The source and receiver were both placed within the surface duct, and a frequency of 30 Hz was selected. Only 27 modes were trapped in the water column. Of these, only one mode (mode 11) contributed to the transmission loss in the surface duct. Modes which uniquely contribute to surface duct propagation are referred to as 'virtual modes' (Labianca, 1973); thus, mode 11 is also referred to as virtual mode 1. Modes 13–27 accounted for the transmission loss associated with the convergence zones.

4.4.5 Range dependence

As noted previously in Fig. 4.1, normal mode models assume range independence. Extensions to range dependence are accomplished either by mode coupling, which considers the energy scattered from a given mode into other modes, or by invoking the adiabatic assumption, which assumes that all energy in a given mode transfers to the corresponding mode in the new environment, provided that the environmental variations in range are gradual. One recent development investigated downslope propagation based on coupled wedge modes, which are the intrinsic normal modes in a wedge (sloping-bottom) coordinate system (Primack and Gilbert, 1991).

Three-dimensional (3D) propagation modeling using normal mode theory has been attempted using two different approaches. The first approach employs the $N \times 2D$ technique, in which the 3D problem is solved using N horizontal radials (or sectors) in conjunction with range-dependent (2D) adiabatic mode theory along each radial. The resulting quasi-3D propagation fields can then be contoured on polar plots, for example. The second approach directly includes the effects of horizontal refraction through use of the lateral wave equation. One specific implementation (Kuperman *et al.*, 1988; Porter, 1991) employs Gaussian beam tracing (Porter and Bucker, 1987) to solve the lateral wave equation. In essence, the horizontal field of each modal wavenumber is translated into a horizontal sound speed field which then defines the Gaussian beam environment for each mode. A new development, called WRAP (Wide-Area Rapid Acoustic Prediction), precomputes local acoustic eigenvalues and normal modes for complex 3D ocean environments comprising a number of distinct local environments (Perkins *et al.*, 1990); for mild horizontal variability, the full 3D acoustic field is constructed by adiabatic mode computations. Another newly developed 3D model called CMM3D includes horizontal refraction and radial mode coupling (Chiu and Ehret, 1990, 1994).

4.4.6 High-frequency adaptations

Normal mode approaches tend to be limited to acoustic frequencies below about 500 Hz due to practical (numerical) computational considerations. Specifically, the number of modes required to generate a reliable prediction of transmission loss increases in proportion to the frequency. By invoking some simplifying assumptions with regard to the complexity of the ocean environment, upper frequency limits in the multikilohertz regions are achievable (Ferla *et al.*, 1982).

4.5 MULTIPATH EXPANSION MODELS

Multipath expansion techniques expand the acoustic field integral representation of the wave equation (equation (4.13)) in terms of an infinite set of integrals, each of which is associated with a particular ray path. This method has also been referred to as the WKB method because a generalized WKB (Wentzel, Kramers and Brillouin) approximation is used to solve the depth-dependent equation derived from the normal mode solution (equation (4.11)). Each normal mode can then be represented by corresponding rays. Multipath expansion models do not allow for environmental range dependence.

The WKB approximation (also referred to as the WKBJ or Liouville–Green approximation) facilitates an asymptotic solution of the normal mode equation by assuming that the speed of sound varies gradually as a function of depth. Advanced versions of the WKB method provide connection formulas to carry the approximation through 'turning points' (i.e. depths where an equivalent ray becomes horizontal). Unlike ray-theoretical solutions, however, the WKB method normally accounts for first-order diffraction effects and caustics.

The specific implementation of this approach is accomplished by directly evaluating the infinite integral of equation (4.13) over a limited interval of the real ξ-axis. Thus, only certain modes are considered. By using a restricted number of modes, one can, in effect, produce an angle-limited source. The resulting acoustic pressure field ϕ in equation (4.13) is then expressed as a sum of finite integrals, where each integral is associated with a particular ray path. As implemented, this approach is particularly applicable to the modeling of acoustic propagation in deep water at intermediate and high frequencies. Multipath expansion models thus have certain characteristics in common with ray models. Moreover, the pressure field is properly evaluated in caustics and shadow zones. Weinberg (1975) has provided a brief summary of the historical development of this technique.

This approach is explored in more detail in section 4.8 where the RAYMODE model (which is based in part on the multipath expansion approach) is described.

4.6 FAST-FIELD MODELS

In underwater acoustics, fast-field theory is also referred to as wavenumber integration; in seismology, it is commonly referred to as the reflectivity method or discrete-wavenumber method. In fast-field theory, the wave equation parameters are first separated according to the normal mode approach. Then, the Hankel function expression in equation (4.13) is replaced by the first term in the asymptotic expansion (DiNapoli and Deavenport, 1979):

$$\mathrm{H}_0^{(1)}(\xi r) \approx \sqrt{\frac{2}{\pi \xi r}}\, \mathrm{e}^{i\xi r} \qquad \text{for } \xi r \gg 1$$

Equation (4.13) may now be written as

$$\phi = \int_{-\infty}^{\infty} \sqrt{\frac{2\xi}{\pi r}}\, \mathrm{G}(z, z_0; \xi)\, \mathrm{e}^{i\xi r}\, \mathrm{d}\xi \tag{4.17}$$

The infinite integral is then evaluated by means of the Fast Fourier Transform (FFT), which provides values of the potential function ϕ at n discrete range points for a given source–receiver geometry

Evaluation of the Green's function can be simplified be approximating the sound speed profile by exponential functions. Such an approximation facilitates the matrizant solution, but complicates the specification of the sound speed profiles.

Models based on fast field theory do not yet fully allow for environmental range dependence. However, two recent developments have introduced the possibility of range-dependent calculations of transmission loss. First, Gilbert and Evans (1986) derived a generalized Green's function method for solving the one-way wave equation exactly in an ocean environment that varied discretely with range. They obtained an explicit marching solution in which the source distribution at any given range step was represented by the acoustic field at the end of the previous step. Gilbert and Evans (1986) further noted that their method, referred to as the Range-Dependent FFP (RDFFP) model, was computationally intensive. Second,

Seong (1990) used a hybrid combination of wavenumber integration and Galerkin boundary element methods, referred to as the SAFRAN model, to extend the fast-field theory technique to range-dependent ocean environments (see also Schmidt, 1991). At present, the experimental nature of these methods imposes uncertain restrictions on their application to sonar performance modeling in range-dependent ocean environments. However, further research may prove these methods to be useful in modeling range-dependent elastic wave propagation, including the effects of compressional-to-shear conversion and elastic mode coupling.

The fast-field program (FFP) approach has been modified to accommodate acoustic pulse propagation in the ocean by directly marching the formulation in the time domain (Porter 1990). Applications include specification of arbitrary source time series instead of the more conventional time-harmonic sources used in frequency-domain solutions of the wave equation.

4.7 PARABOLIC EQUATION MODELS

Use of the parabolic approximation in wave propagation problems can be traced back to the mid-1940s when it was first applied to long-range tropospheric radio wave propagation (Keller and Papadakis, 1977, pp. 282–4). Subsequently, the parabolic approximation method has been successfully applied to microwave waveguides, laser beam propagation, plasma physics and seismic wave propagation. The first application to underwater acoustic propagation problems was reported by F. D. Tappert and R. H. Hardin. (Hardin and Tappert, 1973; Spofford, 1973b, pp. 14–16).

4.7.1 Basic theory

The parabolic equation (or approximation) approach replaces the elliptic reduced wave equation (equation (4.3b)) with a parabolic equation (PE). The PE is derived by assuming that energy propagates at speeds close to a reference speed – either the shear speed or the compressional speed, as appropriate (Collins, 1991). The following derivation is adapted from that presented by Jensen and Krol (1975).

The basic equation of acoustic propagation, equation (4.3a), can be rewritten as

$$\nabla^2\phi + k_0^2 n^2 \phi = 0 \tag{4.18}$$

where k_0 = reference wavenumber (ω/c_0)
ω = source frequency
c_0 = reference sound speed
c = sound speed; $c = c(r, \theta, z)$; θ is the azimuthal angle

n = refraction index (c_0/c)
ϕ = velocity potential
∇^2 = Laplacian operator

Equation (4.18) can be rewritten in cylindrical coordinates as

$$\frac{\partial^2\phi}{\partial r^2}+\frac{1}{r}\frac{\partial\phi}{\partial r}+\frac{\partial^2\phi}{\partial z^2}+k_0^2n^2\phi=0 \tag{4.19}$$

where azimuthal coupling has been neglected, but the index of refraction retains a dependence on azimuth. Further, assume a solution of the form

$$\phi=\Psi(r,z).S(r) \tag{4.20}$$

and obtain

$$\Psi\left[\frac{\partial^2 S}{\partial r^2}+\frac{1}{r}\frac{\partial S}{\partial r}\right]+S\left[\frac{\partial^2\Psi}{\partial r^2}+\frac{\partial^2\Psi}{\partial z^2}+\left(\frac{1}{r}+\frac{2}{S}\frac{\partial S}{\partial r}\right)\frac{\partial\Psi}{\partial r}+k_0^2n^2\Psi\right]=0 \tag{4.21}$$

Using k_0^2 as a separation constant, separate equation (4.21) into two differential equations, as follows:

$$\left[\frac{\partial^2 S}{\partial r^2}+\frac{1}{r}\frac{\partial S}{\partial r}\right]=-Sk_0^2 \tag{4.22}$$

and

$$\left[\frac{\partial^2\Psi}{\partial r^2}+\frac{\partial^2\Psi}{\partial z^2}+\left(\frac{1}{r}+\frac{2}{S}\frac{\partial S}{\partial r}\right)\frac{\partial\Psi}{\partial r}+k_0^2n^2\Psi\right]=\Psi k_0^2 \tag{4.23}$$

Rearrange terms and obtain

$$\frac{\partial^2 S}{\partial r^2}+\frac{1}{r}\frac{\partial S}{\partial r}+K_0^2\,S=0 \tag{4.24}$$

which is the zero-order Bessel equation, and

$$\frac{\partial^2\Psi}{\partial r^2}+\frac{\partial^2\Psi}{\partial z^2}+\left(\frac{1}{r}+\frac{2}{S}\frac{\partial S}{\partial r}\right)\frac{\partial\Psi}{\partial r}+k_0^2n^2\Psi-k_0^2\Psi=0 \tag{4.25}$$

The solution of the Bessel equation (4.24) for outgoing waves is given by the zero-order Hankel function of the first kind:

$$S=\mathrm{H}_0^{(1)}(k_0r) \tag{4.26}$$

For $k_0r \gg 1$ (far-field approximation)

$$S\approx\sqrt{\frac{2}{\pi k_0r}}\exp\left(i[k_0r-\pi/4]\right) \tag{4.27}$$

which is the asymptotic expansion for large arguments. The equation for

$\Psi\,(r,z)$ (equation (4.25)) can now be simplified to

$$\frac{\partial^2 \Psi}{\partial r^2} + \frac{\partial^2 \Psi}{\partial z^2} + 2ik_0 \frac{\partial \Psi}{\partial r} + k_0^2(n^2 - 1)\Psi = 0 \qquad 4.28)$$

Further assume that

$$\frac{\partial^2 \Psi}{\partial r^2} \ll 2k_0 \frac{\partial \Psi}{\partial r} \qquad (4.29)$$

which is the paraxial approximation. Then, equation (4.28) reduces to

$$\frac{\partial^2 \Psi}{\partial z^2} + 2ik_0 \frac{\partial \Psi}{\partial r} + k_0^2\,(n^2 - 1)\Psi = 0 \qquad (4.30)$$

which is the parabolic wave equation. In this equation, n depends on depth (z), range (r) and azimuth (θ). This equation can be numerically solved by 'marching solutions' when the initial field is known (e.g. Tappert, 1977). Typically, a Gaussian field or a normal mode solution is used to generate the initial solution. Additional methods for generating accurate starting fields have been described by Greene (1984) and by Collins (1992).

The computational advantage of the parabolic approximation lies in the fact that a parabolic differential equation can be marched in the range dimension. Alternatively, the elliptic reduced wave equation must be numerically solved in the entire range–depth region simultaneously.

4.7.2 Numerical techniques

Existing parabolic equation, or PE, models employ one of four basic numerical techniques: (1) split-step Fourier algorithm; (2) implicit finite-difference (IFD); (3) ordinary differential equation (ODE); or (4) finite element (FE). The original split-step algorithm developed by Tappert (1977) solves the parabolic wave equation by imposing an artifical zero bottom-boundary condition and pressure-release surface condition. While the split-step algorithm is an efficient method for solving a pure initial-value problem (Perkins *et al.*, 1982), several difficulties arise when there is significant interaction with the sea floor (Bucker, 1983). These are due to: density and compressional sound speed discontinuities at the water–sediment interface; strong gradients of the compressional sound speed in the sediment layers; and rigidity in the sediment layers that produces shear waves. Thus, when bottom interaction is strong, a more general-purpose solution is desirable. For this reason, implicit finite-difference (IFD) schemes (Lee *et al.*, 1981; Robertson *et al.*, 1989) and ordinary differential equation (ODE) methods (Lee and Papadakis, 1980) have also been developed to solve the parabolic wave equation. Lee and McDaniel (1987) provide an in-depth development of the implicit finite-difference (IFD) technique. Their development is made more useful by the inclusion of benchmark test examples together with a listing of the computer program (comprising approximately 1700 lines of FORTRAN code). A microcomputer implementation of the IFD model has been reported by Robertson *et al.* (1991), who included a listing of their computer program. Collins (1988a) described a finite-element solution for the parabolic equation.

Recent modeling developments utilizing parabolic approximations have been directed at refining and expanding the capabilities of existing techniques. Useful summaries of this progress are available in the literature (Lee, 1983; Scully-Power and Lee, 1984). Actual comparisons of computer model results for a set of four ocean acoustic test cases have been documented by Davis *et al.* (1982). A more recent comparison of PE models involving seven test cases has been described by Chin-Bing *et al.* (1993b). One of the most interesting test cases involved a leaky surface duct. So-called 'ducted precursors' are generated by modal energy in the surface duct that leaks out and travels as convergence zone or bottom bounce paths, or both, before coupling back into the surface duct down range. The importance of this problem is that a small phase error in the refracted (leaky) path can produce large changes in the predicted sound level in the duct beyond the first convergence zone (Porter and Jensen, 1993). This pathology was particularly evident in some wide-angle, split-step PE models. Yevick and Thomson (1994) explored this problem further and formulated a new propagation operator that retains the computational efficiency of the split-step algorithm, but which is more accurate.

Traditionally, PE models have been applied to anelastic ocean-bottom regions by treating the shear waves as an additional loss mechanism. This approximation breaks down when anelastic propagation becomes significant, as in shear conversion due to backscattered acoustic fields. In general, PE models propagate the acoustic field only in the forward direction, thus excluding backscatter. A two-way PE model (Collins and Evans, 1992) has been developed to improve the computation of backscattered energy for use in reverberation simulations (e.g. Schneider, 1993). A modified version of this two-way PE model, which now treats backscattering problems in three-dimensional geometries, is referred to as spectral PE (Orris and Collins, 1994).

By using operator formalism, the parabolic equation can be derived in another fashion (Lee and McDaniel, 1987, pp. 314–15). By assuming commutation of operators,

$$\frac{\partial}{\partial r}\frac{\partial}{\partial z}\Psi = \frac{\partial}{\partial z}\frac{\partial}{\partial r}\Psi \tag{4.31}$$

the equation

$$\frac{\partial^2 \Psi}{\partial r^2} + \frac{\partial^2 \Psi}{\partial z^2} + 2ik_0\frac{\partial \Psi}{\partial r} + k_0^2(n^2 - 1)\Psi = 0 \tag{4.32}$$

can be factored as

$$\left[\frac{\partial}{\partial r} + ik_0 - i\sqrt{k_0^2 + k_0^2(n^2-1) + \frac{\partial^2}{\partial z^2}}\right]$$
$$\times\left[\frac{\partial}{\partial r} + ik_0 + i\sqrt{k_0^2 + k_0^2(n^2-1) + \frac{\partial^2}{\partial z^2}}\right]\Psi = 0 \quad (4.33)$$

By considering only the outgoing wave, equation (4.33) reduces to

$$\left[\frac{\partial}{\partial r} + ik_0 - i\sqrt{k_0^2 + k_0^2(n^2-1) + \frac{\partial^2}{\partial z^2}}\right]\Psi = 0 \quad (4.34)$$

which can be rewritten as

$$\frac{\partial}{\partial r}\Psi = i(\sqrt{Q} - k_0)\Psi \quad (4.35)$$

where

$$Q = k_0^2 + k_0^2(n^2-1) + \frac{\partial^2}{\partial z^2} \quad (4.36)$$

The square-root operator (Q) can be approximated using a rational functional representation

$$\sqrt{Q} = k_0\left(\frac{A+Bq}{C+Dq}\right) \quad 4.37)$$

where

$$q = \left(\frac{Q}{k_0^2}\right) - 1 \quad (4.38)$$

Using equation (4.37), equation (4.35) can be rewritten as

$$\frac{\partial}{\partial r}\Psi = ik_0\left(\frac{A+Bq}{C+Dq} - 1\right)\Psi \quad (4.39)$$

For $[A\ B\ C\ D] = [1\ \frac{1}{2}\ 1\ 0]$, equation (4.39) reverts back to that of Tappert (equation (4.30)).

4.7.3 Wide-angle and 3D adaptations

Tappert's (1977) original split-step PE model handles small-angle ($\leqslant 15°$) propagation paths, a restriction imposed by the basic paraxial assumption. Unless the split-step algorithm is modified for wide angles (e.g. Thomson and Chapman, 1983), large phase errors may be introduced into the solution. In order to extend the maximum angle of half-beamwidth propagation in the PE model, alternate forms of the square-root operator have been explored. For split-step PE formulations, the Thomson–Chapman

(1983) wide-angle approximation has been used. The Claerbout (1976, p. 206) wide-angle approximation is used in some finite-difference PE formulations with values of $[A\ B\ C\ D] = [1\ \frac{3}{4}\ 1\ \frac{1}{4}]$ in equation (4.39). Other approximations of the square-root operator have been explored, including higher-order Padé forms that have been incorporated in finite-difference and finite-element PE formulations to achieve near 90° half-beamwidth propagation (Chin-Bing *et al.*, 1993b). Thomson and Wood (1987) investigated a post-processing method for correcting phase errors in the parabolic approximation approach. This method was later extended by Thomson and Mayfield (1994) to include an exact, nonlocal boundary condition at the sea floor. Since strong boundary (surface and bottom) interactions are associated with wide-angle propagation, the accurate treatment of irregular interfaces and rough boundaries assumes greater importance in PE models (Bucker, 1983; Lee and McDaniel, 1983; Dozier, 1984; Collins and Chin-Bing, 1990). Adaptations to under-ice environments are also of interest to sonar modelers.

Extensions to three dimensions are generally implemented in an approximate manner (e.g. $N \times 2D$), although the primary parabolic wave equation is fully three-dimensional (Perkins *et al.*, 1983; Siegmann *et al.*, 1985; Lee and Siegmann, 1986). Such extensions are frequently coupled with wide-angle modifications in order to achieve maximum utility from the models (Botseas *et al.*, 1983; Thomson and Chapman, 1983).

4.7.4 Range-refraction corrections

Range-refraction corrections are introduced to accommodate propagation through strong oceanic fronts. The standard parabolic approximation to the acoustic wave equation is known to have intrinsic phase errors which will degrade the accuracy of any PE solution for long-range propagation in the ocean (Tappert and Lee, 1984; Jensen and Martinelli, 1985). The accuracy can be improved using updated mean phase speeds as a function of range. Tolstoy *et al.* (1985) have suggested a simple transformation of the sound speed profile to compensate for the inability of the PE approach to correctly locate the range of the signal turning points (e.g. Brock *et al.*, 1977).

A model called LOGPE (Berman *et al.*, 1989) uses a logarithmic expression for the index of refraction in the standard parabolic equation in order to closely approximate solutions of the more exact Helmholtz equation in weakly range-dependent ocean environments.

Schurman *et al.* (1991) developed an energy-conserving PE model that incorporated range refraction.

4.7.5 High-frequency adaptations

At frequencies higher than about 500 Hz, PE models become impractical due to excessive execution times. Computational intensity is proportional to the number of range-interval steps. As the frequency increases, the step size decreases and more steps are thus required to achieve the desired prediction range. High-frequency approximations can be obtained by introducing a hybrid approach that combines aspects of PE and ray theory models (Tappert *et al.*, 1984). A special-purpose microcomputer model (PESOGEN) has also been developed to compute the acoustic field in a fully range-dependent environment over a wide range of frequencies for both deep- and shallow-water regions with greatly reduced execution times (Nghiem-Phu *et al.*, 1984).

The so-called calculation frequency method (CFM) can be used to extend PE solutions to high frequencies with execution speeds typically associated with those at low frequencies (Moore-Head *et al.*, 1989). The CFM produces range-averaged transmission loss information by substituting a (low) calculation frequency for a (high) prediction frequency. This substitution is accomplished by sacrificing volume attenuation and boundary-loss phase information under the assumption that diffraction and interference effects are not important at the desired (high) prediction frequency.

4.7.6 Time-domain applications

The parabolic equation approach has been adapted to consider time-domain applications. Specifically, the acoustic pressure is advanced in time by the so-called progressive wave equation (PWE). However, the PWE is, for practical, purposes, limited to narrow-angle propagation paths. Collins (1988b) has derived an inverse Fourier transform of the wide-angle parabolic equation which he terms the time-domain PE (TDPE). The TDPE advances the acoustic pressure field in range. Collins (1988b) used the TDPE to investigate the effects of sediment dispersion on pulse (or broadband) propagation in the ocean.

The progressive wave equation has also been adapted to model the propagation of a nonlinear acoustic pulse which is subject to refraction and diffraction. McDonald and Kuperman (1987) examined the problem of acoustic propagation of finite-amplitude pulses and weak shocks in the ocean where refraction can lead to caustic formation. Their result was a first-order nonlinear progressive-wave equation (NPE) which is the nonlinear time-domain counterpart of the linear frequency-domain parabolic equation (PE). When the nonlinear term is omitted, the NPE reduces to the linear frequency-domain PE. The simplicity of their formulation suggests additional applications to broadband linear acoustic problems in the ocean.

Orchard *et al.* (1992) described the development of a time-domain PE model appropriate for use in 3D ocean acoustic propagation problems.

4.8 THE RAYMODE MODEL – A SPECIFIC EXAMPLE

The RAYMODE model, originally developed by Leibiger (1968), numerically solves the wave equation for underwater acoustic propagation using the multipath expansion approach (section 4.5). RAYMODE is actually a hybrid approach involving elements of both ray theory and wave theory. RAYMODE is a stand-alone propagation model intended for application to passive sonar performance prediction problems in range-independent ocean environments, and it has been designated the US Navy standard model for such applications.

The RAYMODE model is configuration managed, meaning that all updates and modifications are strictly controlled and disseminated by the US Navy (Naval Oceanographic Office, 1993). Accordingly, prospective users should consult the latest version of the model's physics before implementation.

The model accommodates beam patterns at both the source and receiver. Boundary interaction losses are also included. Sea-surface losses are computed by a user-supplied table or by internal algorithms. Surface-scattering losses assume two mechanisms: (1) a high-frequency, large-roughness loss (Beckmann and Spizzichino, 1963, p. 89; Clay and Medwin, 1964); and (2) a low-frequency loss (Marsh and Schulkin, 1962a). Sea-floor losses are computed through identification of bottom province types in order to access stored values, or by a user-supplied table. The volume attenuation expression of Thorp (1967) is used. Transmission loss is calculated both coherently and incoherently.

The basic assumptions incorporated in RAYMODE are:

(1) range-independent sound speed and bathymetry;
(2) sound speed profile fit with segments such that the index of refraction squared is a linear function of depth:
(3) multiple reflections due to sound speed discontinuities are ignored except in surface duct situations;
(4) plane wave reflection coefficients are assumed;
(5) only a finite number of ray cycles are considered in the multipath evaluation of pressure integrals;
(6) a harmonic source is assumed;
(7) a water density of unity is assumed.

The specific implementation of the multipath expansion technique is explained as follows. The infinite integral in equation (4.13) is expressed as a sum of finite integrals, each of which is associated with a particular ray path. In equation (4.13), ϕ is taken to represent the acoustic field pressure and the integration is performed piecewise over particular ray path regions defined by the limiting rays (R.L. Deavenport, 1978, unpublished manuscript).

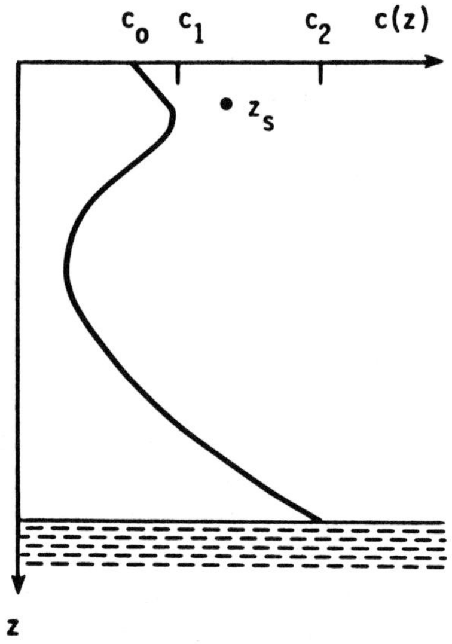

Fig. 4.12 Typical environment for the RAYMODE propagation model; sound speed profile with source located at depth z_s.

$$\int_{-\infty}^{\infty} \simeq \int_{\xi_3}^{\xi_0} = \underset{\text{(SD)}}{\int_{\xi_1}^{\xi_0}} + \underset{\text{(CZ)}}{\int_{\xi_2}^{\xi_1}} + \underset{\text{(BB)}}{\int_{\xi_3}^{\xi_2}} \tag{4.40}$$

where $\xi_i = \omega/c_i$, c_i = sound speed at vertexes, SD = surface duct region, CZ = convergence zone region, and BB = bottom bounce region.

Figure 4.12 illustrates a typical geometry. Here, z_s is the source depth, c_0 is the sound speed at the sea surface, c_1 is the sound speed at the sonic layer depth, and c_2 is the sound speed at the sea floor (at the base of the water column). The particular sound speed profile and source depth geometry illustrated here would support all three ray paths described (i.e. surface duct, convergence zone and bottom bounce). Note that

$$\xi_3 = \left[\frac{\omega}{c_s}\right] \cos \theta_s$$

where θ_s is the maximum ray angle leaving source and c_s is the sound speed at source depth.

The integral for each region is expanded into four parts:

$$\phi = \sum_{i=1}^{4} \phi_i$$

Each ϕ_i corresponds to one of the four ray types associated with the upgoing and downgoing rays at both the source and the receiver. The path geometries and associated sign conventions are prescribed in the RAYMODE L-path chart illustrated in Fig. 4.13. These paths may be compared with those described previously in Fig. 4.2. The cycle counter (q) represents the number of bottom reflections. The sign convention is the same as that described earlier in connection with arrival structure diagrams: a positive sign indicates an upgoing ray while a negative sign indicates a downgoing ray. Thus, a receiver directed upward receives downgoing (or negative) rays while a source directed upward transmits upgoing (or positive) rays.

Each segment is then evaluated by either wave theory or ray theory. A special option is available to the user for surface ducts in order to incorporate the effects of leakage (diffraction effects). This option is exercised only for that wavenumber segment for which it is appropriate. If the special surface duct option is not exercised, then the segment is evaluated using the normal mode approach. However, if the number of normal modes for a given segment exceeds ten, then a multipath expansion of the four parts of the integral is employed. If the segment contains bottom bounce paths, then ray-theoretical methods are used; however, at frequencies below 45 Hz, a normal mode solution is used to evaluate propagation through the sediments. There is also a separate high-frequency routine that is used for frequencies greater than 1 kHz. The key parameters controlling the decision for the method of solution are the acoustic frequency and the physical environment. Consequently, slight variations in the problem statement may result in unexpectedly abrupt changes in transmission loss brought about by changes in the method of solution.

The resulting acoustic pressure field is modified by a beam pattern attenuation factor characterizing the off-axis beam position of an equivalent ray.

The real and imaginary pressure components (P_i) are then used to form incoherent and coherent intensities for each range point and each ξ-partition. Transmission loss (TL), relative to unit intensity at unit distance from the source, is calculated from the incoherent and coherent intensity sums as follows:

$$\text{TL (incoherent)} = -10\log\left\{\sum_{i=1}^{4}\left[(\text{Re}\,\overset{(\text{SD})}{P_i})^2 + (\text{Im}\,\overset{(\text{SD})}{P_i})^2 + (\text{Re}\,\overset{(\text{BB})}{P_i})^2 + (\text{Im}\,\overset{(\text{BB})}{P_i})^2 + (\text{Re}\,\overset{(\text{CZ})}{P_i})^2 + (\text{Im}\,\overset{(\text{CZ})}{P_i})^2\right]\right\} + \alpha r \tag{4.41}$$

$$\text{TL (coherent)} = -10\log\left\{\left(\sum_{i=1}^{4}\left[\text{Re}\,\overset{(\text{SD})}{P_i} + \text{Re}\,\overset{(\text{BB})}{P_i} + \text{Re}\,\overset{(\text{CZ})}{P_i}\right]\right)^2 + \left(\sum_{i=1}^{4}\left[\text{Im}\,\overset{(\text{SD})}{P_i} + \text{Im}\,\overset{(\text{BB})}{P_i} + \text{Im}\,\overset{(\text{CZ})}{P_i}\right]\right)^2\right\} + \alpha r \tag{4.42}$$

RESOLUTION* OF SIGN OF SOURCE AND RECEIVER ANGLES IN RAYMODE:

GENERAL RANGE $\bar{R} = qR_C \pm R_S \pm R_d$ **OR** $qR_C + P_2R_S + P_1R_d$ **WHERE**

q = CYCLE 0, 1, 2,...λ
R_C = CYCLE RANGE
R_S = SOURCE TERM
R_d = RECEIVER TERM

PATH L	P_2	P_1	EQUATION $\bar{R}$	q = 0 (D.P.)	q = 1 (EX. BB)	q = 2	SOURCE ANGLE	RECEIVER ANGLE
1	−	+	$qR_C - R_S + R_d$	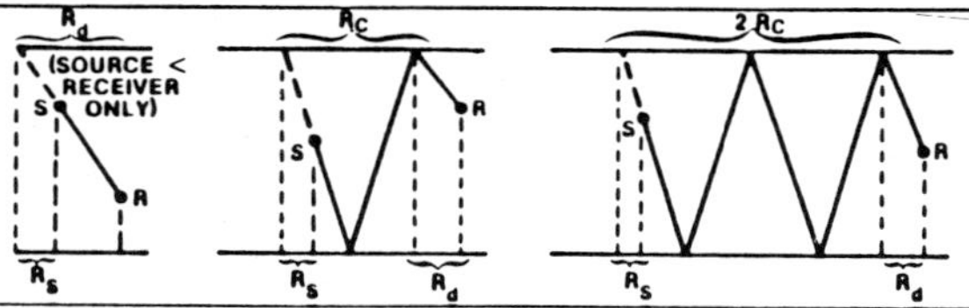			+ DOWN	− UP
2	+	+	$qR_C + R_S + R_d$	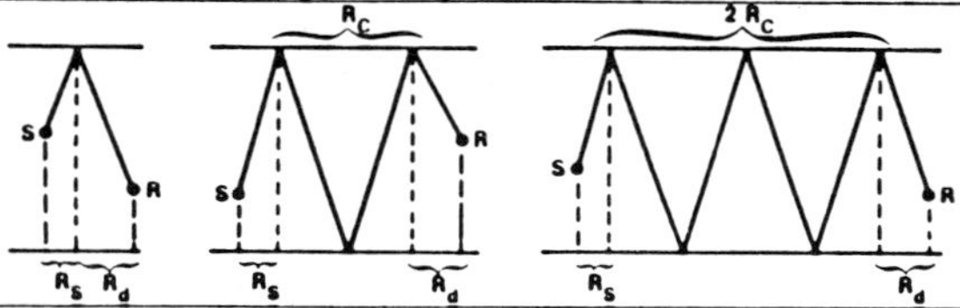			− UP	− UP
3	−	−	$qR_C - R_S - R_d$	(DOES NOT EXIST)	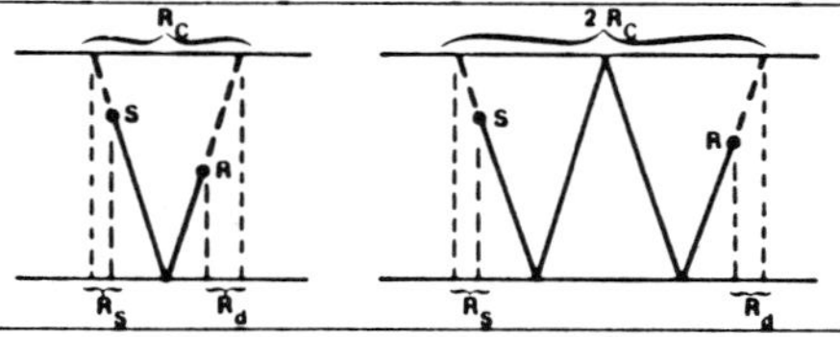		+ DOWN	+ DOWN
4	+	−	$qR_C + R_S - R_d$	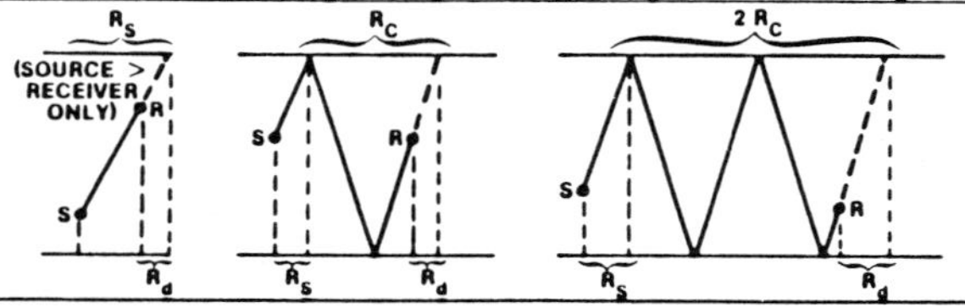			− UP	+ DOWN

*SIGN OF ANGLE IS OPPOSITE SIGN OF RANGE ADJUSTMENT TERM FOR BOTH SOURCE AND RECEIVER

Fig. 4.13 RAYMODE L-path chart illustrating sign resolution of source and receiver angles (Yarger, 1982).

where α is Thorp's (1967) attenuation coefficient.

In the RAYMODE model, user-supplied input information consists of:

(1) sound speed profile
(2) source and receiver depths
(3) frequency
(4) horizontal range determinats
*(5) bottom loss
*(6) surface loss
*(7) source and receiver beam patterns
#(8) program controls
#(9) output options

where * = only if required, otherwise default values are used; and # = examples include: number of ray cycles to be processed; ray angle limits; mode cutoff values; calculation method options.

Graphical outputs from RAYMODE include:

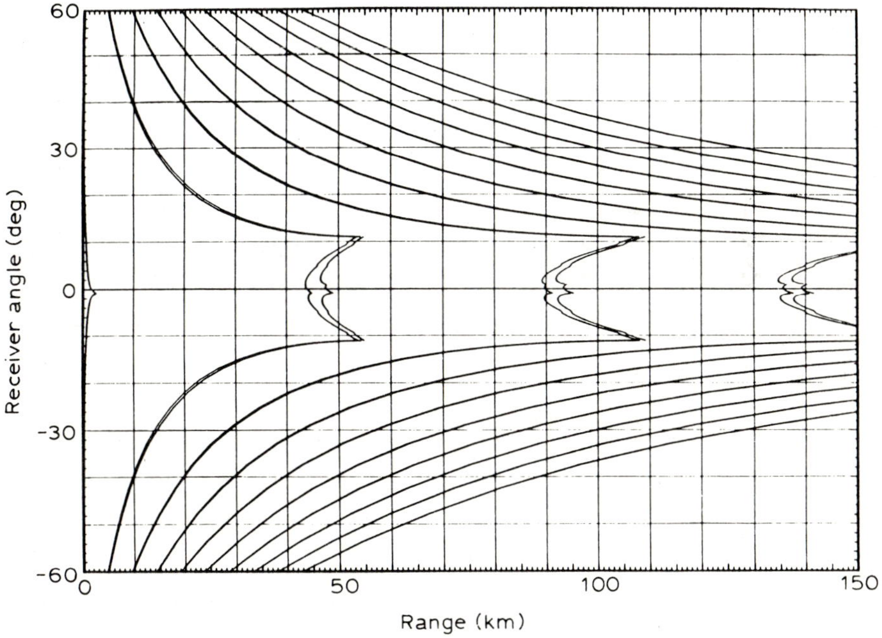

Fig. 4.14 RAYMODE arrival structure (Yarger, 1982). Graph shows receiver angle versus horizontal range for selected rays representing bottom bounce and convergence zone (CZ) paths. The CZ paths are associated with the near-axis features located at 45, 90 and 135 km.

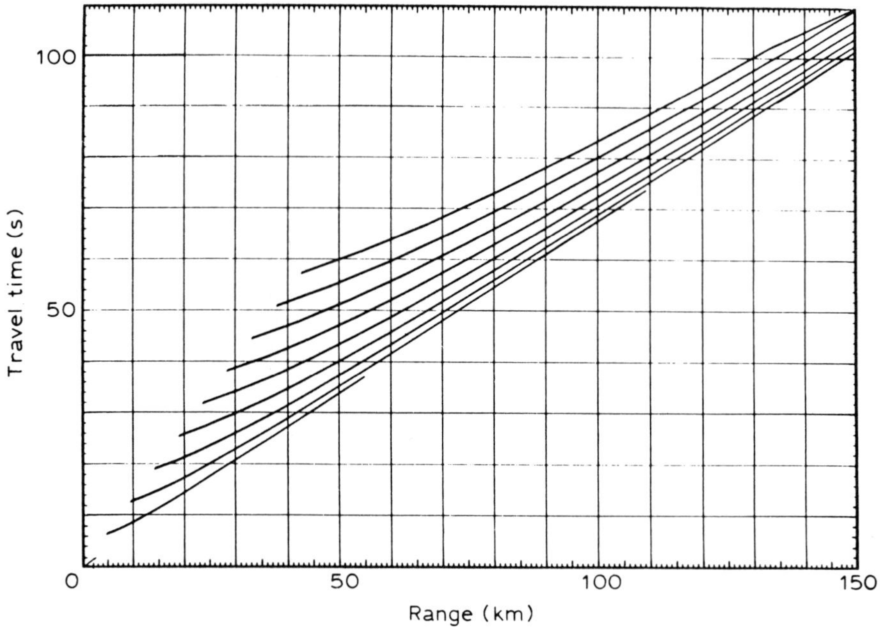

Fig. 4.15 RAYMODE travel time versus horizontal range for the ray paths selected in Fig. 4.14 (Yarger, 1982).

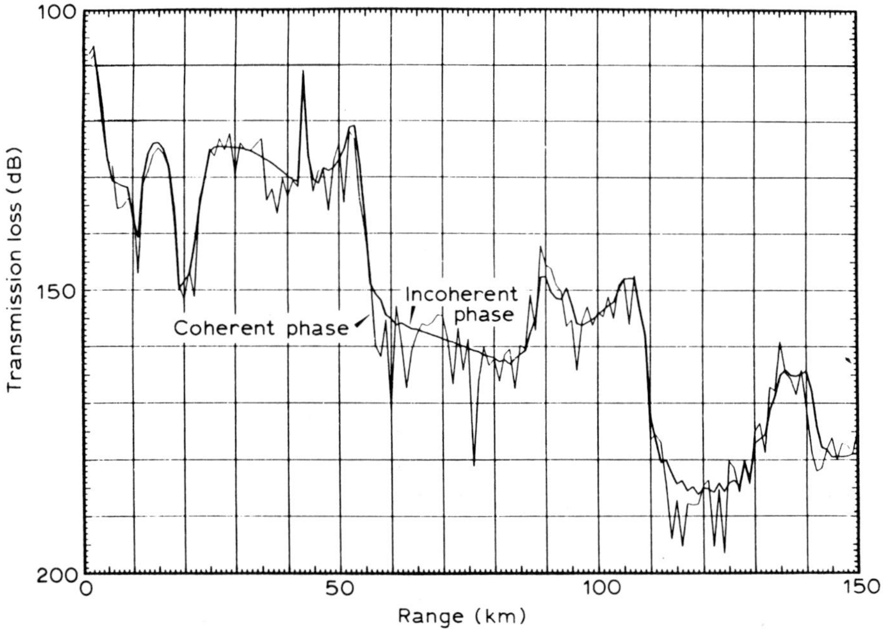

Fig. 4.16 RAYMODE transmission loss versus range for both coherent and incoherent phase summation (Yarger, 1982).

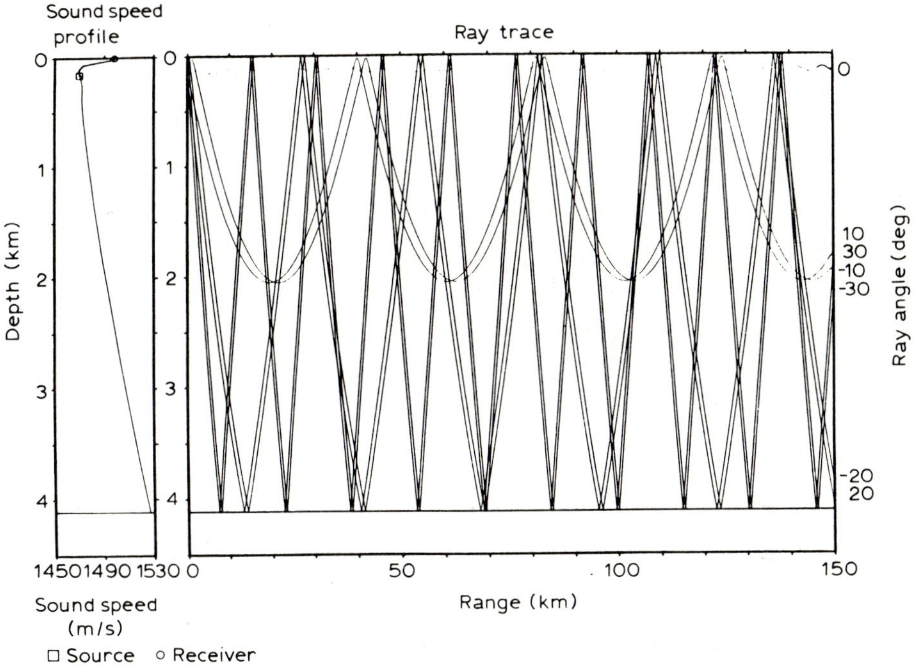

Fig. 4.17 RAYMODE ray trace diagram (Yarger, 1982).

(1) sound speed profile plot, and plots of bottom loss, surface loss and beam deviation loss;
(2) source and receiver angle versus horizontal range for selected ray paths (Fig. 4.14);
(3) travel time versus horizontal range for the ray paths in item (2) above (Fig. 4.15);
(4) transmission loss versus range for coherent or incoherent combination of multipaths with beam pattern attenuation (Fig. 4.16);
(5) ray trace diagram (Fig. 4.17).

Additional references of RAYMODE are cited in the model summary section of this chapter. Operational versions of RAYMODE can be interfaced directly with the US Navy standard databases (section 10.5).

4.9 NUMERICAL MODEL SUMMARIES

This section presents a summary of available underwater acoustic propagation models.These models are applicable to research in support of sonar

MODEL TYPE	APPLICATIONS							
	SHALLOW WATER				DEEP WATER			
	LOW FREQUENCY		HIGH FREQUENCY		LOW FREQUENCY		HIGH FREQUENCY	
	RI	RD	RI	RD	RI	RD	RI	RD
RAY THEORY	○	○	◑	●	◑	◑	●	●
NORMAL MODE	●	◑	●	◑	●	◑	◑	○
MULTIPATH EXPANSION	○	○	◑	○	◑	○	●	○
FAST FIELD	●	○	●	○	●	○	◑	○
PARABOLIC EQUATION	◑	●	○	○	◑	●	◑	◑

LOW FREQUENCY (< 500 HZ)

HIGH FREQUENCY (> 500 HZ)

RI: RANGE-INDEPENDENT ENVIRONMENT

RD: RANGE-DEPENDENT ENVIRONMENT

● MODELING APPROACH IS BOTH APPLICABLE (PHYSICALLY) AND PRACTICAL (COMPUTATIONALLY)

◑ LIMITATIONS IN ACCURACY OR IN SPEED OF EXECUTION

○ NEITHER APPLICABLE NOR PRACTICAL

Fig. 4.18 Domains of applicability of underwater acoustic propagation models (adapted from Jensen, 1982; Proc. MTS/IEEE Oceans 82 Conf., pp. 147–54; copyright by IEEE).

design and sonar operation. As is true of any summary of this nature, the information represents a snap-shot in time of the present state of the art. While comprehensive, this summary does not claim to be exhaustive.

In order to optimize the utility of this information, the models are arranged in categories reflecting the basic modeling technique employed (i.e. the five canonical approaches) as well as the ability of the model to handle environmental range dependence. Such factors define what is termed 'domains of applicability'. Strict categorization is occasionally compromised by hybrid models, and some arbitrariness has been allowed in the classification process. The environmental range dependence considers variations in sound speed or bathymetry. Other parameters may be considered to be range-dependent by some of the models, though they are not explicitly treated in this summary.

The specific utility of these categories is further explained below. In sonar design and operation problems, the analyst is normally faced with a decision matrix involving water depth (deep versus shallow), frequency (high versus low) and range dependence (range-independent versus range-dependent ocean environments). Jensen (1982, 1984) has developed a very useful classification scheme for optimizing this decision logic against the available modeling approaches (and their domains of applicability). Jensen's scheme is slightly modified here to accommodate the five modeling approaches utilized in this book (Fig. 4.18).

The following assumptions and conditions were imposed in the construction of Fig. 4.18:

1. Shallow water includes those water depths for which the sound can be expected to significantly interact with the sea floor.
2. The threshold frequency of 500 Hz is somewhat arbitrary, but it does reflect the fact that above 500 Hz many wave-theoretical models become computationally intensive; also, below 500 Hz the physics of some ray-theoretical models may become questionable.
3. A solid circle indicates that the modeling approach is both applicable (physically) and practical (computationally). Distinctions based on speed of execution will continually change as progress is made in computational capabilities.
4. A partial circle indicates that the modeling approach has some limitations in accuracy or in speed of execution.
5. An open circle indicates that the modeling approach is neither applicable nor practical.

Eighty-three stand-alone passive propagation models have been identified in Table 4.1, where they are categorized according to the five modeling approaches discussed earlier. They are further segregated according to their ability to handle range-dependent environments. The numbers in parentheses following each model refer to appropriate documentation. In

Table 4.1 Summary of underwater acoustic propagation models

Technique	*Range independent*		*Range dependent*[a]	
Ray theory	CAPARAY [1] FACT [2] FLIRT [3] GAMARAY[c] [4] ICERAY[b] [5] INSIGHT [6]	PLRAY/VLA [7] RANGER [8]	ACCURAY [9] FACTEX [10] GRASS [11] HARPO [12] MEDUSA [13] MPC [14]	MPP [15] Pedersen [16] RAYWAVE [17] RP-70 [18] SHALFACT [19] TRIMAIN [20]
Normal mode	AP-2/5 [21] COMODE [22] DODGE [23] FNMSS[b] [24] NEMESIS [25] NLNM [26] NORMOD3 [27]	NORM2L [28] PROTEUS[c] [29] SHEAR2 [30] Stickler [31]	ADIAB [32] ASERT [33] ASTRAL 4.1[a,b] [34] CENTRO [35] CMM3D [36] COUPLE [37] Kanabis [38]	KRAKEN [39] MOATL [40] PROLOS [41] SNAP [42] WEDGE [43] WRAP [44] 3D Ocean [45]
Multipath expansion	FAME [46] MULE[c] [47] NEPBR [48] RAYMODE 8.05[a] [49]			
Fast field	FFP [50] Kutschale FFP[b] [51] MSPFFP[b] [52] OASES [53] Pulse FFP[c] [54] SAFARI [55]		RDFFP [56] SAFRAN [57]	
Parabolic equation			Corrected PE [58] DREP [59] FEPE [60] FEPES [61] FOR3D [62] HAPE [63] HYPER [64] IFD Wide Angle [65] LOGPE [66] MaCh1 [67] MIPE [68] MOREPE [69] OS2IFD [70]	PAREQ [71] PE 3.4[a,b] [72] PE-FFRAME [73] PESOGEN [74] Spectral PE [75] TDPE[c] [76] Two-Way PE [77] ULETA [78] UMPE [79] UNIMOD [80] 3DPE [81] 3DPE [82] 3D TDPA[c] [83]

[a] US Navy standard model.
[b] Under-ice capability.
[c] Broadband capability.

(Continued)

(Continued)

References to propagation models: [1] Westwood and Vidmar (1987); [2] Spofford (1974), Baker and Spofford (1974), Jacobs (1982), McGirr *et al.* (1984), Holt (1985), King and White (1986a, 1986b); [3] McGirr and Hall (1974); [4] Westwood (1992); [5] Stotts and Bedford (1991), Stotts *et al.* (1994); [6] Harrison *et al.* (1990), Packman *et al.* (1992); [7] Bartberger (1978a); [8] Foreman (1977); [9] Westwood (1989b); [10] Garon (1975); [11] Cornyn (1973a, 1973b); [12] Jones (1982), Jones *et al.* (1982, 1986), Weickmann *et al.* (1989), Newhall *et al.* (1990), Georges *et al.* (1990), Harlan *et al.* (1991a, 1991b), Jones and Georges (1991); [13] Foreman (1982, 1983); [14] Miller (1982, 1983); [15] Spofford (1973a), Jacobs (1974); [16] Pedersen *et al.* (1962), Gordon (1964); [17] Bucker (1971), Watson and McGirr (1975); [18] Colilla (1970); [19] Garon (1976); [20] Roberts (1974); [21] Bartberger (1978b); [22] Gilbert *et al.* (1983); [23] Ward (1989); [24] Gordon and Bucker (1984); [25] Gonzalez and Payne (1980); [26] Gordon (1979); [27] Blatstein (1974); [28] Ellis (1980); [29] Gragg (1985); [30] Ellis and Chapman (1985); [31] Stickler (1975), Stickler and Ammicht (1980), Ammicht and Stickler (1984); [32] Etter *et al.* (1984) – secondary reference; [33] Lukas *et al.* (1980a); [34] Spofford (1979), Blumen and Spofford (1979), White *et al.* (1988), Dozier and White (1988), White (1992), White and Corley (1992a, 1992b); [35] Arvelo and Überall (1990); [36] Chiu and Ehret (1990, 1994); [37] Evans (1983, 1986); [38] Kanabis (1975, 1976); [39] Porter and Reiss (1984, 1985), Porter (1991); [40] Miller and Ingenito (1975), Ingenito *et al.* (1978), Miller and Wolf (1980); [41] Ellis (1985); [42] Jensen and Ferla (1979); [43] Primack and Gilbert (1991); [44] Perkins *et al.* (1990), Kuperman *et al.* (1991); [45] Weinberg and Burridge (1974); [46] Weinberg (1981); [47] Weinberg (1985a); [48] McCammon and Crowder (1981); [49] Leibiger (1968), Yarger (1976, 1982), Medeiros (1982a, 1985a), Almeida and Medeiros (1985), Naval Oceanographic Office (1991a, 1991c); [50] DiNapoli (1971), DiNapoli and Deavenport (1980); [51] Kutschale (1973); [52] Kutschale (1984); [53] Dozier and Cavanagh (1993) – secondary reference; [54] Porter (1990); [55] Schmidt and Glattetre (1985), Schmidt (1988); [56] Gilbert and Evans (1986); [57] Seong (1990); [58] Perkins and Baer (1978); [59] Thomson and Chapman (1983), Thomson (1990); [60] Collins (1988a); [61] Collins (1993a, 1993b); [62] Botseas *et al.* (1987), Lee *et al.* (1988, 1992); [63] Greene (1984); [64] Tappert *et al.* (1984); [65] Lee and Botseas (1982), Botseas *et al.* (1983); [66] Berman *et al.* (1989); [67] Brown (1994); [68] Chin-Bing *et al.* (1993b) – secondary reference; [69] Schurman *et al.* (1991); [70] Robertson *et al.* (1991); [71] Jensen and Krol (1975); [72] Brock (1978), Holmes (1988), Holmes and Gainey (1992a, 1992b, 1992c); [73] Chin-Bing and Murphy (1988), Murphy and Chin-Bing (1988); [74] Nghiem-Phu *et al.* (1984); [75] Orris and Collins (1994); [76] Collins (1988b); [77] Collins and Evans (1992); [78] Palmer *et al.* (1988b); [79] Smith and Tappert (1993); [80] Etter *et al.* (1984) – secondary reference; [81] Perkins and Baer (1982); [82] Collins and Chin-Bing (1990); [83] Orchard *et al.* (1992).

---

some cases, documentation is either unavailable or is inaccessible (i.e. unpublished); reference is then made to secondary references which provide some information on the models. It should be noted that model documentation can range from informal programming commentaries to journal articles to detailed technical reports containing a listing of the actual computer code. Those propagation models designated as standards by the US Navy are noted by [a]; models suitable for use in ice-covered regions are noted by [b]; and models capable of broadband (or pulse) computations are noted by [c]. Abbreviations and acronyms are defined in Appendix A.

Taken together, Fig. 4.18 and Table 4.1 provide a useful mechanism for selecting a subset of candidate models once some preliminary information is available concerning the intended applications. Although Table 4.1 identifies two range-dependent models in the fast-field category, the experimental nature of these methods imposes uncertain restrictions on their application to sonar performance modeling in range-dependent ocean environments (see section 4.6). Therefore, Fig. 4.18 recommends that application of fast-field models be limited to range-independent environments.

There are many cases in which two different modeling approaches are suitable (as indicated by a solid circle in Fig. 4.18). The user should be encouraged to use candidates from both categories in order to assess any intermodel variations. Even when only one modeling approach is suitable, more than one candidate model should be tried. In the event of divergences in model predictions (a very likely event), the envelope defining spreads in model outputs can be used to determine the degree of uncertainty. Moreover, this envelope will likely bracket the correct result.

5

Propagation II. Mathematical models (Part Two)

5.1 BACKGROUND

Mathematical models of underwater acoustic propagation include both numerical models and empirical models. Chapter 4 addressed the theoretical development of numerical models and summarized their availability. This chapter addresses the development of empirical models applicable to special propagation paths such as surface ducts, shallow water and Arctic half-channels. Where appropriate, comparisons are made with predictions generated by numerical models. Data support requirements for mathematical models of propagation are discussed and a select number of special applications are described in order to highlight promising areas for future research and development.

5.2 SURFACE DUCT MODELS

Properties of the surface duct were discussed previously in Chapter 3. Both ray-theoretical and wave-theoretical solutions can be applied to propagation in the surface duct.

5.2.1 Ray theory models

An expression for transmission loss in a surface duct may be obtained through simple ray-theoretical considerations (Urick, 1983). In Fig. 5.1, let a nondirectional source of sound be located at P in a surface duct (or mixed layer). Also, let c_0 denote the reference sound speed in the duct. Of all the rays leaving the source, only those within a certain limiting angle 2θ remain in the duct. At a distance of 1 m, the power contained in this ray bundle is distributed over a portion of the spherical surface A_1. At a long

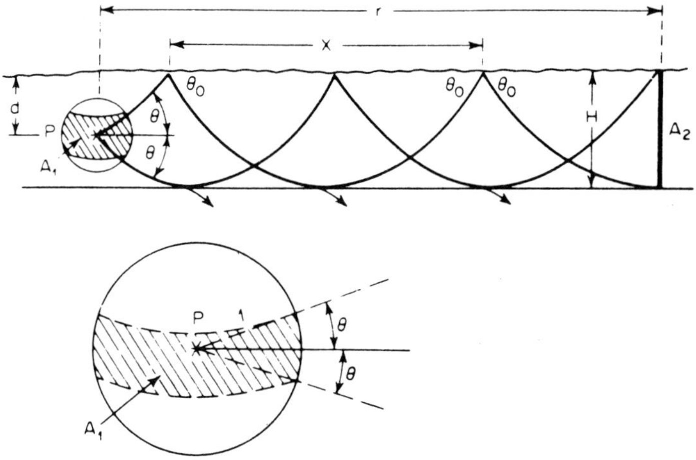

Fig. 5.1 Propagation geometry in a surface duct (Urick, 1983; *Principles of Underwater Sound*, 3rd edn; reproduced with permission of McGraw-Hill Publishing Company).

distance r, this same amount of power (in the absence of leakage and absorption) is distributed over a cylindrical surface A_2. Because the power crossing areas A_1 and A_2 is conserved, the transmission loss to range r, averaged over the duct thickness H, is

$$\mathrm{TL} = 10 \log \frac{A_2}{A_1}$$

By geometry

$$A_2 = 2\pi r H \quad \text{and} \quad A_1 = 2\pi \int_{-\theta}^{\theta} \cos\theta \, \mathrm{d}\theta = 4\pi \sin\theta$$

Therefore, for the surface duct

$$\mathrm{TL} = 10 \log \frac{rH}{2\sin\theta} = 10 \log r r_0 = 10 \log r_0^2 \frac{r}{r_0}$$

where $r_0 = H/(2\sin\theta)$ and θ is the inclination at source of maximum trapped ray. The quantity $r_0^2 r/r_0$ indicates that the propagation to range r may be viewed as the result of spherical spreading out to a transition range r_0, followed by cylindrical spreading from r_0 to r. The range r_0 also corresponds to the condition in which the vertical extent of the beam subtended by 2θ equals the duct thickness (H); ranges greater than r_0 are generally considered to represent far-field conditions. When the attenuation due to absorption and leakage is added, the duct-transmission-loss expression becomes

$$\mathrm{TL} = 10 \log r_0 + 10 \log r + (\alpha + \alpha_L) r \times 10^{-3} \tag{5.1}$$

where α is the absorption coefficient (dB/km), α_L is the leakage coefficient (dB/km), which expresses the rate at which acoustic energy leaks out of the duct, and r is the range (m). This expression is applicable to all ducts.

For a duct with a constant sound speed gradient (g), in which the rays are arcs of circles with radius of curvature $R = c_0/g$, the following relationships result from geometry for the conditions of large $R (R \gg H)$ and small θ_0 ($\sin\theta_0 \ll 1$) at source depth d (refer to Fig. 5.1):

$$R = c_0/g = \text{radius of curvature of rays}$$

$$x = \sqrt{8RH} = \text{skip distance of limiting ray}$$

$$\theta_0 = \sqrt{\frac{2H}{R}} = \text{maximum angle of limiting ray (rad)}$$

$$\theta = \sqrt{\frac{2(H-d)}{R}} = \text{angle of limiting ray at source depth (rad)}$$

$$r_0 = \sqrt{\frac{RH}{8}}\sqrt{\frac{H}{H-d}} = \frac{x}{8}\sqrt{\frac{H}{H-d}} = \text{transition range}$$

Kinsler *et al.* (1982, pp. 402–6) also present an insightful development of these relationships.

Empirical surface duct equations have been developed by Marsh and Schulkin (1955) based on an extensive set of data collected during Project AMOS (Acoustic, Meteorological and Oceanographic Survey) in 1953–54. These equations have been incorporated into many ray-theoretical models to handle the special case of surface duct propagation. Graphical summaries of the AMOS data were presented in Chapter 3. Other empirically derived formulas have been reviewed by Urick (1982, Chapter 6).

5.2.2 Wave theory models

Pedersen and Gordon (1965) adapted Marsh's (1950) normal mode approach to short ranges for a bilinear-gradient (positive gradient overlying a negative gradient) model of a surface duct (Fig. 5.2). This now classic bilinear surface duct model has been incorporated into some ray-theoretical models to augment their capabilities.

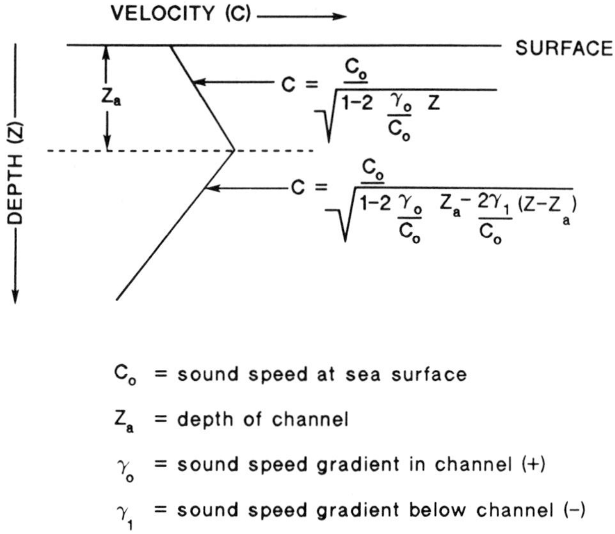

Fig. 5.2 Geometry of bilinear-gradient model (Pedersen and Gordon, 1965).

The transmission loss (TL) is given as a function of range (assuming leakage due to surface roughness can be neglected) by

$$\mathrm{TL} = -10 \log \left| \sum_{1}^{N} \mathrm{H}_0^{(2)}(\lambda_n r) u_n(t) u_n(t_0) \right|^2 - 20 \log \pi + \alpha r \tag{5.2}$$

where N = number of modes included in the computation
$H_0^{(2)}$ = zero-order Hankel function of the second kind
λ_n = complex wavenumber
n = mode counter index ($n = 1, 2, \cdots, N$)
r = range
$u_n(\,)$ = depth function; compare with the normalized mode functions of equation (4.14)
t = ratio of receiver depth to depth of channel
t_0 = ratio of source depth to depth of channel
α = absorption coefficient

Qualitatively, the relationship in equation (5.2) predicts cylindrical spreading as a function of range combined with the effects of modal interactions. These modes may be thought of as damped sinusoidal waves. There are two basic factors that determine the degree to which a particular mode contributes to the result. First, the product $|u_n(t)u_n(t_0)|$ depends on the source and receiver depths, but not on range. Specifically, an exponential damping factor implicit in the formulation causes the relative contribution of a mode to decrease with increasing range and increasing mode number.

By invoking simplifying approximations, Pedersen and Gordon (1965) were able to generate analytical solutions using a variation of equation (5.2) that agreed favorably with both ray-theoretical solutions and with experimental data.

There are three important aspects of this approach that differ from other normal mode solutions: (1) the solution is valid for short ranges because the branch-cut integral is zero for this model; (2) the modes are damped since the wavenumbers are complex; and (3) there are no cutoffs in the frequency domain. Thus, higher-order modes are highly damped and only the lower-order modes need be considered for most practical problems.

Gabrielson (1982) has further examined the application of normal mode models to leaky ducts.

5.2.3 Oceanographic mixed layer models

New developments in numerical modeling in oceanography now permit the depth of the mixed layer to be forecast in both time and location. These predicted values can then be incorporated into surface duct models to predict the corresponding acoustic transmission loss. Such interfacing of models is referred to as coupled ocean acoustic modeling (Mooers *et al.*, 1982).

Models of the mixed layer have largely been restricted to the treatment of one-dimensional (1D) approximations which have proven useful when horizontal advection can be neglected. In many ocean areas, however, the 1D approximations appear to be inappropriate for estimating mixed layer depths (Garwood, 1979).

Mixed layer models are of two basic types: differential and bulk. Differential models use the equations for the conservation of momentum, heat, salt and turbulent kinetic energy (TKE) in their primitive form, and are not integrated over the mixed layer. The mixed layer for these models is defined by the region where the local TKE is large enough to provide a certain minimum level of vertical mixing. Bulk, or integrated, models assume that the mixed layer is a well-defined layer that is uniform in temperature and salinity. The governing equations for these models are obtained by integrating the primitive equations over the depth of the mixed layer.

Mixed layer models respond to three basic types of forcing conditions: wind deepening, heating and cooling. Wind deepening is defined to occur when the mixed layer deepens due to the erosion of the stably stratified region at its base by wind-generated turbulence; the depth of mixing is governed by the strength of the density stratification and the wind speed. Under conditions of heating, the depth of the mixed layer is governed by a balance between the stabilizing effect of surface heating, or a positive surface buoyancy flux, and the effect of mixing due to wind-generated turbulence. This balance governs the mixed layer depth during periods when the mixed layer is shallowing. Under conditions of cooling, a net surface heat loss, or a negative surface buoyancy flux, causes the mixed layer to deepen due to convection. Convection usually occurs at night and is the dominant mechanism for deepening the mixed layer in fall and winter, especially under conditions of reduced solar heating and increased evaporative cooling due to increased wind speeds.

A number of mixed layer models have been developed and some have been coupled with underwater acoustic transmission loss models to provide input (and feedback) of the parameters important for prediction of sound propagation in the surface duct. One system which is operational for fleet applications is the Thermodynamical Ocean Prediction System, or TOPS (Clancy and Martin, 1979). TOPS is classified as a differential mixed

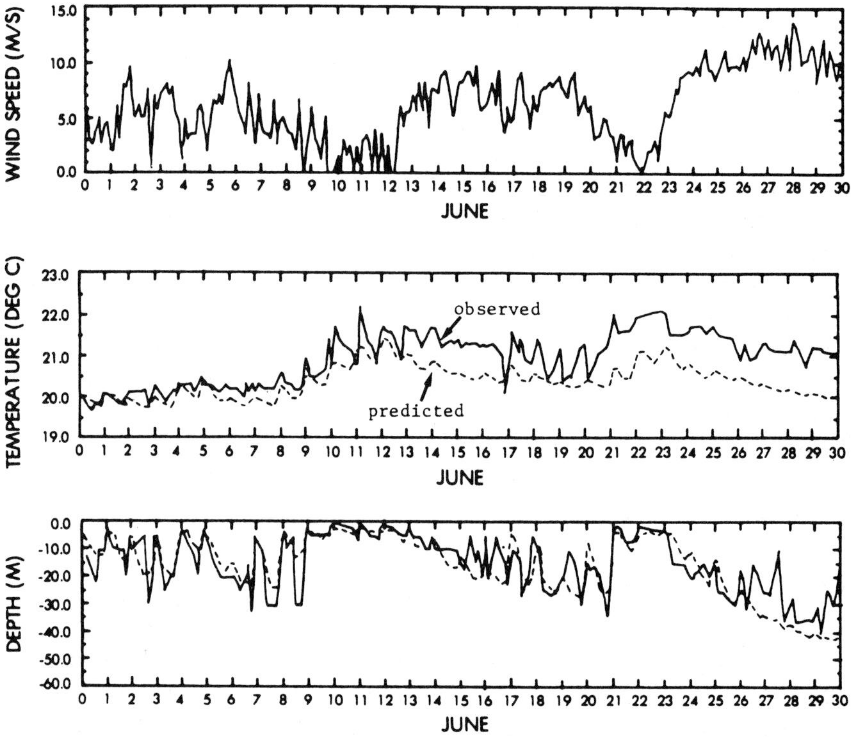

Fig. 5.3 Observations and predictions at Ocean Station November (in the North Pacific Ocean) during June 1961 (Clancy *et al.*, 1981). Upper panel: observed wind speed. Middle panel: observed (solid line) and predicted (dashed line) sea-surface temperatures. Bottom panel: observed (solid line) and predicted (dashed line) mixed layer depths.

layer model. Forecasts appear to agree well with measurements in those ocean areas where the layer depth is dependent primarily on local conditions and not on advection from neighboring regions. A comparison of observations and predictions generated by TOPS (Fig. 5.3) shows reasonable agreement. The top panel is the observed-wind speed. The two lower panels compare the observed (solid lines) and predicted (dashed lines) mixed layer temperatures and depths over the period.

Clancy and Pollack (1983) have advocated coupling the TOPS synoptic mixed layer model to an objective ocean thermal analysis system in order to produce a continuously updating, real-time, analysis-forecast-analysis system. The forecast component (TOPS) employs the Mellor and Yamada (1974) Level-2 turbulence parametrization scheme. It includes advection by instantaneous wind drift and climatologically averaged geostrophic currents, and is forced by surface fluxes supplied by atmospheric models.

The Diurnal Ocean Surface Layer (DOSL) model, which predicts diurnal patterns of the sea-surface temperature (SST) field, has been developed to support high-frequency sonar operations against shallow targets (Clancy *et al.*, 1991b; Hawkins *et al.*, 1993). Diurnal variability in the mixed layer dominates short-term changes in the acoustic behavior of the surface duct; the well-known 'afternoon effect' discovered during World War II denotes the loss of a surface duct due to the creation of a shallow, transient thermocline by local heating. Diurnal SST changes also affect the use and interpretation of satellite data. Specifically, satellites measure the skin temperature of the ocean, and these surface measurements may or may not be characteristic of the bulk mixed layer temperature (the desired parameter for naval operations) depending on the amplitude and phase of the SST cycle at the time of measurement. Since the diurnal SST response is a strongly nonlinear function of the wind speed, smooth variations in the synoptic wind field can produce sharp horizontal SST gradients that might be misinterpreted as the thermal signature of ocean frontal features.

5.3 SHALLOW-WATER DUCT MODELS

5.3.1 Shallow-water propagation characteristics

Acoustic propagation in shallow water is dominated by repeated interactions with the sea floor. Generally, shallow water is restricted to consideration of the continental shelves with depths less than 200 m. Detection ranges in shallow water are severely limited both by the high attenuation which results from interaction with the bottom and by the limited water depth, which will not support the long-range propagation paths available in deep water. Moreover, determination of source location (bearing, range and depth) can be affected by horizontal refraction caused by the repeated boundary reflections over a sloping bottom. Doolittle *et al.* (1988) have

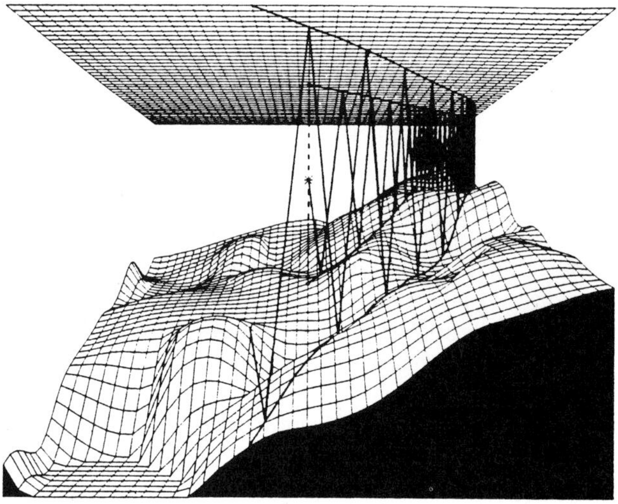

Fig. 5.4 Typical beam trace in a shallow-water region (Bucker, 1994).

experimentally confirmed the horizontal refraction of CW acoustic radiation from a point source in a wedge-shaped ocean environment. A striking graphical presentation of a 3D ray trace in a complicated, wedge-shaped ocean environment is illustrated in Fig. 5.4 (Bucker, 1994). This ray trace displays the effects of horizontal refraction caused by a sloping bottom boundary. The source (denoted by an asterisk) appears in the background.

5.3.2 Optimum frequency of propagation

Understanding and predicting acoustic sensor performance in shallow water is further complicated by the relatively high temporal and spatial variability of the environment. Degradation in sensor performance can result from improper optimization of the frequency or depth of operation at a particular location and time of year.

In shallow-water environments, the optimum frequency of propagation is often the result of competing propagation and attenuation mechanisms at either end of the frequency spectrum. Jensen and Kuperman (1979, 1983) investigated this problem and concluded that the optimum frequency is strongly dependent on water depth, is somewhat dependent on the particular sound speed profile, and is only weakly dependent on the bottom type. Jensen and Kuperman (1983) also noted that shear waves in the bottom were important in determining the optimum frequency of propagation and the actual transmission loss levels at lower frequencies. A major loss mechanism for low-frequency acoustic propagation in shallow water

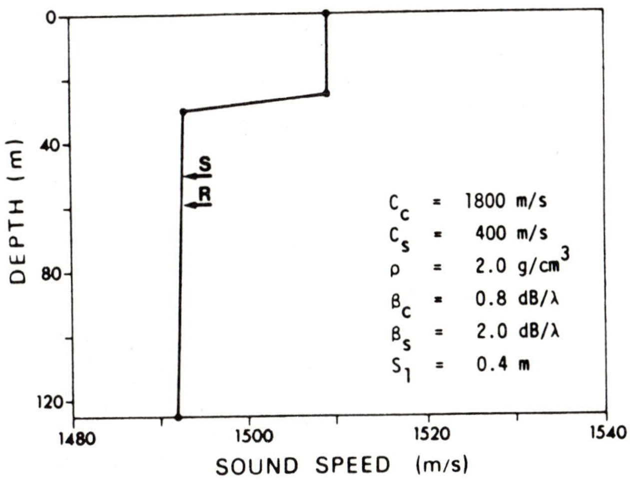

Fig. 5.5 Sound speed profile and bottom properties for a shallow-water area located in the eastern North Atlantic Ocean: c = sound speed, ρ = density, β = attenuation, S_1 = rms bottom roughness. Subscripts 'c' and 's' refer to compressional and shear waves, respectively. Source (*S*) and receiver (*R*) depths are indicated on the sound speed profile (Jensen and Kuperman, 1979).

is the attenuation in ocean sediments. Research results (e.g. Focke, 1984) indicate that variations in attenuation as a function of sediment depth have a significant impact on propagation.

An important analytical tool used by Jensen and Kuperman (1979, 1983) was a frequency-range representation of transmission loss, as illustrated in Figs 5.5 and 5.6. For the particular environment studied (Fig. 5.5), a normal mode model was used to generate repeated transmission loss runs versus frequency. In Fig. 5.6(b), the optimum frequency predicted by the model is about 200 Hz, as evidenced by the elongated axis of low-loss values. These predicted results compared favorably with experimental results (Fig. 5.6 (a)).

Shallow-water modeling techniques include both numerical models and empirical models. Eller (1984b, 1986) has summarized important developments in shallow-water acoustic modeling.

5.3.3 Numerical models

Both ray-theoretical and wave-theoretical approaches have been used to numerically model sound propagation in shallow water. Since shallow water environments are best approximated by range-dependent geometries (e.g. sloping bottom and high spatial variability of water column and sediment properties), attention will be focused on the range-depend-

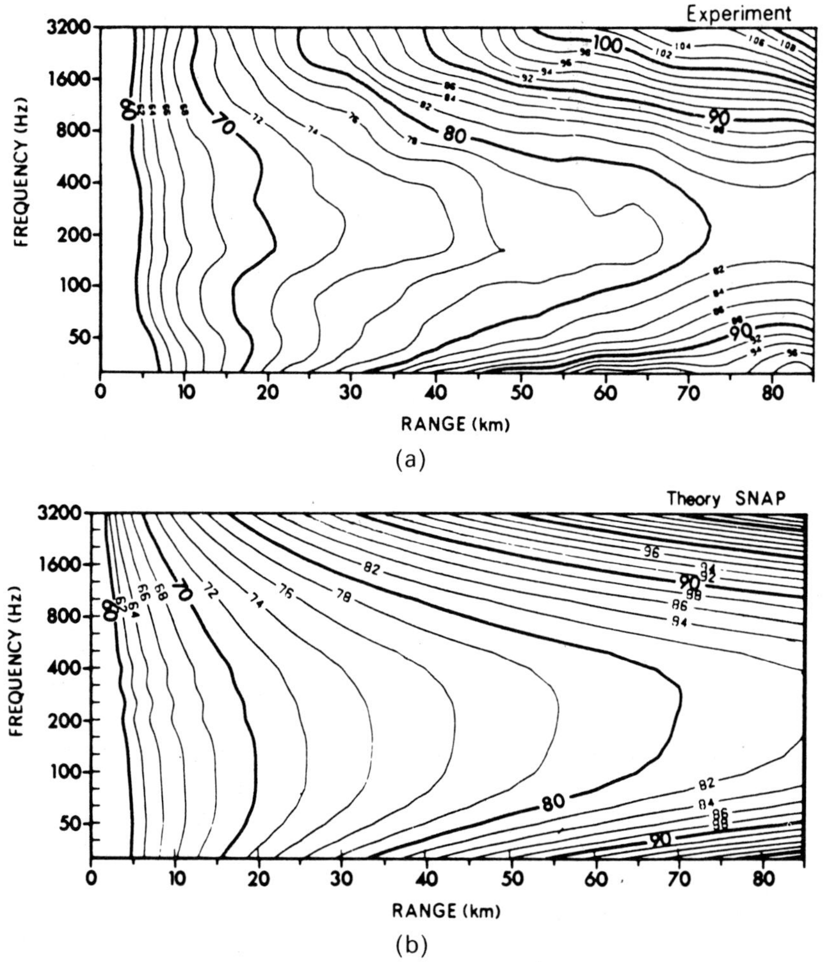

Fig. 5.6 Transmission loss contours (in 2 dB intervals) for the environment presented in Fig. 5.5: (a) experimental measurements and (b) predictions generated by the SNAP propagation model (Jensen and Kuperman, 1979).

ent modeling approaches. Moreover, the high-angle boundary interactions encountered in shallow water have traditionally limited consideration to two basic approaches: ray theory and normal mode solutions (with adiabatic approximations or mode coupling). More recently, appropriately modified parabolic equation (PE) models have also been utilized successfully in shallow-water environments (Jensen, 1984; Jensen and Schmidt, 1984).

Emphasis has recently been placed on modeling of sound propagation over a sloping bottom. This geometry is commonly referred to as a wedge problem and involves both upslope and downslope propagation. The direction of propagation (i.e. upslope or downslope) considerably alters the observed propagation characteristics. Consequently, this problem is of great practical interest to sonar operations in shallow water. This geometry has also been used as a benchmark problem in model evaluation (Chapter 11).

The basic mechanisms involved in acoustic propagation in a horizontally stratified waveguide are: spreading loss, attenuation due to bottom interaction effects, and intermode phasing effects. In a range-variable waveguide, an additional mechanism must be considered. This mechanism is related to changes in the acoustic energy density that occur with bathymetric changes and is often referred to as renormalization loss, or megaphone effects. The term renormalization is used because the so-called megaphone effects are manifested as changes in the normalization of the normal mode depth function due to changes in the waveguide depth (Koch *et al.*, 1983). These megaphone effects produce a gain in upslope propagation and a loss in downslope propagation.

The processes involved in upslope propagation can be better understood by using the ray-mode analogy. This analogy is a heuristic concept which states that any given mode trapped in the water column can be associated with upgoing and downgoing rays corresponding to specific grazing angles at the bottom (e.g. Urick, 1983, pp. 174–6; Boyles, 1984, pp. 197–204). As sound propagates upslope, the horizontal wave number associated with each mode decreases; in the ray analogy, the grazing angle at the bottom increases. For each mode, then, a point on the slope will be reached at which the grazing angles of the analogous rays will approach the critical angle at the bottom. At this point, the energy essentially leaves the water column and enters the bottom. In the ray analogy, the bottom reflection losses associated with those rays become very large. In the wave analogy, the modes transit from the trapped (waterborne) to the continuous (bottom propagating) spectrum. This point is called the cutoff depth for the equivalent modes. Upslope propagation is then said to exhibit a transition from a trapped to a radiative state (e.g. Arnold and Felsen, 1983; Jensen, 1984). In the case of downslope propagation, the transition is from a radiative to trapped state.

Conventional modal formulations fail to adequately explain circumstances in which a mode suddenly disappers from the water column with its energy being radiated into and dissipated within the sea floor (Pierce, 1982). Thus, the particular case of mode coupling where discrete modes (trapped in the water column) couple into continuous modes (which propagate in the bottom) has been further explored as a matter of practical interest (e.g. Miller *et al.*, 1986).

Evans and Gilbert (1985) developed a stepwise-coupled-mode method which overcomes the failure of previous coupled mode techniques to properly conserve energy over sloping bottoms. Their method of stepwise coupled modes avoids the problems associated with sloping bottoms by using only horizontal and vertical interfaces. The full solution thus includes both forward and backscattered energy (Jensen and Ferla, 1988).

Collins (1990a) suggested using a rotated PE in those wedge geometries involving complicated bottom boundary conditions. Specifically, by rotating the coordinate system, the PE can be marched parallel to the sea floor. The sea surface is now a sloping boundary with simplistic boundary conditions (pressure-release surface) that can be approximated by a sequence of range-independent regions in which the surface is specified as a series of stair steps.

Primack and Gilbert (1991) investigated downslope propagation using coupled wedge modes, which are the intrinsic normal modes in a wedge coordinate system.

Eigenray formulations can sometimes be useful in determining significant propagation paths and mechsnisms in a wedge geometry. One particular phenomenon of interest in the upslope problem is that of backscattered eigenrays. These rays have paths that travel up the slope, past the receiver and then back down the slope before arriving at the receiver (Westwood, 1990). The method of images has also been used to construct ray path solutions in shallow-water environments with a sloping bottom (Macpherson and Daintith, 1967).

Jensen (1984) examined both upslope and downslope propagation using an appropriately modified parabolic equation (PE) model. These results will be described below. Other researchers have investigated the wedge problem utilizing normal mode solutions (Evans, 1983) and ray theory (Arnold and Felsen, 1983; Westwood, 1989c).

(*a*) *Upslope propagation*

For the wedge problem involving upslope propagation, Jensen (1984) considered the particular environment in Fig. 5.7. The water–bottom interface is indicated on the contour plot by the heavy line starting at 350 m depth, then inclining towards the sea surface beyond a range of 10 km. The bottom slope is 0.85°. The frequency is 25 Hz, and the source is located at a depth of 150 m. The sound speed in constant at 1500 m/s. The bottom has a sound speed of 1600 m/s, a density of 1.5 g/cm^3 and an attenuation coefficient of 0.2 dB/wavelength. Shear waves were not considered.

In Fig. 5.7, Jensen displayed transmission loss contours between 70 and 100 dB in 2 dB intervals. Thus, high-intensity regions (where the loss is <70 dB) appear as black areas while low-intensity regions (loss >100 dB) appear as white areas. The PE solution was started by a Gaussian initial field, and there are four propagating modes. The high intensity within the bottom at ranges under 10 km corresponds to the radiation of continuous modes into the bottom. As sound propagates upslope, four well-defined beams (numbered 1–4) are seen, one corresponding to each of the four modes. This phenomenon of energy leaking out of the water column as discrete beams has been confirmed experimentally by Coppens and Sanders (1980). These points correspond to the cutoff depths associated with each mode.

(b) Downslope propagation

For the wedge problem involving downslope propagation, Jensen (1984) considered the environment illustrated in Fig. 5.8. The initial water depth is 50 m, and the bottom slope is 5°. The sound speed is constant at 1500 m/s. The sound speed in the bottom is 1600 m/s and the attenuation coefficient is 0.5 dB/wavelength. The density ratio between the bottom and the water is 1.5. The contour plot is for a source frequency of 25 Hz. The initial field for the PE calculations was supplied by a normal mode model, and only the first mode was propagated downslope. Shear waves were not considered.

Figure 5.8 shows that some energy propagates straight into the bottom at short ranges; that is, it couples into the continuous spectrum. Beyond the near field, however, propagation within the wedge is adiabatic, with the one mode apparently adapting well to the changing water depth. At a range of 20 km, the energy is entirely contained in the local first mode, even though as many as 21 modes can exist in a water depth of 1800 m.

5.3.4 Empirical models

Two noteworthy empirical algorithms have been developed for use in predicting transmission loss in shallow water, both of which provide depth-averaged estimates for range-independent ocean environments. One model (Rogers, 1981) is derived from theoretical considerations. The second model (Marsh and Schulkin, 1962b; Schulkin and Mercer, 1985), also known as Colossus, is based on field measurements from a small number of geographic areas.

(a) Rogers model

Rogers (1981) found that virtually all shallow-water transmission loss (TL) curves can be described by an equation of the form

$$\mathrm{TL} = 15 \log R + AR + B + CR^2 \tag{5.3}$$

where R is the range, and A, B and C are coefficients.

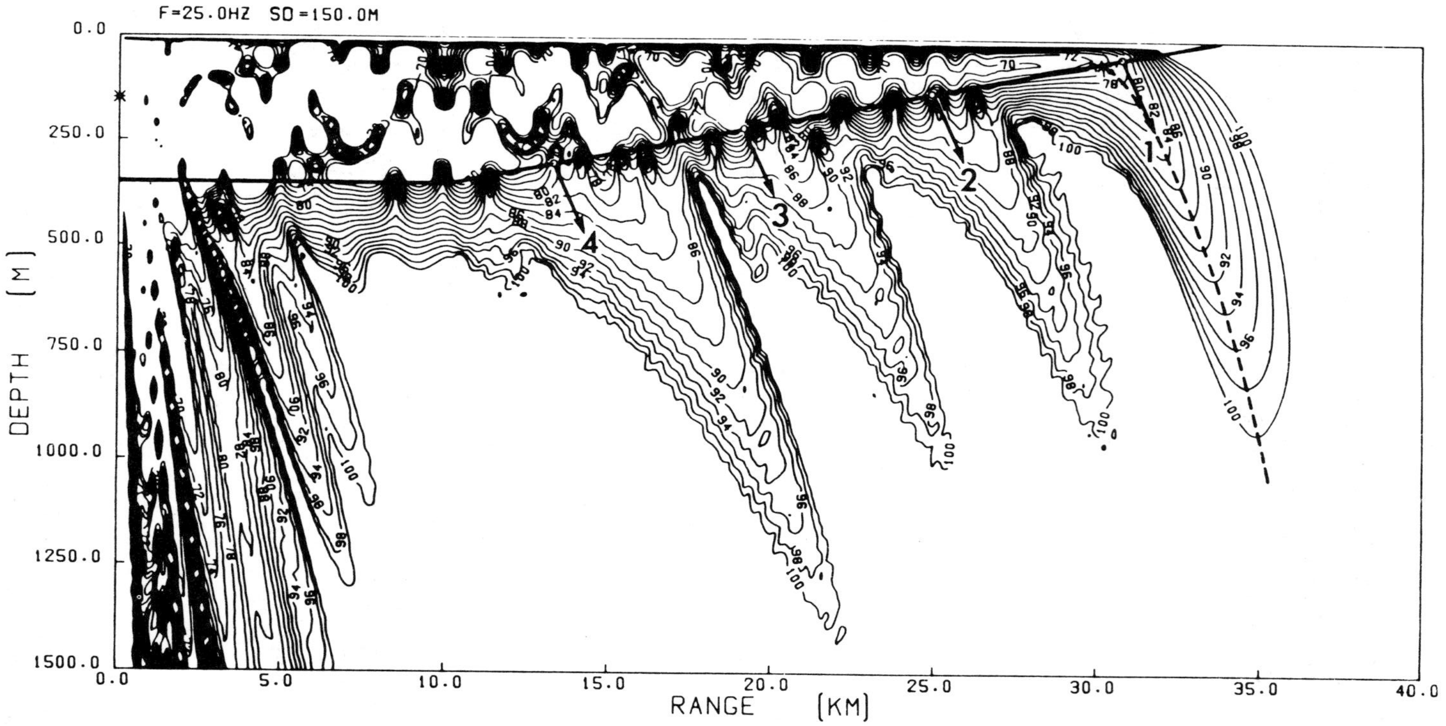

Fig. 5.7 Upslope propagation showing discrete mode cutoffs. Results were generated using a parabolic equation (PE) propagation model (Jensen and Schmidt, (1984).

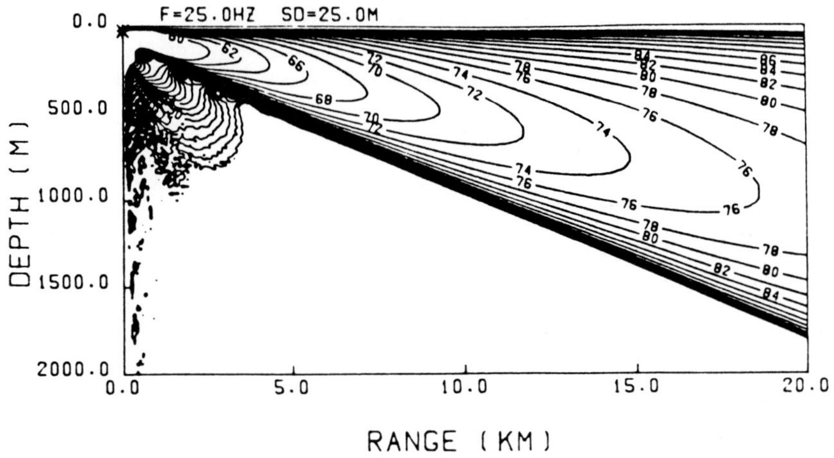

Fig. 5.8 Downslope propagation over a constant 5° slope (Jensen and Schmidt, 1984). Results were generated using a parabolic equation (PE) propagation model.

For the case of a negative sound speed gradient (i.e. sound speed decreases with increasing depth), Rogers obtained the following equation:

$$\mathrm{TL} = 15 \log R + 5 \log (H\beta) + \frac{\beta R \theta_L^2}{4H} - 7.18 + \alpha_w R \tag{5.4}$$

where R = range (m)
H = water depth (m)
β = bottom loss (dB/rad)
θ_L = limiting angle (rad)
α_w = absorption coefficient of sea water

The term 15 log R represents the spreading loss for the mode-stripping regions; thereafter, the spreading loss corresponds to 10 log R.

The limiting angle (θ_L) is the larger θ_g or θ_c, where θ_g is the maximum grazing angle for a skip distance (i.e. the maximum refracted-bottom-reflected, or RBR, ray) and θ_c is the effective plane-wave angle corresponding to the lowest propagating mode:

$$\theta_g = \sqrt{\frac{2Hg}{c_w}} \text{ (rad)} \tag{5.5}$$

$$\theta_c = \frac{c_w}{2fH} \text{ (rad)} \tag{5.6}$$

where g is the magnitude of the negative sound speed gradient (1/s), c_w is the maximum (sea surface) sound speed (m/s), and f is the frequency (Hz).

The bottom loss (β) is derived from the theoretical expression for the Rayleigh reflection coefficient for a two-fluid lossy interface, for small grazing angles. For most cases of interest (small values of θ_L), the bottom loss can be approximated as

$$\beta \approx \frac{0{\cdot}477 M_0 N_0 K_s}{[1 - N_0^2]^{3/2}} \text{(dB/rad)} \tag{5.7}$$

where $N_0 = c_w/c_s$
$M_0 = \rho_s/\rho_w$
c_s = sound speed in sediment
ρ_w = density of sea water
ρ_s = sediment density
K_s = sediment attenuation coefficient (dB/m∗kHz)

For example, at a fixed frequency of 200 Hz, Rogers (1981) considered eight different sound speed profiles for which sound speed decreased monotonically with depth (Fig. 5.9). Using equation (5.3), the coefficients, *A*, *B* and *C* were determined for a number of test cases between 5 and 100 km. The maximum deviation between equation (5.3) and the actual transmission loss curves generated by a normal mode model was also reported (Table 5.1). The sediment properties were based on Hamilton (1980). The intent was to demonstrate the importance of the sound speed profile (versus the average gradient in the duct) in determining depth-averaged transmission loss.

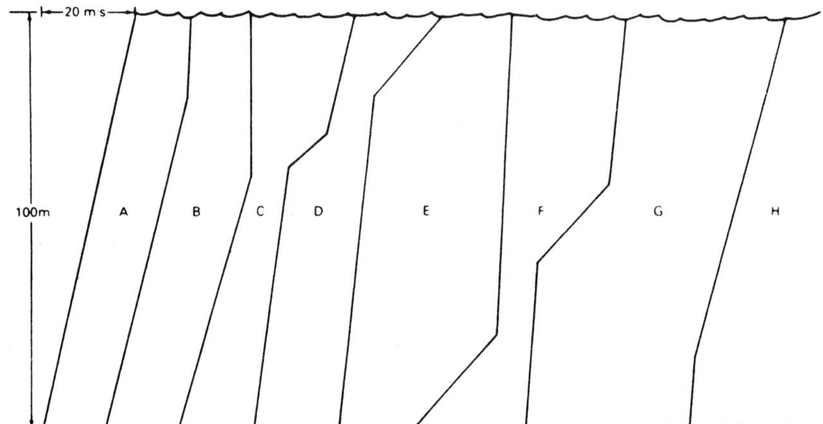

Fig. 5.9 Eight different sound speed profiles considered by Rogers (1981). In each profile, sound speed decreases with depth with an average gradient of 0.2/s.

Table 5.1 Coefficients for Rogers' (1981) empirical formula for shallow-water transmission loss (TL): $TL = 15 \log R + AR + B + CR^2$ (dB)

Sediment	*Profile type*	*Three-parameter fit*			
		A	*B*	$C \times 10^4$	*Maximum deviation*
Fine sand	A	0.18	49.18	+1.098	0.017
Fine sand	B	0.22	49.04	+1.772	0.043
Fine sand	C	0.27	49.02	+0.461	0.042
Fine sand	D	0.149	49.04	+0.020	0.047
Fine sand	E	0.0729	49.61	−0.801	0.042
Fine sand	F	0.146	49.76	−0.587	0.051
Fine sand	G	0.252	48.8	−4.36	0.046
Fine snad	H	0.173	49.14	+0.09	0.021

Range (R) is measured in km; the coefficients A, B and C are valid for a frequency of 200 Hz for the sound speed profiles (A–H) illustrated in Fig. 5.9, over a fine-sand bottom.

(*b*) *Marsh–Schulkin model*

The Marsh and Schulkin (1966b) model, also referred to as Colossus, is an empirical model for predicting transmission loss in shallow water (e.g. Podeszwa, 1969). Colossus employs several concepts: (1) refractive cycle, or skip distance; (2) deflection of energy into the bottom at high angles by scattering from the sea surface; and (3) a simplified Rayleigh two-fluid model of the bottom for sand or mud sediments. With a few free parameters (including water depth), about 100 000 measurements spanning the frequency range 100 Hz–10 kHz were fitted within stated error bounds.

Along with the concepts of skip distance and bottom loss, the Colossus model used the AMOS results (Marsh and Schulkin, 1955) for a deep-water isothermal surface duct, but with an average thermocline appropriate for shallow water. It also used measurements of actual transmission losses in shallow water off the east coast of the United States as a function of frequency, categorized by bottom type (sand or mud) and by season. Two other mechanisms characteristic of shallow-water processes were also included in the Colossus model: a 'near-field anomaly' correction in the direct radiation zone that included the gain due to multiple bottom and surface bounces, and an energy conservation rule that was used to establish the effective shallow-water attenuation coefficient (α_t) which included the additional loss due to the coupling of energy from the wind-roughened sea surface to the bottom.

The bilinear gradient used in the Colossus model is composed of two constant, linear segments drawn toward the surface and toward the bottom from the depth of maximum sound speed (or temperature). The sound ray cycles have one upward radius of curvature (positive sound speed gradient) for surface bounces and one downward radius of curvature (negative gradient) for bottom bounces. Based on the depth of the surface layer and the water depth, a single effective skip distance (H) is formulated. Multiples of this effective skip distance are used to define a zone of direct ray paths (20 log R, where R -is the range), a zone of mode stripping (15 log R), and a zone of single-mode control (10 log R). The mode-stripping process was found to be complete at a range equal to $8H$, where H is the skip distance.

In the Marsh–Schulkin (or Colossus) model, transmission loss is a function of sea state (or wave height), bottom type, water depth, frequency, and the depth of the positive gradient layer. The skip distance (H) is used as a reference to define regions where wavefront spreading follows square, three-halves and first-power laws as a function of range (R). Accordingly, three equations were developed to provide for the gradual transition from spherical spreading in the near field to cylindrical spreading in the far field:

Short range ($R < H$)

$$\mathrm{TL} = 20 \log R + \alpha R + 60 - K_{\mathrm{L}} \tag{5.8}$$

Intermediate range ($H < R < 8H$)

$$\mathrm{TL} = 15 \log R + \alpha R + \alpha_{\mathrm{t}}[(R/H) - 1] + 5 \log H + 60 - K_{\mathrm{L}} \tag{5.9}$$

Long range ($R > 8H$)

$$\mathrm{TL} = 10 \log R + \alpha R + \alpha_{\mathrm{t}}[(R/H) - 1] + 10 \log H + 64.5 - K_{\mathrm{L}} \tag{5.10}$$

where $H = [(L + D)/3]^{1/2}$ = skip distance (km)

L = mixed layer depth (m)
D = water depth (m)
R = range (km)
α = absorption coefficient (dB/km)
α_t = effective shallow-water attenuation coefficient (dB/bounce)
K_{L} = near-field anomaly (dB)

The absorption coefficient (α) can be estimated from Fig. 3.14. The effective shallow-water attenuation coefficient (α_{t}) and the near-field anomaly (K_{L}) are functions of frequency, sea state and bottom composition. Values of α_{t} range from about 1 dB/bounce to about 8 dB/bounce. Typical values of K_{L} range from about 1 dB to about 7 dB.

According to Schulkin and Mercer (1985), the model's chief criticisms have been that it could not be adjusted for arbitrary negative sound speed gradients (it uses the same constant gradient in all cases), and that it uses empirical bottom loss values. Consequently, the Colossus model has recently been extended for the purpose of treating arbitrary negative and bilinear sound speed gradients. These extensions use new general expressions for the skip distance, the near-field anomaly, and the reflection coefficients. The extended model (Schulkin and Mercer, 1985) and the model by Rogers (1981) are found to give about the same predictions when the same inputs are used.

5.4 ARCTIC MODELS

5.4.1 Arctic environmental models

Basic descriptors of the Arctic marine environment which require specialized algorithms for use in propagation models are limited to absorption and surface (under-ice) scattering (Etter, 1987c; Ramsdale and Posey, 1987). The generation of other parameters, such as sound speed and bottom scattering, appear to be adequately supported by existing algorithms which are valid over a wide range of oceanic conditions.

Absorption is regionally dependent, due mainly to the pH dependence of the boric acid relaxation. In the Arctic, the pH range is roughly 8.0–8.3 (versus 7.7–8.3 for nominal sea water), but the greatest variability occurs much closer to the sea surface than in other ocean areas. Formulas appropriate for use in Arctic regions have been presented by Mellen *et al.*(1987c).

Existing under-ice scattering loss models appropriate for inclusion in mathematical models of acoustic propagation have been reviewed and evaluated by Eller (1985). Chapter 3 described physical models of under-ice roughness.

5.4.2 Arctic propagation models

Developments in acoustic propagation modeling for the Arctic Ocean have been very limited. Much of the past effort on characterizing propagation in the Arctic has been devoted to gathering acoustical data and developing empirical models based on that data. While these models tend to be site and season specific with little generality, they do provide basic information on the frequency and range dependence of propagation.

In general, there are four factors peculiar to the Arctic environment which complicate the modeling of acoustic propagation: (1) the ice keels present a rapidly varying surface; (2) the reflection, transmission and scattering properties at the pack ice interface are not well known; (3) the measurement of under-ice contours is difficult; and (4) the diffraction around ice obstacles may be important. In the Arctic half-channel, ray theory may provide a useful predictive method, whereas its utility in temperate water surface ducts may actually be quite limited. The utility of ray theory in the Arctic derives from the fact that the Arctic half-channel is between one and two orders of magnitude greater in gradient (and much greater in depth) than the temperate water surface ducts. The strong positive thermocline and halocline produce an exceptionally strong positive sound speed gradient in the subsurface layer. This markedly shortens the ray loop length, causing many surface reflections for rays with small grazing angles (Mobile Sonar Technology, 1983, unpublished manuscript).

The open ocean region in the Arctic environment also poses potential problems as far as existing propagation models are concerned. Surface reflection losses, which may not be significant in deep ocean propagation, become more important in the upward-refracting environment of the Arctic since multiple surface reflections play a dominant role. Moreover, multiple ducts are prevalent in this region and are not adequately treated by many propagation models.

Two basic modeling approaches are currently being pursued: the application of ice-scattering coefficients to existing numerical models of acoustic propagation and the development of empirical models. These two approaches are discussed in more detail below.

5.4.3 Numerical models

Numerical models of underwater acoustic propagation specifically designed for ice-covered regions are limited. Since the Arctic half-channel acts as a low-pass filter (discriminating against higher-frequency components), and since bottom interaction is not as important as in other ocean regions (because of the upward-refracting nature), most modeling applications in the Arctic have employed either normal mode (Gordon and Bucker, 1984) or fast-field (Kutschale, 1973, 1984) approaches. These models are considered most appropriate for prediction of low-frequency (<350 Hz) propagation at long ranges (>25 nm (nautical miles)). Other modeling techniques (e.g. ray theory and parabolic equation) might also be suitable for calculation of mean transmission losses if the needed descriptions of the ice cover were available (e.g. Chin–Bing and Murphy, 1987). In fact, Stotts *et al.* (1994) developed a ray-theoretical propagation model called ICERAY that is valid for under-ice environments, and at least one implementation of the PE technique has been modified to include the effects of ice scattering (see Table 4.1).

5.4.4 Empirical models

Empirical models are inherently limited by the databases from which they were derived. Attempts to fit results from a large data set with simplistic curves generally imply large errors in the model results. Comparisons of model results with data are clearly quite limited by a lack of comprehensive data sets. A recent intercomparison of available empirical techniques applicable to the Arctic has revealed large, unresolved discrepancies (Deavenport and DiNapoli, 1982).

Two of the better known empirical models are described below.

(a) Marsh–Mellen model

The Marsh–Mellen Arctic Transmission Loss Model (Marsh and Mellen, 1963; Mellen and Marsh, 1965) is based on observations made during the summers of 1958 and 1959 between Arctic drift stations separated by 800–1200 km. The measured arrivals were found to consist of a dispersive, quasi-sinusoidal wave train in the 10–100 Hz frequency range. These features were explained by a half-channel model in which the higher frequencies were attenuated by under-ice scattering. Using these and other experimental data, long-range, low-frequency (<400 Hz) transmission loss data in the Arctic were fitted with an equation of the form

$$\mathrm{TL} = 10 \log r_0 + 10 \log R + \alpha_s N_s \tag{5.11}$$

where r_0 is the skip distance for limiting ray, N_s is the number of surface reflections, R is the range in m ($R = r_0 N_s$), and α_s is the loss per bounce. The wind-generated ocean wave spectrum (Marsh, 1963) was taken to approximate the ice roughness.

(b) Buck model

The Buck Arctic Transmission Loss Model (Buck, 1981) consists of a short-range (10–100 nm) and a long-range (100–1000 nm) model for low-frequency (<100 Hz) transmission loss in that part of the Arctic Ocean deeper than 1000 m. These preliminary empirical models represent linear regression fits to winter data collected in the Beaufort Sea in 1970, in the Fram Strait in 1977, and in an intermediate area in 1979 for a source depth of 244 m and a receiver depth of 30 m:

Short range (10–100 nm)

$$\mathrm{TL} = 62.4 + 10 \log R + 0.032f + 0.065R + 0.0011fR \tag{5.12}$$

Long range (100–1000 nm)

$$\mathrm{TL} = 68.5 + 10 \log R + 0.07f - 0.0015\,sR + 0.000487fsR \tag{5.13}$$

where f is the frequency (Hz), R is the range (nm) and s is the standard deviation ice depth (m), also referred to as under-ice roughness parameter. A chart of the standard deviation ice depth (s) suitable for use in equation (5.13) is presented in Fig. 5.10 (from Buck, 1985).

There is some doubt at this time as to the applicability of equations (5.12) and (5.13) to summer conditions and to other source–receiver depth combinations. Also refer to Fig. 3.28 for average curves of Arctic transmission loss versus frequency based on measured data (Buck, 1968).

5.5 DATA SUPPORT REQUIREMENTS

The development of numerical models requires data with which to support model initialization and model evaluation.

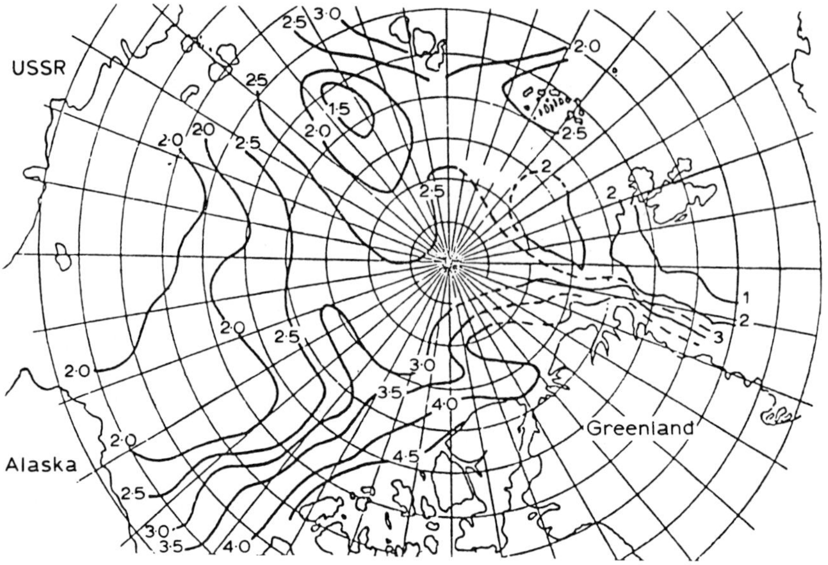

Fig. 5.10 Estimates of under-ice roughness in standard deviation (in meters) about the mean ice depth (Buck, 1985); derived in part from nuclear submarine sonar data analyzed L.A. LeSchack.

Initialization of propagation models requires various descriptors of the ocean environment including the water column, the sea surface and the sea floor. Wave-theoretical models tend to be more demanding of bottom sediment information than do the ray-theoretical models.

Evaluation of propagation models requires transmission loss measurements which are keyed to descriptions of the prevailing ocean environment. Such coordinated measurements are necessary to ensure that the model is initialized to the same environment for which the transmission loss data are valid (e.g. Hanna, 1976).

Data support requirements are further complicated by the fact that long-range transmission loss can rarely be considered to occur in a truly range-independent ocean environment. With few exceptions, changes in sound speed or water depth (among other parameters) can be expected, not only as a function of range but also as a function of azimuth (or bearing). Consequently, data management in support of model development and operation can be formidable, and is often a limiting factor in the proper employment and evaluation of numerical models.

When working with a variety of propagation models (either different models within one category or models from different categories, as illustrated previously in Fig. 4.1), it becomes obvious very quickly (sometimes painfully so) that each model not only requires somewhat different parameters but sometimes requires varying format specifications for any given parameter.

In response to this situation, two related developments have occurred. First, gridded databases (i.e. those organized by latitude and longitude) have been established which contain all the required ocean environmental parameters in a standardized format with automated update and retrieval mechanisms. Second, model-operating systems (see Chapter 10) have been created to automatically interface the standard databases with the various models represented in the operating system, and to accommodate the format specifications peculiar to each of the models.

A graphic example of a retrieval from a gridded ocean database is presented in Fig. 5.11. This presentation of range-dependent sound speed and bathmetry data is suitable for use in many propagation models. Chapter 10 presents a summary of available oceanographic and acoustic databases.

One of the newest automated ocean data products is the Generalized Digital Environmental Model (GDEM), developed by Dr Tom Davis at the

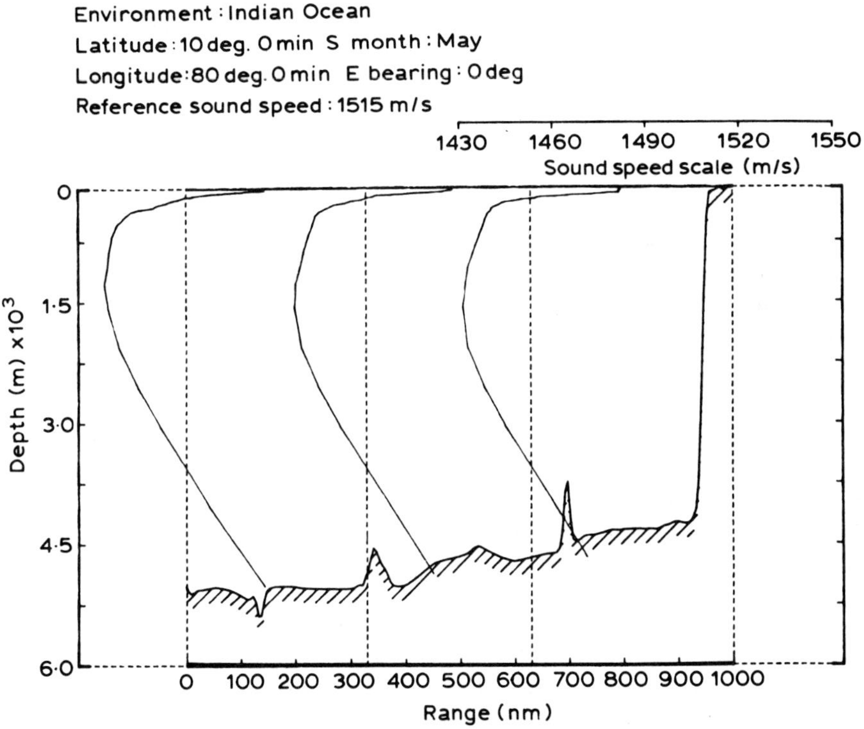

Fig. 5.11 Example of a range-dependent ocean environment extracted from a gridded database appropriate for use in many propagation models.

US Naval Oceanographic Office (Wells and Wargelin, 1985). GDEM uses a digitized database of major oceanic features together with their climatic location to display temperature, salinity and sound speed profiles on a 30-min latitude and longitude grid. A simple parabolic fit is used to describe the deep portion of the sound speed profile. The more variable near-surface fields of temperature, salinity and sound speed are constructed using orthogonal polynomials between the surface and a depth of 800 m. GDEM has been rigorously compared with reliable climatologies (Teague *et al.*, 1990). Refer to Tables 10.4 and 10.5 for further descriptions of GDEM.

The specification of sound speed profiles for initialization of propagation models is very important. The following sections will eleborate on this topic.

5.5.1 Sound speed profile synthesis

For application to propagation models, the sound speed profile is usually synthesized from a finite number of sound speed values at discrete depths. The functional representation of the synthesized profile is of great importance to the type of modeling technique employed. There are basically two different approaches in use, each with its own advantages and disadvantages, as summarized in Fig. 5.12.

(*a*) *Segmented constant gradient*

The sound speed profile can be constructed by connecting the discrete points with straight-line segments. Because of discontinuities in the first derivative of the resulting function, transmission losses calculated on the basis of ray-tracing techniques are undefined at certain ranges. These regions are referred to as false caustics since they are false regions of infinite intensity.

In some propagation models, the sound speed profile is fitted with segments in which the inverse of the sound speed squared varies linearly with depth. This often permits a more efficient mathematical solution in which the mode functions are expressed in terms of Airy functions.

(*b*) *Curvilinear or continuous gradient*

Curvilinear profile approximations that preserve the continuity of slope as well as sound speed have been developed by Pedersen and Gordon (1967). The sound speed profile can also be fitted with quadratic equations to within specified tolerances (Weinberg, 1969, 1971).

Cubic splines can be used to approximate the sound speed profile for application to ray-theoretical techniques (Solomon *et al.*, 1968; Moler and

SOUND SPEED PROFILE CURVE FITTING TECHNIQUES

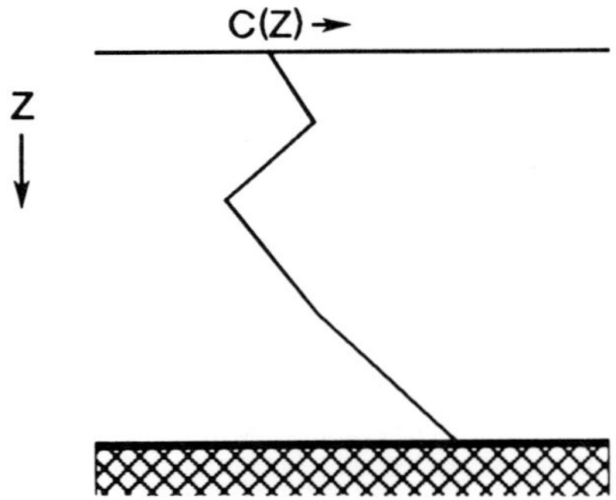

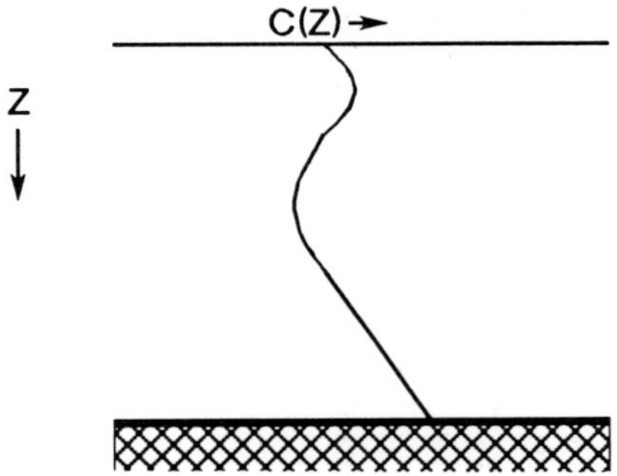

Segmented Constant–Gradient

- **Discontinuous first derivative leads to false caustics in ray tracing techniques.**
- **Line segments in which $1/C^2$ varies linearly with depth permits mode functions to be expressed in terms of Airy functions in normal mode techniques.**

Curvilinear or Continuous Gradient

- **Quadratic equations – fit data points to within specified tolerances.**
- **Cubic splines – do not result in closed–form solutions for ray path equation in ray tracing techniques; continuous 1st and 2nd derivatives.**
- **Conic sections and hyperbolic cosines – result in closed–form solutions for ray path equation using elliptical and hyperbolic ray paths in ray tracing techniques.**
- **Exponential forms – facilitate matrizant solution in Fast Field Theory techniques.**

Fig. 5.12 Summary of sound speed profile curve-fitting techniques.

Solomon, 1970). However, this method does not result in closed form solutions for the rays. Approximating profiles with conic sections and hyperbolic cosines results in closed form expressions for the ray path equations when using elliptical and hyperbolic ray paths (Flanagan *et al.*, 1974).

Exponential forms are sometimes used in those modeling techniques employing fast-field theory to facilitate a matrizant solution. To simplify the mathematical treatment of long-range sound propagation in the ocean, investigators (Munk, 1974; Flatté, 1979) have introduced what is termed a canonical model of the sound speed profile. This model has an exponential form which is valid in the vicinity of the deep sound channel axis:

$$c(z) = c_1[1 + \varepsilon(\eta + \mathrm{e} - \eta - 1)]$$
$$\eta = \frac{2(z - z_1)}{B}, \quad \varepsilon = \frac{B\gamma_A}{2} \tag{5.14}$$

where $c(z)$ = sound speed as a function of depth
c_1 = sound speed at channel axis (z_1)
η = dimensionless distance beneath channel axis
B = scale depth
ε = perturbation coefficient
γ_A = fractional sound speed gradient for adiabatic ocean

Munk (1974) used the following typical values: $c_1 = 1492$ m/s, $B = 1.3$ km, $z_1 = 1.3$ km, $\gamma_A = 1.14 \times 10^{-2}$ km^{-1} and $\varepsilon = 7.4 \times 10^{-3}$.

5.5.2 Earth curvature corrections

In problems involving acoustic propagation over great distances (say, greater than 30 nm, or 56 km), the fact that our reference plane is situated on a spherical surface can no longer be ignored.

Curvature of the Earth's surface is roughly equivalent to a linear sound speed profile. A linear surface duct is closely analogous to whispering gallery.

First-order spherical Earth curvature corrections are usually applied to the sound speed profile before any curve-fitting techniques are employed. These corrections are typically of the form (Watson, 1958; Hoffman, 1976)

$$c_i = c'_i(1 + z'_i/R_0) \tag{5.15}$$

$$z_i = z'_i(1 + z'_i/2R_0) \tag{5.16}$$

where R_0 = radius of Earth (~6370.9 km)
c'_i = input sound speed at depth z'_i
c_i = corrected sound speed at corrected depth z_i

The effect of curvature is more than just a matter of chords versus arcs of great circles. Rather, the important feature is that the contours of constant sound speed (as represented on a range–depth plane) are actually concentric spheres instead of flat, parallel planes. The differences in ray angles as calculated by Snell's law produce effects of potential importance to sonar performance predictions. A detailed comparison between state-of-the-art ray theory and experimental transmission loss data (Pedersen *et al.*, 1984) revealed that Earth curvature corrections to the sound speed profile produced model predictions that were in better agreement with the data than would have otherwise been obtained. The effect of curvature was to reduce the range to the leading edge of the first convergence zone by about 1%. Errors might be expected to increase with increasing ranges.

5.5.3 Merging techniques

Propagation models generally require sound speed data from the surface of the ocean to the bottom. When the local water depth exceeds the limits of available sensing devices such as the XBT, the required sound speed profiles are often derived from a combination of *in-situ* data in the near-surface (or shallow) layer together with climatology in the deep layer (Fisher and Pickett, 1973).

A number of algorithms have been developed for automatically merging data from these two layers. Statistical relationships based on oceanographic analyses are typically employed to associate the shallow-layer data with the most appropriate deep-layer climatology. This practice is usually satisfactory when the historical databases are characterized by high spatial and temporal resolutions. However, merging algorithms may introduce errors into the derived surface-to-bottom sound speed profile as a result of curve-fitting artifacts or incomplete data near the juncture of the two layers. Additional errors may result from measurement and digitization processes.

5.6. SPECIAL APPLICATIONS

This section will review specialized applications of underwater acoustic propagation models. Progress in these particular areas is advancing rapidly and the following discussions are intended to identify areas of particular interest at the present time as well as areas appropriate for future research.

5.6.1 Stochastic modeling

Methods by which to investigate and model the stochastic (versus deterministic, or completely predictable) nature of acoustic propagation in the ocean have been refined as part of ongoing, long-term projects. Such efforts are concerned with probabilistic predictions of the environmental limits to sonar aperture designs (Perkins *et al.*, 1984). A workshop conducted at the US Naval Research Laboratory addressed the model and database requirements for current and new highly complex sonar systems (Spofford and Haynes, 1983). Among the topics addressed were estimates of both the first and second moments (variances) or spreads in space, time and frequency required in a variety of ocean environments to support the operation of current sonar systems and the design of advanced sonar systems.

A review of stochastic signal-modeling efforts (Wood and Papadakis, 1985) concluded that spread function models are efficient for computing lower-order statistics although with limited capability. General purpose Monte Carlo models are desirable for their higher-order capabilities, despite higher computing costs.

In the so-called model-based approach to acoustic signal-processing in the ocean, mathematical models of physical phenomena and measurement processes are incorporated into the processor. The inclusion of a propagation model in the signal-processing scheme introduces environmental information in a self-consistent manner. Furthermore, stochastic properties of the oceanic medium can be included in the model. Solutions using state–space techniques employ two sets of equations: the state equation and the measurement equation. The state equation describes the evolution in space of the modal and range functions while the measurement equation relates these states to the hydrophone array measurements. One specific implementation (Candy and Sullivan, 1992) cast the normal mode propagation model into state–space form, extended the formulation to a Gauss–Markov model and applied the results to an ocean-acoustic signal-processing problem in the context of a horizontally stratified ocean with a known source position. A related implementation investigated the inverse reconstruction of a sound speed profile from hydrophone measurements (Candy and Sullivan, 1993).

The use of finite element techniques for computing underwater acoustic propagation has been studied as part of a computationally intensive probabilistic acoustics program, the major objective of which was to develop models for propagation of the moments of the acoustic field in regions where the ocean boundaries are random surfaces (Goldstein, 1984). The possibility of using exact techniques such as the finite element method (FEM) for solving the wave equation has been made more attainable by the augmentation of computers with array processors.

Recent developments in finite element modeling include ISVRFEM (Pack, 1986), the Finite Element Parabolic Equation (FEPE) (Collins, 1988a), the Finite Element Ocean Acoustic Model (FOAM) (Murphy and Chin–Bing, 1989) and the Seismo–Acoustic Finite Element (SAFE) model (Murphy and Chin–Bing, 1991).

5.6.2 Broadband modeling

Broadband (also referred to as pulse or wideband) propagation modeling is concerned with simulating the effects associated with the transmission of a signal characterized by a frequency spectrum (versus a single-frequency continuous wave). When considering such signals, the most simplistic approach has been to calculate the geometric mean frequency (f_M), defined as $f_M = (f_1 f_2)^{1/2}$, where f_1 and f_2 are the lower and upper limits, respectively, of the frequency band. The propagation of an equivalent signal at frequency f_M is then simulated using one of the available propagation models. When the bandwidth is small, this approach probably results in a reasonable approximation; otherwise, this approach may lead to substantial errors, particularly when the spectrum is not flat.

Beyond this simplistic approach, there are two more sophisticated techniques: Fourier synthesis and time domain methods.

In Fourier synthesis, multiple executions of an existing propagation model are performed over the frequency range (f_1, f_2) at a number of discrete frequencies at intervals of Δf, where Δf might be 1 Hz, for example. The resulting transmission losses for each frequency in the bandwidth are then combined through an appropriate weighting and averaging process (i.e. interpolation postprocessor) to arrive at the transmission loss corresponding to the bandwidth. Examples of such multi-frequency extensions include the ray theoretical model GAMARAY (Westwood, 1992), the normal mode model PROTEUS (Gragg, 1985) and the Multilayer Expansion Model, MULE (Weinberg, 1985a).

Alternatively, the method developed by McDonald and Kuperman (1987) for modeling the propagation of a broadband linear pulse in a waveguide is one example of a broader class of techniques referred to as time domain methods (Kuperman, 1985). In principle, the frequency domain wave equation (valid for a single-frequency continuous wave signal) can treat broadband signals by Fourier synthesis of the individual CW solutions over the frequency spectrum. In the presence of non-linearities, however, interactions among frequency components invalidate the frequency-domain approach. In the time domain, the wave equation can be formulated using methods which remove such pathological limitations from the numerical solutions.

In related developments Porter (1990) developed a time-marched fast-field program (Pulse FFP) for modeling acoustic pulse propagation in the ocean, Collins (1988b) used the TDPE model to investigate the effects of sediment dispersion on pulse propagation and Orchard *et al.* (1992) developed the 3D TDPA model for simulating pulse propagation in 3D ocean geometries.

Jensen (1988) provided a summary of wave-theoretical techniques suitable for practical modeling of low-frequency acoustic pulse propagation in the ocean. Jensen emphasized the computational efficiency of pulse propagation predictions using Fourier synthesis of existing CW propagation models based on normal mode and parabolic equation approaches. Jensen (1993) further explored these issues by placing particular emphasis on propagation in leaky surface ducts.

5.6.3 Matched field processing

The use of matched field-processing techniques in underwater acoustics has been explored in recent years by a number of investigators; comprehensive overviews have been provided by Tolstoy (1992, 1993) and Baggeroer *et al.* (1993). This technique correlates the acoustic pressure field from a submerged source, as detected at each receiver in a hydrophone array, with the field modeled at the array by assuming a particular source position and ocean environment. Consequently, a high degree of correlation between the experimental and modeled pressure fields indicates a high probability that the source is located at the estimated position in the range-depth plane. The basic components of matched field processing are illustrated in Fig. 5.13 (Tolstoy, 1993). Accordingly, this technique shows great promise as a high-resolution localization tool. This technique can also be used to reconstruct prevailing oceanic conditions when the source and receiver positions are known *a priori.*

A variety of mathematical estimator functions have been utilized to perform the comparisons between the experimental and modeled pressure fields including: linear; minimum variance; multiple constraint; matched mode; eigenvector; maximum entropy; approximate orthogonal; variable coefficient likelihood; and optimum uncertain field processors (Fizell, 1987; Tolstoy, 1993). The outputs of these processors are often presented graphically on so-called 'performance surfaces' to facilitate identification of probable source positions; the ambiguity surface illustrated in Fig. 5.13 is one example. The matched field technique has been found attractive for application to shallow-water geometries. For example, Bucker (1976) used a conventional cross correlation of the experimental and modeled pressure fields while Del Balzo *et al.* (1988) used a maximum-likelihood estimator. The maximum-likelihood estimator has also been applied to deep water Arctic environments by Fizell and Wales (1985).

The use of broadband signals in matched field processing has been investigated by Westwood (1992). The use of matched mode localization as an alternative to matched field processing has been investigated by Bogart and Yang (1992), whose preliminary results indicated that the matched mode ambiguity surface showed equal or improved (i.e. lower) sidelobes compared with that of matched field processing and was easier to compute. Dosso *et al.* (1993) used matched field inversion to estimate ocean-bottom properties.

5.6.4 Transmutation approaches

A transmutation approach has recently been applied to underwater acoustic propagation modeling by Duston *et al.* (1986). As described by Gilbert and Wood (1986), who also provided historical references to the literature, transmutation theory allows one to find an integral operator that transforms the solutions of a simpler partial differential equation into solutions of another, more complicated, partial differential equation. A third partial differential equation exists that the kernel of the integral operator must then satisfy. The advantage gained is that more freedom exists in assigning useful initial or boundary conditions to the kernel. Thus, instead of separating the Helmholtz equation into two ordinary differential equations, it is separated into two related partial differential equations, one of which is solved analytically and the other by a hybrid of symbolic and numerical computations.

5.6.5 Chaos

Chaos in underwater acoustic modeling has been explored by Tappert *et al.* (1988) and by Palmer *et al.* (1988a). They observed that ray path solutions exhibit 'classical chaos', that is, unpredictable and stochastic behavior. The phenomenon of chaos in underwater acoustics is presumed to be caused by the exponential proliferation of catstrophes (in the form of caustics) due to the loss of control implied by the nonseparability of variables in the eikonal equation. Consequently, there exists a prediction horizon which cannot be exceeded even when the ocean environment is known exactly.

The application of chaotic concepts to underwater acoustic propagation modeling has been further investigated by Smith *et al.* (1992a, b), Tappert *et al.* (1991) and Brown *et al.* (1991a, b). These studies have emphasized the practical importance of chaotic ray trajectories as limiting factors in generating deterministic predictions of acoustic propagation, particularly in the presence of meso-scale ocean structure. Collins and Kuperman (1994) suggested that the computational difficulties associated with chaos may be overcome by solving eigenray problems with boundary value techniques as opposed to initial value techniques.

The deliberate use of chaotic waveforms as sonar signals has also been investigated. One theoretical study (Alapati *et al.*, 1993) used chaotic metrics (e.g. Rasband, 1990) and conventional ambiguity functions to evaluate the performance of several nonlinear waveforms after convolution with realistic ocean impulse–response functions. These impulse–response (or Green's) functions were generated using the Generic Sonar Model (section 10.4.4). Ziomek (1985, pp. 4–5) discusses the mathematical formalism of impulse–response functions in a signal-processing context. Because of their amenability to signal enhancement (or noise reduction) techniques, chaotic waveforms have been suggested for use in bistatic active sonar systems operating in shallow-water regions (section 9.5).

5.6.6 Three-dimensional modeling

Modeling of underwater acoustic propagation in three dimensions (3D), sometimes referred to as volume acoustic field modeling (Chin–Bing *et al.*, 1986), has assumed greater importance as sonar systems have become more complex (Jones, 1983). Such modeling is generally, but not always, accomplished by extending the capabilities of existing range-dependent (2D) techniques such as PE, normal mode and ray theory models to form composite 3D pictures. Such approaches are commonly termed $N \times 2D$ since the models are sequentially executed for N adjacent range-dependent (2D) radials (or sectors). Ray theory and parabolic equation techniques can theoretically treat 3D propagation, although such implementations are rarely accomplished in practice. Advanced computational techniques to support true 3D modeling have also been investigated (e.g. Johnson, 1984).

Three-dimensional propagation modeling has been used to simulate acoustic interactions with seamounts (Medwin *et al.*, 1984) and with mesoscale oceanographic features (Kuperman *et al.*, 1987). Coupled with appropriate color graphic displays, 3D modeling promises to be a powerful analytical and predictive tool.

In the case involving seamounts, Medwin *et al.* (1984) were able to show that: (1) depending on the roughness of the sea surface, diffraction over the crest of the seamount can be the strongest contributor to the sound in the shadow region; (2) the diffracted signal always arrives before the multiply reflected sound; (3) the diffraction loss is proportional to the square root of the acoustic frequency; and (4) a 2D model produces excessive diffraction and excessive multiple reflection signal compared to a more realistic 3D model. The 3D model used by Medwin *et al.* is a hybrid solution formed by combining the range-independent FACT model (to calculate acoustic propagation to and from the seamount) with 3D physical models of the frequency-dependent wave interaction at the seamount. Comparisons with experimental data obtained from the Dickins Seamount in the northeastern Pacific Ocean (Chapman and Ebbeson, 1983; Ebbeson and Turner, 1983) verified the accuracy of this model in the frequency range 50–500 Hz. A ray trace illustrating interaction with a seamount is presented in Fig. 5.14.

The 3D Hamiltonian ray-tracing model, HARPO, has been used in tomographic studies of the ocean (Jones *et al.*, 1991; Newhall *et al.*, 1990). Both HARPO (Lynch *et al.*, 1994) and the 3D coupled-mode model CMM3D (Chiu and Ehret, 1994) have been interfaced with a sound speed field generated by the Harvard Open Ocean Model (HOOM).

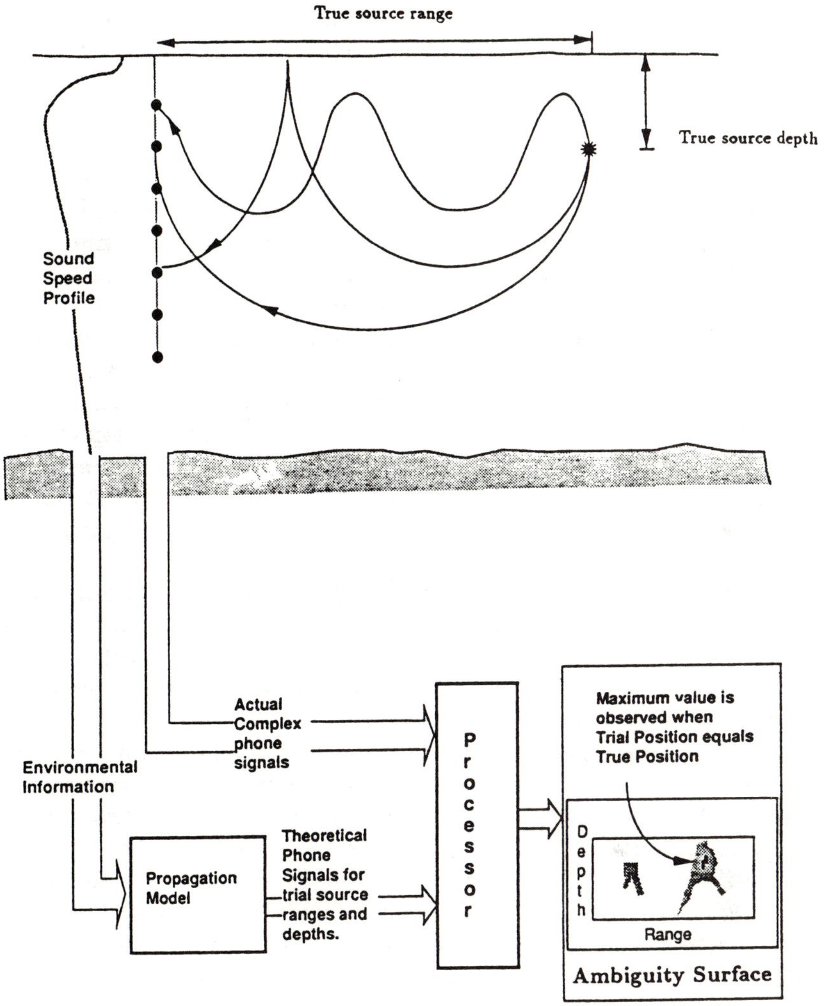

Fig. 5.13 Illustration of the basic components of matched field processing (Tolstoy, 1993; *Matched Field Processing for Underwater Acoustics*; reproduced with permission of World Scientific Publishing).

Perkins *et al.* (1993) modeled the ambient noise field in 3D ocean environments.

5.6.7 Ocean fronts and eddies

The distribution of ocean fronts and eddies was previously discussed in Chapter 2. The effects of eddies on acoustic propagation were investigated

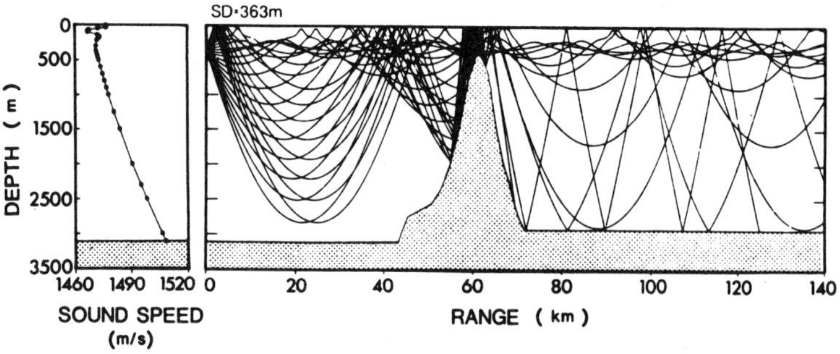

Fig. 5.14 Ray-trace representation of sound propagation across a seamount (Jensen, 1988; *IEEE J. Oceanic Engr.*, **13**, 186–97; copyright by IEEE; based on data from Chapman and Ebbeson, 1983).

by Vastano and Owens (1973), among others, using ray-theoretical models. More sophisticated wave-theoretical models have since been employed in more comprehensive studies. For example, Baer (1981) utilized a primary three-dimensional version of the parabolic equation to study propagation through an eddy, and Hall and Irving (1989) used an adiabatic normal mode technique to investigate propagation through Baer's example of an eddy. All investigators indicate that the eddies significantly modify major characteristics of the acoustic field, relative to those obtained in the absence of the eddies. Such effects include the shifting of convergence zones, altered multipath arrival sequences and horizontal refraction (Munk, 1980), among others. In a comprehensive study of warm-core eddies, Browning *et al.* (1994) found that the greatest acoustic impact was obtained for a shallow source and receiver configuration.

As an illustration, the results of Vastano and Owens (1973) can be used to quantify the effects of a cold-core eddy (surrounded by warmer Sargasso Sea water) on acoustic propagation. Assuming a frequency of 100 Hz, a source depth (S) of 200 m, a receiver depth (R) of 300 m, and a Figure of Merit (FOM) of 90 dB (see Chapter 10), the resulting transmission loss curves in Fig. 5.15(a) would result in the detection zones in Fig. 5.15(b) for a sonar located both inside and outside the eddy. The shaded areas denote detection opportunities under the stated conditions. The shifts of the convergence zones are readily apparent as one effect of the eddy. The particular situation illustrated in Fig. 5.15 could arise, for example, if a sonar were located on the outer edge of an eddy, with the eddy on the left and the Sargasso Sea waters on the right.

To further explore the effects of eddies (or rings, as they are sometimes called) on sonar detection, the case of a surface ship located in the Sargasso Sea trying to detect a submarine located inside an eddy will be examined. Figure 5.16 shows contours of constant sound speed through a typical Gulf

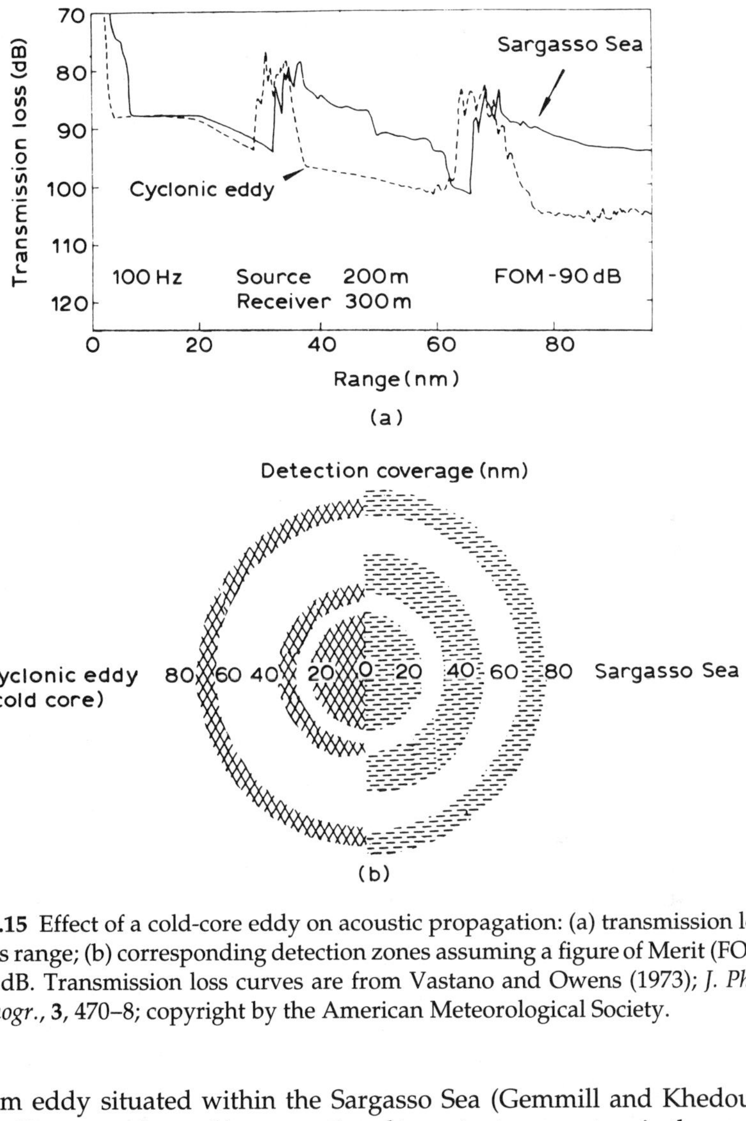

Fig. 5.15 Effect of a cold-core eddy on acoustic propagation: (a) transmission loss versus range; (b) corresponding detection zones assuming a figure of Merit (FOM) of 90 dB. Transmission loss curves are from Vastano and Owens (1973); *J. Phys. Oceanogr.*, **3**, 470–8; copyright by the American Meteorological Society.

Stream eddy situated within the Sargasso Sea (Gemmill and Khedouri, 1974). Since sound speed is proportional to water temperature in the upper layers of the ocean, this figure also demonstrates this to be a cold-core eddy since the low sound speed values in the center reflect lower water temperatures relative to the surrounding waters. Also shown (by a dashed line) is the SOFAR channel axis which indicates the depth of minimum sound speed. Next, consider Fig. 5.17 which schematically portrays the corresponding isotherms. A surface ship located in the warmer Sargasso Sea is frustrated in its attempts to passively detect the sound energy emitted by the submarine located in the colder interior of the eddy since

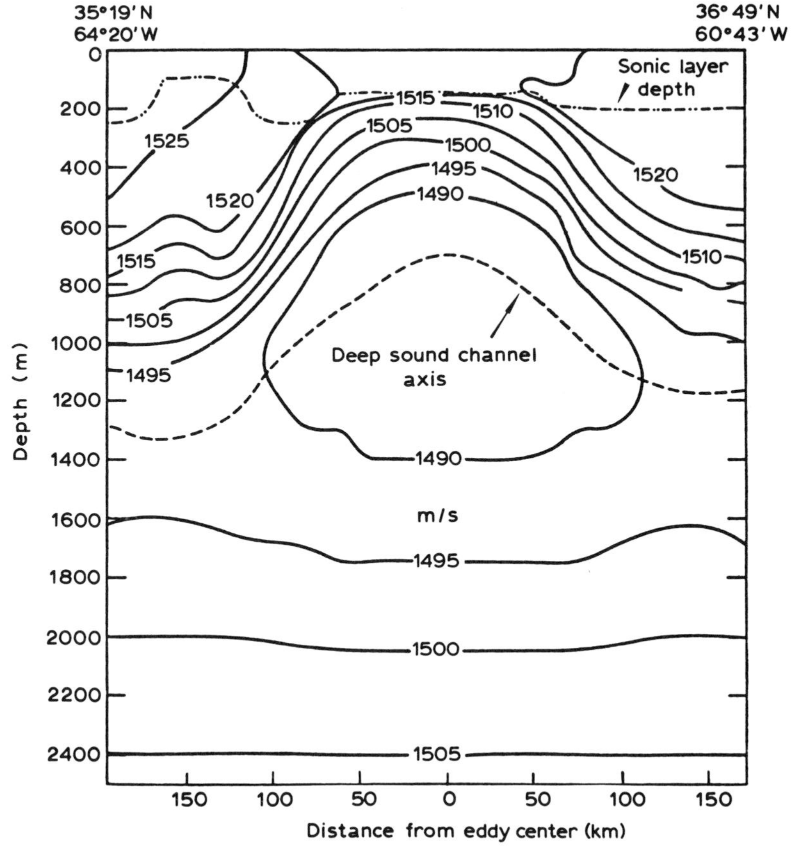

Fig. 5.16 Sound speed contours through a cold-core eddy (Gemmill and Khedouri, 1974).

the sound waves are refracted downward (away from the higher temperatures and higher sound speeds) and away from the ship's passive listening device. The submarine thus can successfully avoid detection by positioning itself inside known eddy locations.

Heathershaw *et al.* (1991) domonstrated the feasibility of coupling the range-dependent propagation model GRASS with a 3D eddy-resolving ocean model to study acoustic propagation through frontal systems.

TIhe Optimum Thermal Interpolation System (OTIS) (Clancy *et al.*, 1991a) will form the basis for an upgraded Tactical Ocean Thermal Structure (TOTS) system (Hawkins, 1992) designed for use aboard ships equipped with the Tactical Environmental Support System (TESS). OTIS assimilates real-time observations from multiple sources into a complete three-dimensional (3D) representation of the oceanic thermal field. A

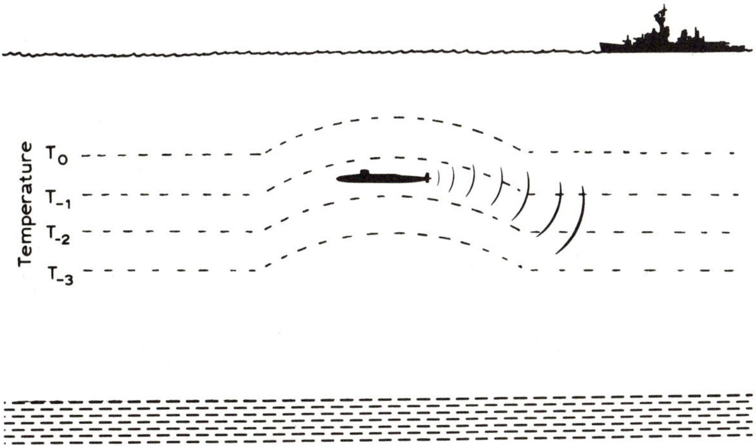

Fig. 5.17 Schematic illustration of a submarine in a cold-core eddy avoiding detection by a surface ship with a passive sonar. Sound is refracted downward away from the ship by the effects of the colder water in the interior of the eddy.

corresponding 3D representation of the salinity field is derived using empirical techniques. To compensate for sparse subsurface measurements, OTIS supplements actual observations with synthetic data derived from an empirical orthogonal function (EOF) representation of water masses, from ocean feature models, and from subjective interpretation of satellite imagery. The resulting product provides an accurate 3D representation of the temperature and salinity fields together with a realistic depiction of fronts and eddies. Such information is useful in organizing ASW assets as well as in planning field exercises. OTIS has also been used to generate high-quality data for the evaluation of oceanographic forecasting models (Lai *et al.*, 1994).

5.6.8 Coupled ocean–acoustic modeling

Oceanographers have recently developed synoptic forecasting techniques appropriate for predicting the locations of frontal features such as currents and eddies (Mooers *et al.*, 1982; Robinson *et al.*, 1984; Robinson, 1987; Peloguin, 1992). These forecasts can be used by forces afloat to facilitate the efficient allocation of naval resources during ASW operations.

These same dynamical ocean models can be used in conjuction with underwater acoustic propagation models to generate timely forecasts of sonar performance in the vicinity of highly variable frontal features. When used in this fashion, the ocean models generate input variables necessary for initialization of the acoustic models. This synergistic arrangement is referred to as coupled ocean–acoustic modeling. Such coupling has been successfully demonstrated, for example, by Botseas *et al.* (1989), who interfaced an implicit-finite-difference PE model with ocean forecasts generated by the Harvard open ocean model.

Coupled ocean–acoustic forecast systems comprise three basic components: an oceanic forecast scheme, a coupling scheme and an ocean acoustic propagation scheme (Robinson *et al.*, 1994). These systems can also be used to generate nowcasts and hindcasts. Nowcasts are estimates of the present state of a system; they are based on a combination of observations and dynamical modeling. Hindcasts are *a posteriori* forecasts; they are useful in evaluating modeling capabilities based on historical benchmark data (e.g. Martin, 1993).

Requirements for oceanographic data to support coupled ocean–acoustic forecast systems often exceed observational capabilities. Therefore, data assimilation, which introduces data generated by feature models, is used to achieve accurate synoptic realizations. Feature models are statistical representations of common synoptic structures in the ocean such as fronts and eddies (Robinson *et al.*, 1994).

The coupling of oceanographic mixed layer models with surface duct propagation models was discussed earlier in this chapter.

The subject of marine modeling, which properly embraces the numerical modeling of ocean physics, geology, chemistry and biology, has been summarized by Goldberg *et al.* (1977) and by Kraus (1977).

5.6.9 Acoustic tomography

Ocean acoustic tomography (Munk and Wunsch, 1979) is an inverse technique which measures perturbations in travel time between fixed acoustic sources and receivers (see Fig. 5.18). This technique is analogous to the medical procedure called tomography (from the Greek word for 'slice' or 'cut'), and also has elements in common with conventional seismology (Menke, 1989).

This technique has been proposed for large-scale monitoring of ocean basins. Specifically, the perturbations in travel times between sources and receivers can be used to deduce sound speed (and thus water density) fluctuations in the interior of the oceans. Moreover, if reciprocal transmissions are used between sources and receivers, the differences in travel times can be used to compute the mean ocean currents along the acoustic path (Spiesberger, 1989). Sea surface spectra can also be estimated using acoustic tomography (Miller *et al.*, 1989).

The temporal resolution required to distinguish multipath arrivals has been estimated at 50 ms. Existing low-frequency (<200 Hz) broadband sources appear to satisfy these requirements. The real limiting factor will probably be the lack of spatial resolution owing to practical limitations on the number of sources and receivers. Consequently, there will usually be

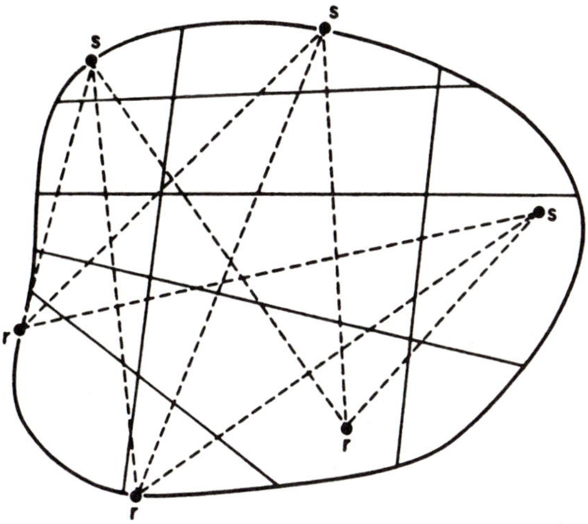

Fig. 5.18 Schematic illustration of an ocean tomographic geometry. Perturbations in acoustic travel time from any source *s* to any receiver *r* (dashed paths) are used to estimate sound speed perturbations in an arbitrary grid area formed by the solid lines (Munk and Wunsch, 1979; *Deep-Sea Res.*, **26**, 123–61; copyright by Pergamon Press PLC).

insufficient information with which to determine a unique solution to the inverse problem.

A field test of ocean–acoustic tomography was conducted in the North Atlantic Ocean in 1981 (Cornuelle *et al.*, 1985). The test was conducted over a two-month period in a 300 km square centered at 26°N, 70°W using nine acoustic deep-sea moorings with sea-floor transponders. The acoustic sources operated at a center frequency of 224 Hz, with a 5.4 Hz rms bandwidth. The three-dimensional sound speed field obtained by inversion techniques compared favorably with direct observations of sound speed profiles. Tomographic maps constructed at three-day intervals over the test period revealed a pattern of eddy structure in general agreement with that deduced from direct observations. Mapping errors were attributed to noise variance in travel time measurements.

Acoustic tomography experiments have been conducted to detect climatic trends of temperature in the oceans with basin-scale resolution (Munk and Forbes, 1989; Spindel and Worcester, 1990a, 1990b, 1991; Spiesberger and Metgzer, 1991a; Spiesberger *et al.*, 1992; Forbes, 1994). Since acoustic travel times are inversely related to water temperature, tomography provides a method for ascertaining the bulk temperature of the oceans. This bulk temperature is potentially a more reliable metric for global warming than are conventional atmospheric measurements which, for example, can be influenced by the heat island effect that is often

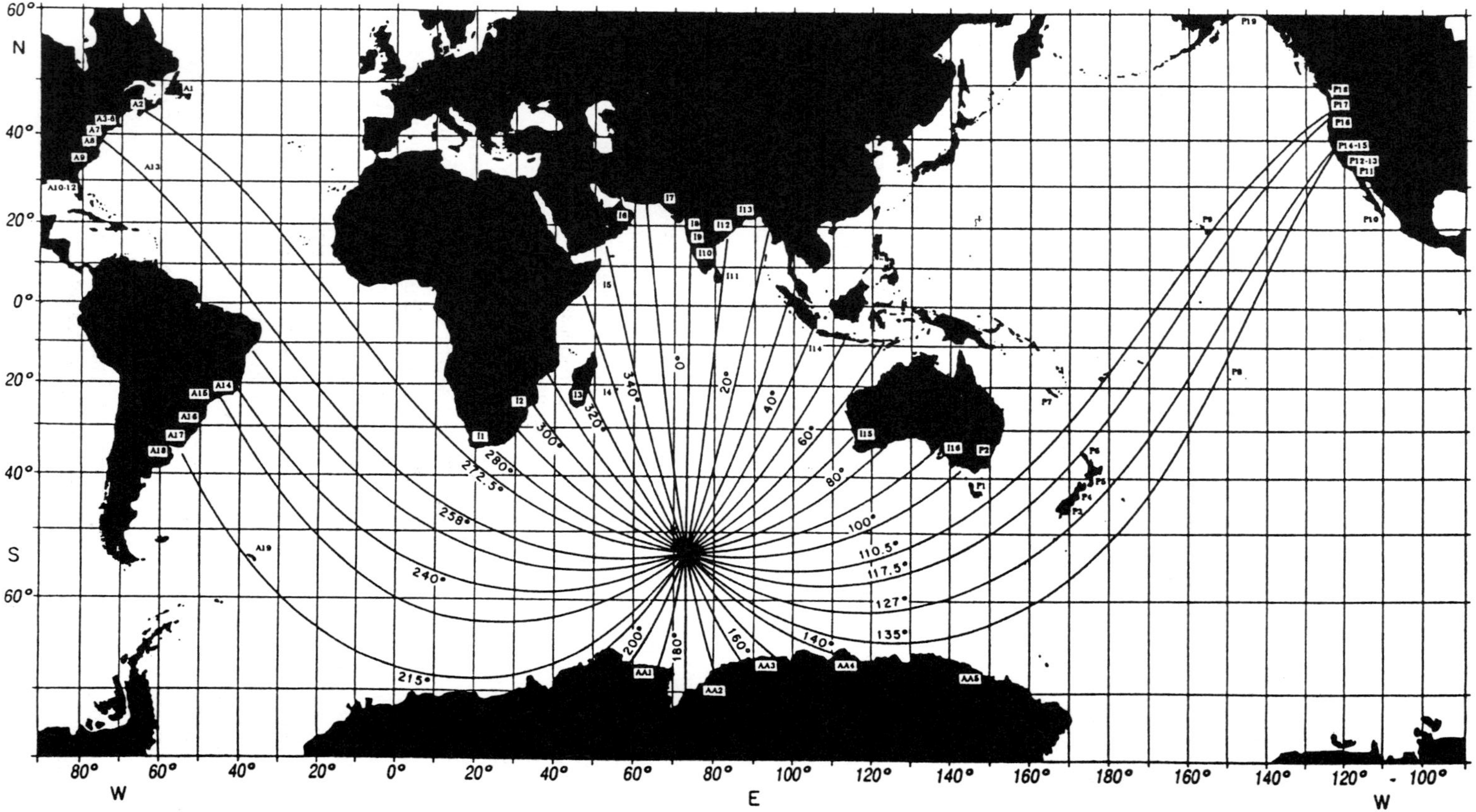

Fig. 5.19 Illustration of bathymetrically unimpeded ray paths emanating from Heard Island with the indicated launch angles. All rays are refracted geodesics along the surface containing the sound channel axis (see Fig. 2.8). White boxes indicate locations of oceanographic research stations capable of receiving acoustic signals (Munk and Forbes, 1989; *J. Phys. Oceanogr.*, **19**, 1765–78; copyright by the American Meteorological Society).

associated with urban temperature records. Moreover, basin-scale averages of water temperature suppress the intense meso-scale eddies that would otherwise dominate the climatic variations.

The Heard Island Feasibility Test (HIFT) established the limits of phase-coded, electrically driven sound sources (instead of explosive sources) in tomographic experiments (Munk, 1994). Figure 5.19 illustrates the ability of an acoustic source situated at Heard Island to sample vast volumes of the oceans over ray paths unimpeded by bathymetric obstructions. These ray paths represent refracted geodesics along the surface containing the sound channel axis (Fig. 2.8). A geodesic path is the shortest distance between two points on the surface of an elliptical (versus spherical) Earth. Goodesics would be the proper ray paths in a uniform sound speed field; however, these paths are altered by the refractive effects of lateral sound speed gradients.

HIFT is a prerequisite to a program for Acoustic Thermometry of Ocean Climate (ATOC). The goals of ATOC are: to verify the acoustically measured changes in travel time by comparison with those inferred from traditional oceanographic measurements along the transmission path in the North Pacific Ocean; to study the coherence of climatic scale variability in the North Pacific gyre; to test long-range acoustic transmission characteristics using 40-element vertical receiver arrays; and to test specially designed ATOC sound sources. The potential effects of acoustic transmissions on marine mammals have also been investigated in related efforts (National Research Council, 1994; Green, 1994). A series of 17 technical papers addressing various aspects of HIFT was recently published; these papers were introduced by Munk and Baggeroer (1994).

Long-range acoustic propagation issues relevant to tomographic investigations include bathymetric interactions, which have been investigated by Munk (1991), Munk and Zachariasen (1991) and Heaney *et al.* (1991). Forward scatter by islands and seamounts was found to produce significant scattered arrivals (Munk and Zachariasen, 1991); such arrivals can complicate interpretation of the recorded acoustic travel times if they are not properly differentiated from the direct arrivals.

Gaillard (1985) demonstrated that mobile sources or receivers can also be used in tomographic experiments, but that this mobility increases the sensitivity to noise. More efficient use of *a priori* knowledge of the ocean reduces this sensitivity by decreasing the number of degrees of freedom.

Moving Ship Tomography (MST) is a method for obtaining high-resolution, quasi-synoptic, three-dimensional maps of the oceanic temperature field over large areas (The Acoustic Mid-Ocean Dynamics Experiment Group, 1994). All previous work in ocean acoustic tomography has been performed with fixed or moored instruments. MST is part of the larger Acoustic Mid-Ocean Dynamics Experiment (AMODE), which has the goal of determining the measurement limits of moving ship tomography. The larger number of acoustic ray paths generated by a moving receiver will improve measurements of advecting fronts and interacting eddies; this increased volumetric resolution will enable more rigorous testing of the predictive capabilities of dynamical ocean circulation models.

As summarized by Spindel (1985), acoustic tomography is particularly attractive for application to the ocean for a number of reasons: (1) images of the ocean interior can be obtained from periphery measurements; (2) large ocean areas can be measured with relatively few instruments; (3) a single source–receiver pair samples the ocean vertically due to multipath propagation effects; and (4) the integrating properties of acoustic propagation produce average quantities not obtainable with point measurements.

6

Noise I. Observations and physical models

6.1 BACKGROUND

Ambient noise is the prevailing, unwanted background of sound at a particular location in the ocean at a given time of the year. It does not include transient sounds such as the noise of nearby ships and marine organisms, or of passing rain showers. It is the background of noise, typical of the time, location and depth against which a signal must be detected. Ambient noise also excludes all forms of self-noise, such as the noise of current flow around the sonar. Thus, ambient noise is the residual sound level remaining after all identifiable, transient noise sources have been removed (Urick, 1983, Chapter 7). Updated summaries of sea surface sound have been provided by Kerman (1988; 1993).

Levels of noise sources are specified as the root-mean-square sound pressure level in a 1 Hz band. A variety of labels are used to express these units: (1) spectrum level, dB re 1 μPa; (2) dB re 1 $\mu Pa^2/Hz$; or (3) dB re 1 $\mu Pa/Hz^{1/2}$ (where 're' is an abbreviation for 'relative to').

6.2 NOISE SOURCES AND SPECTRA

Figure 6.1 is a hypothetical example of the spectrum of ambient noise in the open ocean (Urick, 1983, Chapter 7). This spectrum is composed of segments of different slope each exhibiting a different behavior. A number of frequency bands in the spectrum can be associated with readily identifiable noise sources. Five such frequency bands are indicated in Fig. 6.1. Band I, lying below 1 Hz, is associated with noise of hydrostatic origin (tides and waves) or with seismic activity (Kibblewhite and Ewans, 1985). Valid measurements in this band (and in Band II) are extremely difficult to

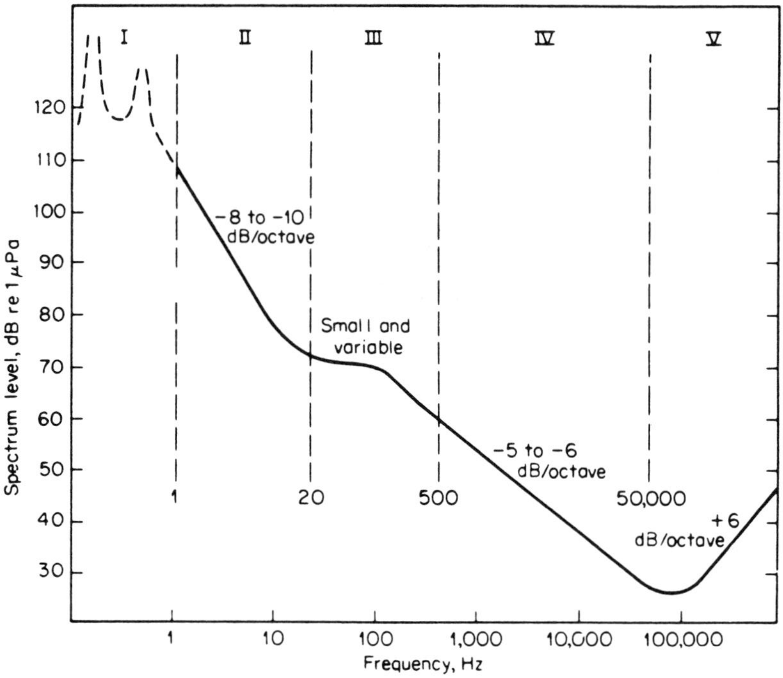

Fig. 6.1 General spectrum of deep-sea noise showing five frequency bands of differing spectral slopes; the slopes are given in decibels per octave of frequency (Urick, 1983; *Principles of Underwater Sound*, 3rd edn; reproduced with permission of McGraw-Hill Publishing Company).

make because of the self-noise of the hydrophone and its supporting structure caused by currents (e.g. cable strumming).

Band II is characterized by a spectral slope of −8 to −10 dB/octave (about −30 dB/decade). The most probable source of noise in deep water appears to be oceanic turbulence. In Band III, the ambient noise spectrum flattens out and the noise appears to be dominated by distant shipping traffic. Band IV contains the Knudsen spectra (Knudsen *et al.*, 1948) having a slope of −5 to −6 dB/octave (about −17 dB/decade) in which the noise originates at the sea surface nearby the point of measurement. Band V is dominated by thermal noise originating in the molecular motion of the sea and is uniquely characterized by a positive spectrum having a slope of +6 dB/octave.

For some prediction purposes, average representative ambient-noise spectra for different conditions are adequate. Such average working curves are shown in Fig. 6.2 for different conditions of shipping and wind speed. These curves are called Wenz curves after Wenz (1962), who first published them. In the infrasonic region below 20 Hz, only a single line is

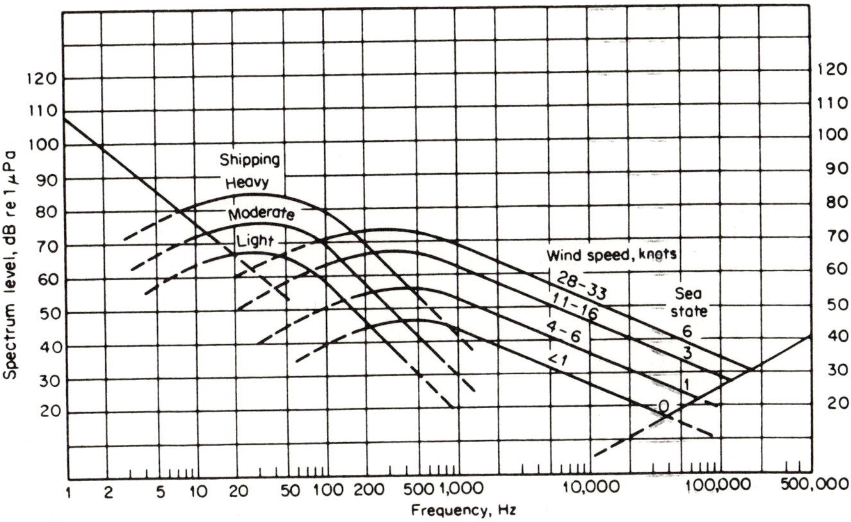

Fig. 6.2 Average deep-sea ambient noise spectra (Urick, 1983; *Principles of Underwater Sound*, 3rd edn; reproduced with permission of McGraw-Hill Publishing Company).

drawn. The ambient noise spectrum at any location and time is approximated by selecting the appropriate shipping and wind curves and connecting them at intermediate frequencies where more than one source is contributing. When more than one source of noise is present, the effective noise background is obtained by summing the intensities of the contributing sources. When using noise levels (specified in units of spectrum level, dB re 1 μPa), this summation process is easily accomplished using the 'power summation' operator, denoted by the symbol $\oplus$, and defined as

$$\oplus = 10 \log \sum_{i=1}^{n} 10^{Li/10} \tag{6.1}$$

where L_i is the level of ith noise source (dB) and n is the number of contributing noise sources. This operation effectively converts the noise levels (L_i) to units of intensity, sums the intensities, and then converts the sum back to units of dB.

For example, in Fig. 6.2 at a frequency of 100 Hz under conditions of moderate shipping and sea state 6, one would obtain noise levels of about 69 dB and 71 dB, respectively. In effect, however, the noise level is obtained from equation (6.1) as

$$69\,\text{dB} \oplus 71\,\text{dB} = 10 \log[10^{6.9} + 10^{7.1}] = 73\,\text{dB}$$

or 2 db higher than the level due to surface weather noise alone.

6.2.1 Seismo-acoustic noise

The term seismo-acoustics is used broadly in reference to low-frequency noise signals originating within the Earth and the ocean. In the frequency range below 3 Hz, Orcutt (1988) has defined three specific frequency bands distinguished by the physics of the noise sources:

1. mocroseism band (80 mHz–3 Hz) – high-level microseismic noise resulting from nonlinear wave–wave interactions;
2. noise-notch band (20–80 mHz) – noise controlled by currents in the turbulent boundary layer near the sea floor; and
3. ultralow-frequency (ULF) band (< 20 mHz) – noise resulting from surface gravity waves.

6.2.2 Shipping noise

Shipping noise can exhibit variability in both space and time. The spatial variability is largely governed by the distribution of shipping routes in the oceans. Temporal variability can be introduced, for example, by the seasonal activities of fishing fleets.

The noise generated by coastal shipping and by high-latitude shipping can contribute to the noise field in the deep sound channel of tropical and subtropical ocean areas. Coastal shipping noise is introduced into the deep sound channel through the process of downslope conversion; high-latitude shipping noise is introduced through the latitudinal dependence of the depth of the sound channel axis.

Wagstaff (1981) used ray-theoretical considerations to illustrate the mechanisms involved in the downslope conversion process. The following hypothetical arrangement was assumed (Fig. 6.3): (1) the continental shelf extends from the coastline to the shelf break (approximately 11.4 km from shore), with an inclination angle of about 1° from the horizontal; (2) the continental slope extends from the shelf break seaward to a depth in excess of 1000 m, with an inclination of about 5°; and (3) the deep sound channel axis is located at a depth of 1000 m. Then, in an 11 km band extending seaward from the shelf break, downward-directed radiated noise from surface ships can enter the sound channel by direct reflection off the bottom.

Kibblewhite *et al.* (1976) demonstrated the importance of the latitudinal dependence of the sound channel axis depth in introducing high-latitude shipping noise into the deep sound channel. In Fig. 6.4, the sound speed structure in the North Pacific Ocean is related to the local water masses; also shown are the critical depth and bathymetry. Figure 6.4 vividly illustrates the shoaling of the sound channel axis as one approaches the Arctic region. Thus, low-frequency noise from shipping sources near the latitude of 50° N will be refracted downward into the (relatively) shallow

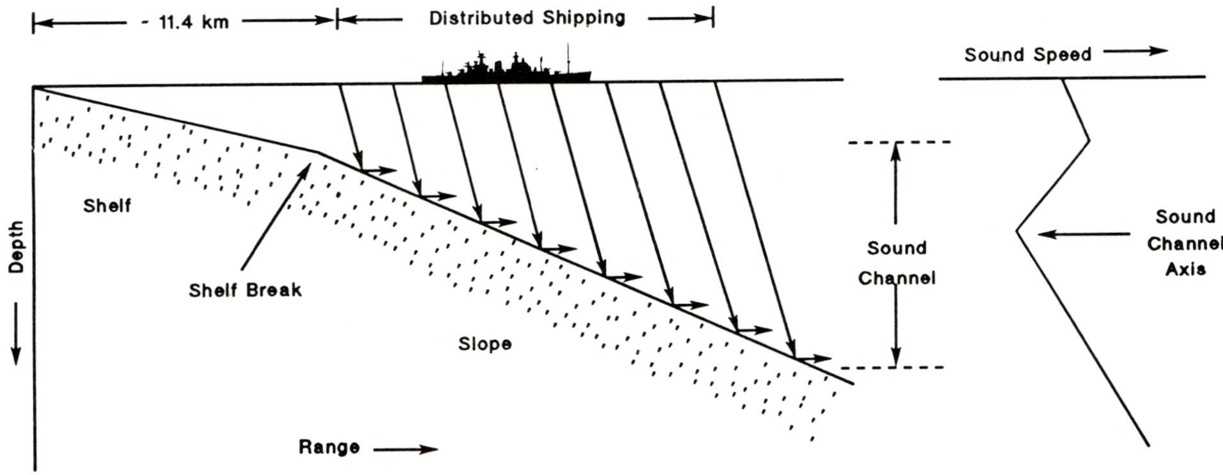

Fig. 6.3 Illustration of the conversion of coastal shipping noise, represented by high-angle rays, to noise in the deep sound channel, represented by horizontal rays (adapted from Wagstaff, 1981).

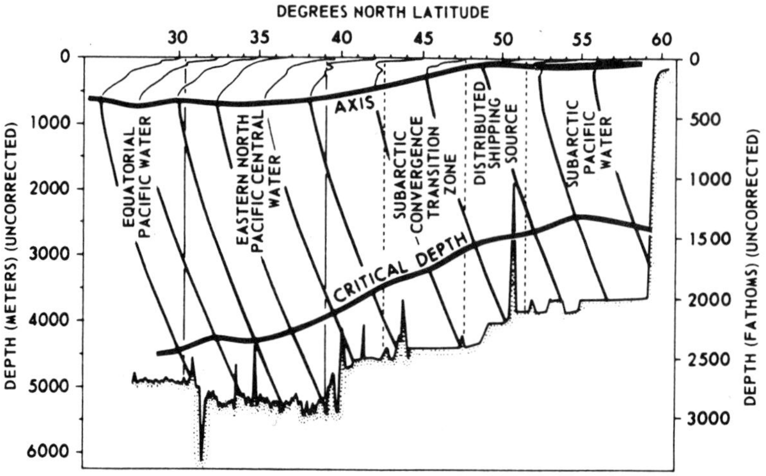

Fig. 6.4 Bathymetric and sound speed structure in the North Pacific Ocean. The noise from distributed shipping sources at high latitudes can enter the sound channel and propagate with little attenuation to lower latitudes. Relationships between the sound speed structure and the prevailing water masses are also illustrated (Kibblewhite *et al.*, 1976).

sound channel and then propagate with little attenuation to lower latitudes.

6.2.3 Bioacoustic noise

Marine bioacoustic signal sources are typically transient in nature and exhibit diverse temporal, spatial and spectral distributions. The main contributors to bioacoustic signals include certain shellfish, fish and marine mammals. Of the marine mammals, whales are the most notable contributors. Cummings and Holliday (1987), Au *et al.* (1987) and Watkins *et al.* (1987) describe whale signal characteristics and distributions. A paper by Watkins and Schevill (1977) includes a phonograph recording of actual whale codas.

6.2.4 Wind and rain noise

The established relationships between surface weather phenomena and ambient noise levels have been used to advantage by oceanographers. Shaw *et al.* (1978) demonstrated that surface wind speeds derived from measured ambient noise spectra can be used to calculate the wind stress over the oceans. Synoptic wind stress information is required for the dynamic modeling of wind-driven ocean currents. Scrimger *et al.* (1987)

and Lemon and Duddridge (1987) described the development and operation of WOTAN (Weather Observation Through Ambient Noise) systems. These systems operate over the frequency range 0.5–30 kHz and have been successfully used to infer wind speeds that were highly correlated with buoy-mounted anemometer measurements.

The ambient noise attributable to rainfall over the oceans has also been used in an inverse fashion to provide estimates of oceanic precipitation (Nystuen, 1986). The underwater ambient noise spectrum generated by rain has a unique spectral shape, distinguishable from other noise sources, which is dominated by a broad peak at about 15 kHz. Moreover, the relationship between spectral level and rate of rainfall is quantifiable. Scrimger *et al.* (1987) made measurements of the underwater noise generated by rain in a lake located on Vancouver Island, British Columbia (Fig. 6.5). These data illustrate the peak in noise levels at 15 kHz. Pumphrey and Crum (1990) determined that the major cause of rain-generated sound is the production of bubbles upon water drop impact at the surface. These bubbles then oscillate with a small amplitude and radiate as a dipole.

The effects of bubbles and rain as noise-generating mechanisms have been investigated further by Buckingham (1991) and Laville *et al.* (1991), respectively. There are two sources of rainfall sound: raindrop impacts on the water surface, and air bubble resonances. Bubble resonances were found to be responsible for the spectral peak near 13–15 kHz, while raindrop impacts were associated with a broadband spectrum having a negative slope (Laville *et al.*, 1991). The underwater sound due to rainfall can be distinguished further according to raindrop diameters (Medwin *et al.*, 1992): small drops (0.8–1.1 mm) radiate primarily from bubble resonances near 15 kHz; mid-size drops (1.1–2.2 mm) radiate only broadband impact sound; and large drops (> 2.2 mm) radiate both impact and resonance sounds. In related investigations, Rohr and Detsch (1992) attributed the effect of films on suppressing high-frequency ambient noise to the opposition of bubble-producing events by these films. Crum *et al.* (1992) and Pumphrey (1994) reviewed the nature of precipitation sounds underwater, while Prosperetti and Oğuz (1993) reviewed the physics of drop impacts on liquid surfaces. Collectively, these studies have improved upon earlier investigations such as those reported by Heindsmann *et al.* (1955), Franz (1959) and Bom (1969).

Indirect measurements of rainfall are important since roughly 80% of the Earth's precipitation occurs over oceans and lakes where only about 10% of the weather stations are located (e.g. islands and buoys). This information allows oceanographers and meteorologists to improve our understanding of the oceanic heat and water budgets (e.g. Etter *et al.*, 1987) which, in turn, can provide measures of both short-term and long-term fluctuations in the global climate.

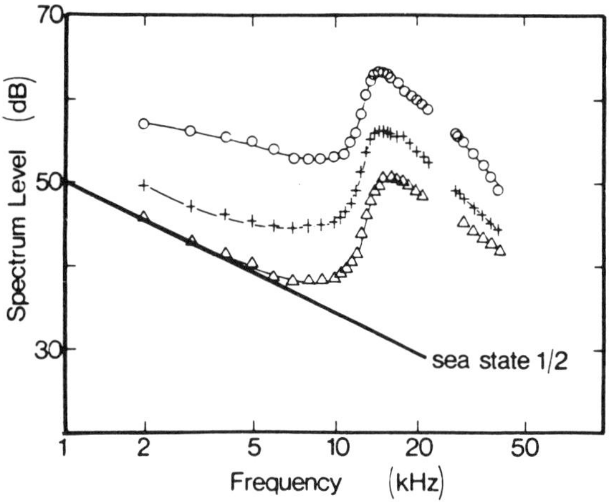

Fig. 6.5 Rain noise spectra observed at moderate wind speeds for rain rates of 0.5 mm/h (Δ), 0.8 mm/h (+) and 1.3 mm/h (○) compared to the Knudsen curve for sea state 1/2 (Scrimger *et al.*, 1987).

Nystuen (1994) developed an Acoustic Rainfall Analysis (ARA) algorithm consisting of: detection of rainfall in the presence of other underwater noise sources; classification of rainfall type based on drop-size distribution; and acoustic quantification of rainfall rate. The classification of rainfall type (e.g. according to stratiform or convective) will permit meteorologists to infer the vertical distribution of atmospheric latent-heat release in support of global climate studies.

Oğuz (1994) developed a theoretical model to predict the ambient noise levels arising from bubble clouds generated by breaking waves and resultant whitecap formation. Cloud geometries were modeled by inverted hemispherical shapes within which solutions to the wave equation could be obtained analytically. Empirical relationships between wind speed and whitecap occurrence permitted calculation of noise levels as a function of frequency and wind speed.

6.3 DEPTH DEPENDENCE

Measurements of the depth dependence of low-frequency ambient noise were made by Morris (1978) in the northeastern Pacific Ocean. Hydrophones were suspended form the research platform FLIP (Floating Instrument Platform). Figure 6.6 shows the sound speed profile, hydrophone depths and average noise profiles in one-third-octave bands for this experiment. There is a decrease of noise with depth at low frequencies, with a smaller decrease with depth at 500 Hz as wind noise overcomes the dominance of shipping noise. Below the critical depth, the fall-off with depth is steeper as

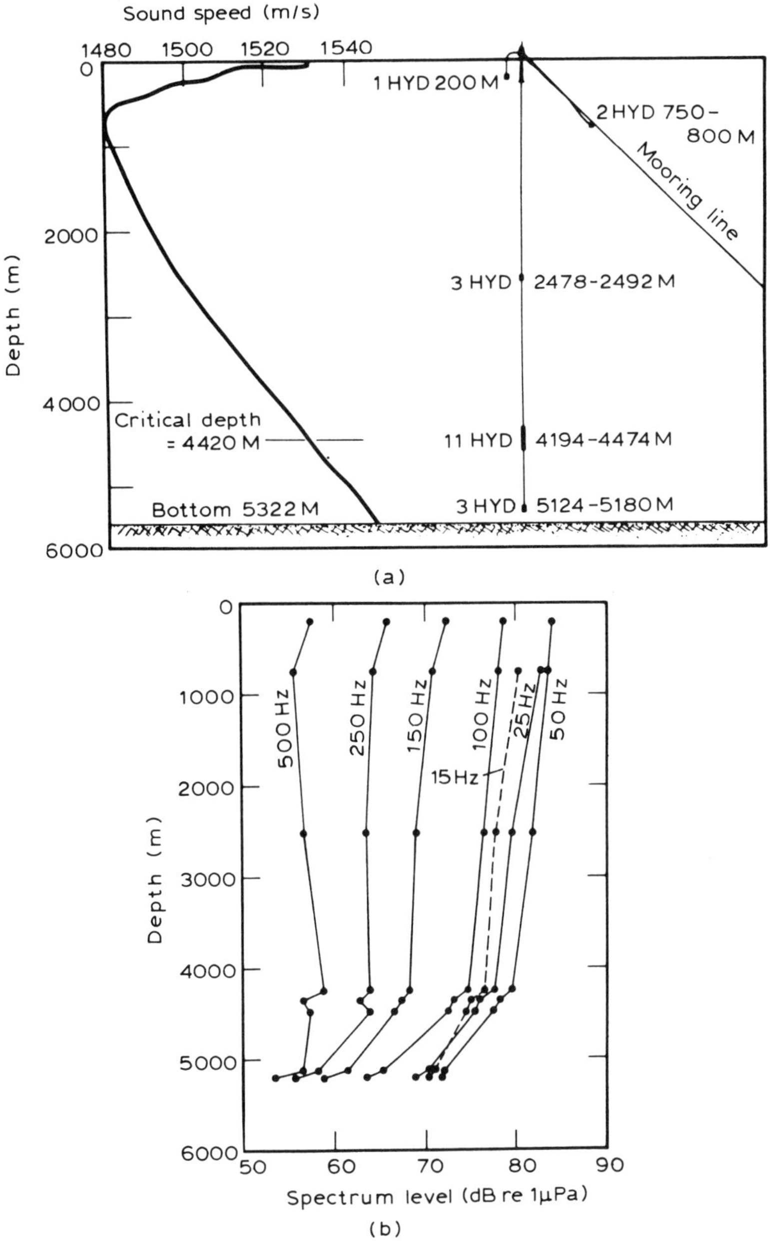

Fig. 6.6 Northeast Pacific Ocean ambient noise measurements (Morris, 1978): (a) sound speed profile and hydrophone depths; (b) measured noise profiles.

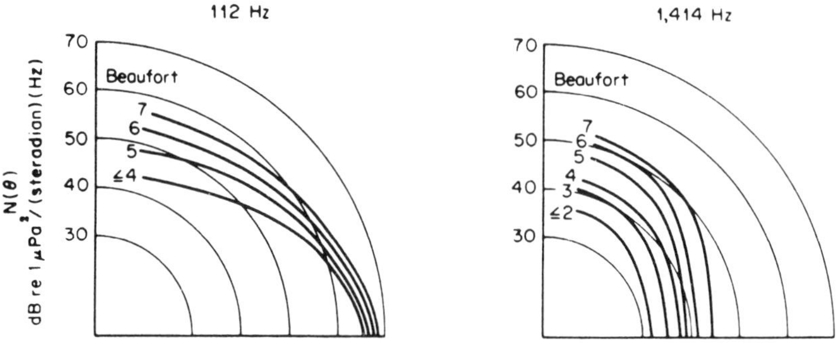

Fig. 6.7 Distribution of ambient noise in the vertical plane at a bottomed hydrophone at two frequencies (Axelrod *et al.*, 1965).

the bottom is approached. This is the result of the loss of refracted sound energy through the effects of bottom interaction.

Urick (1984, Chapter 4) has shown that ambient noise at frequencies greater than 10 kHz is rapidly attenuated with increasing depth due to the effects of absorption.

6.4 DIRECTIONALITY

As a first-order approximation, the noise field of the ocean might be considered to be isotropic in nature, that is, uniform in all directions, both horizontal and vertical. Measurements have shown that this is not the case.

Axelrod *et al.* (1965) made measurements of the vertical directionality of ambient noise at frequencies of 112 and 1414 Hz. Figure 6.7 presents polar plots of the ambient intensity per unit solid angle $N(\theta)$ arriving at a bottomed hydrophone as a function of vertical angle θ. At 112 Hz more noise arrives at the hydrophone from the horizontal than from the vertical; this effect diminishes with increasing wind speed. At 1414 Hz the opposite is true in that more noise arrives from overhead than horizontally; this effect increases with increasing wind speed.

This directional behavior is consistent with the view that low-frequency noise originates at great distances and arrives at the measurement hydrophone principally via horizontal paths, suffering little attenuation, whereas high-frequency noise originates locally at the sea surface overhead (Urick, 1984, Chapter 5).

The horizontal (or azimuthal) directionality of ambient noise can be highly variable, particularly at low frequencies. Shipping traffic is the dominant source of noise at low frequencies, and the temporal and spatial variations in shipping routes and shipping densities explain much of the observed azimuthal variation.

6.5 ARCTIC AMBIENT NOISE

The noise environment under the Arctic ice is different from that of any other ocean area. Shipping noise is extremely low due to the lack of surface traffic. The ice cover itself affects the ambient noise field significantly. It can decouple the water from the effects of the wind and produce ambient noise conditions that are much quieter than a corresponding sea state zero in the open ocean. The ice itself may produce noises as it is acted on by wind, waves and thermal effects (e.g. Milne, 1967). Other sources of noise in the Arctic include seismic and biological activity.

The character of the ice cover is different in areas of shore-fast pack ice, moving pack ice and the Marginal Ice Zone (MIZ). The under-ice noise levels, directionality, spectrum shape and temporal character are very different in each of these regions. Noise originating within the ice stems principally from its state of stress which gives rise to fracturing. Noise measurements under the pack ice span the frequency range from 3 Hz to more than 1000 Hz (Dyer, 1984, 1988).

Pack ice is very dynamic and its characteristics are highly variable in space and time (Makris and Dyer, 1986). Nevertheless, the lack of wind–wave interaction and the absence of local shipping can lead to noise levels 10 dB below those encountered in the open ocean.

Noise levels in the MIZ are typically higher than those in either the pack ice regions or the open ocean regions (Diachok and Winokur, 1974; Diachok, 1980). Figure 6.8 illustrates the variations in median ambient noise levels that occur across a compact ice–water MIZ area. The relative magnitudes of the noise levels generated at the ice–water boundary are a function of the rate of change of ice concentration with distance. Thus, the relative maximum noise level measured at a diffuse ice–water boundary would be smaller than that measured at a compact ice edge. Ambient noise levels in the MIZ also depend on such variables as sea state, water depth and dominant ocean-wave period. The last variable is hypothesized as being related to the efficiency of coupling ocean-wave energy into the ice. Makris and Dyer (1991) demonstrated that surface gravity wave forcing was the primary correlate of ice-edge noise in the MIZ.

Pritchard (1990) developed an ambient noise model to simulate the time history of under-ice noise generated by dynamical ice movement, stress and deformation. The model included noise contributions from local and distant sources as well as a transmission loss module appropriate for the Arctic environment. Time series data were simulated for a frequency of 31.5 Hz.

Lewis and Denner (1988) reported observations of higher-frequency (1000 Hz) ambient noise data in the Arctic Ocean. These noises were attributed to the thermal fracturing of ice.

Related aspects of ice-generated noise have been reviewed by Sagen *et al.* (1990).

6.6 ACOUSTIC DAYLIGHT

Ambient noise in the ocean can be used to form pictorial images of underwater objects. An analogy can be drawn between the natural optical (daylight) field in the atmosphere and the radiating ambient noise field in the ocean: both fields consist of random, incoherent radiation propagating in all directions. Thus, the concept of imaging with ambient noise has been referred to as 'acoustic daylight' (Buckingham *et al.*, 1992). A reasonable operating frequency range for the acoustic daylight system is 5–50 kHz; the lower limit is determined by angular resolution considerations while the upper limit is determined by the onset of thermal noise.

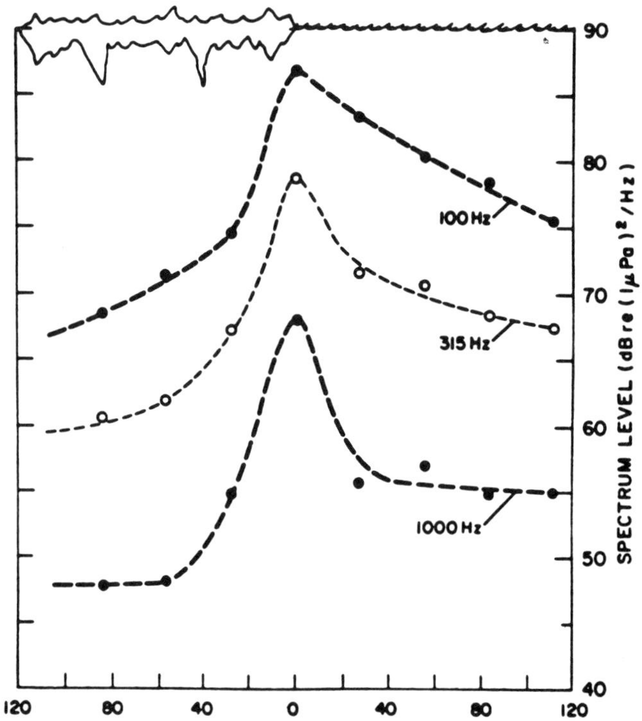

Fig. 6.8 Variation of median ambient noise sound pressure spectrum levels with distance from a compact ice edge for frequencies of 100, 315 and 1000 Hz in sea state 2 (Diachok, 1980).

Buckingham (1993) analyzed the incoherent acoustic imaging of a spherical target in order to quantify the contrast ratio (acoustic contrast) under various degrees of anisotropy. In the case of isotropic noise, the contrast (acoustic visibility) was found to have a maximum value of about 4 dB. This result was consistent with field measurements when the angle of view of the acoustic lens matched the angle subtended by the target at the phase center of the measurement array.

An Acoustic Daylight Ocean Noise Imaging System (ADONIS) is currently under construction. This system will operate over the frequency range 15–75 kHz with a corresponding minimum wavelength of 20 mm, will form 126 individual beams of nominal width of 0.76° (at the upper frequency), and will operate over the spatial range 10–200 m. Simulations were developed to predict the performance of ADONIS (Potter, 1994). These simulations demonstrated that near perfectly reflecting objects are unlikely to be imaged in volume isotropic noise, except perhaps in the near field; however, the ocean is expected to exhibit considerable anisotropy, if only in the vertical direction, thus improving performance over the conjectured isotropic baseline.

7

Noise II. Mathematical models

7.1 BACKGROUND

Chapter 6 discussed aspects of ambient noise measurements in the ocean. In particular, Fig. 6.2 portrayed the average deep-sea ambient noise spectra (in the frequency range 5 Hz–50kHz) as originating from shipping traffic at low frequencies (5–200 Hz) and surface weather at high frequencies (200 Hz–50 kHz). Clearly, in the frequency range 50–500 Hz, both mechanisms (shipping and wind) contribute to the observed noise levels. These dual mechanisms have influenced the direction of noise model developments.

Mathematical models of noise in the ocean can be segregated into two categories: ambient noise models and beam noise statistics models. Ambient noise models predict the mean levels sensed by an acoustical receiver when the noise sources include surface weather, biologics and such commercial activities as shipping and oil drilling. Beam noise statistics models are more specialized in that they predict the properties of low-frequency shipping noise for application to large-aperture, narrow-beam passive sonar systems. The latter models use either analytic (deductive) or simulation (inductive) techniques to generate statistical descriptions of the beam noise. In this context, beam noise is defined as the convolution of the receiver beam pattern with the sum of the intensities from the various noise sources. The analytic models calculate statistical properties directly from the components (e.g. source level, propagation loss), while the simulation models use Monte Carlo techniques.

7.2 THEORETICAL BASIS FOR NOISE MODELING

Mathematical models of noise in the ocean predict both the level and directionality (vertical and horizontal) of noise as a function of frequency, depth, geographic location and time of year.

Noise models are segregated into the categories of ambient noise models and beam noise statistics models. Regardless of type, however, noise models consist of two components: a transmission loss component, and a noise level and directionality component. In principle, the transmission loss can be computed internal to the noise model or it can be input externally from other (stand-alone) model predictions or from field measurements.

Ambient noise models treat noise sources as variable densities distributed over large areas. This represents a fair approximation to the generation of wind noise as well as distant shipping noise. Consequently, the transmission loss calculations can be range averaged, as opposed to point-to-point. This greatly relaxes the accuracy to which transmission loss must be known.

Beam noise statistics models treat noise sources (individual ships) as discrete sources, and thus require a point-to-point representation of transmission loss, as opposed to range averaged.

Various regression formulas can sometimes satisfy elementary modeling requirements. A comprehensive review of regression formulas appropriate for the estimation of average ambient noise spectra below 100 kHz has been assembled by Sadowski *et al.* (1984). Regression formulas have been compiled for noise sources arising from ocean turbulence, shipping traffic, surface weather (both wind and rain) and molecular agitation. Wagstaff (1973) also presented regression formulas which were incorporated into an ambient noise model valid over the frequency range 10–500 Hz. Ross (1976, Chapter 8) reviewed regression formulas appropriate for shipping noise levels. (See also Ross (1993) for a brief history.)

The use of regression formulas to estimate the ambient noise levels, but not directionality, presumes that the noise levels can be considered independently of the sonar system characteristics. Proper noise modeling, however, convolves the system beam pattern (receiver response) with the calculated noise field levels and directionalities (Wagstaff, 1982).

An example of a more elaborate empirical model is the Geophysics Ambient Noise Model (Bradley and Bradley, 1984). This model provides estimates of seasonal deep-water ambient noise levels and azimuthal directionality over the frequency range 25 Hz–15 kHz. The model is based on a comprehensive empirical database of shipping and wind noise, but not noise due to biologics or industrial activity. The model further assumes a nominal receiver depth of 100 m, thus ignoring any depth dependence.

7.3 AMBIENT NOISE MODELS

A simple model of ambient noise in the ocean consists of an infinite layer of uniform water with a plane surface along which the sources of noise are uniformly distributed. In this model, the ambient noise level is essentially independent of depth (Urick, 1983, Chapter 7). A more realistic model, however, would include volume absorption and the effects of refraction and boundary reflections over long-range paths (Talham, 1964).

To compute the low-frequency component of noise due to distant shipping, three inputs are required: (1) the density of shipping as a function of azimuth and range from the receiver; (2) the source level for the radiated noise of each generic type of merchant ship; and (3) the transmission loss as a function of range between the near-surface sources and the depth of the receiver. Then, contributions from successive range rings centered about the receiver can be summed to obtain the level of shipping noise as a function of azimuth at the receiver. The theory behind this kind of modeling, in addition to some observational data on the density of shipping traffic (in the North Atlantic), has been described by Dyer (1973), among others.

The high-frequency component of noise due to surface weather is usually computed on the assumption that it is locally generated and isotropic. Thus, only the weather conditions (sea state or wind speed) prevailing in the immediate vicinity need to be considered in addition to any localized rain showers or biologic activity.

Modeling of the vertical directionality of deep-water ambient noise can be approached by means of a simple model. Consider a bottomless, uniform ocean, without refraction or attenuation, having a surface covered with a dense, uniform distribution of noise sources. Furthermore, let each unit area of the surface radiate with an intensity $I(\theta)$ at one meter. Then, at point P in Fig. 7.1(a) the incremental intensity dI produced by a small circular annulus of area dA at horizontal range r is (Urick, 1983, Chapter 7)

$$\mathrm{d}I = \frac{I(\theta)\mathrm{d}A}{l^2} = \frac{I(\theta)2\pi r\,\mathrm{d}r}{l^2} \tag{7.1}$$

But $r = h\tan\theta$, so that $\mathrm{d}r = h\sec^2\theta\,\mathrm{d}\theta$. Also, $l = h\sec\theta$. On substituting and rearranging, equation (7.1) reduces to

$$\mathrm{d}I = 2\pi I(\theta)\tan\theta\,\mathrm{d}\theta \tag{7.2}$$

If ψ is a solid angle, then $\mathrm{d}\psi = 2\pi\sin\theta\,\mathrm{d}\theta$, so that the noise intensity per unit solid angle, $N(\theta)$, becomes

$$N(\theta) = \frac{\mathrm{d}I}{\mathrm{d}\psi} = I(\theta)\sec\theta \tag{7.3}$$

When $I(\theta) = I_0$, equation (7.3) yields

$$N(\theta) = I_0\sec\theta \tag{7.4a}$$

which is the ambient noise intensity per unit solid angle for a surface distribution of monopole sources. Using a distribution of dipole sources for which the intensity radiated at an angle θ is $I(\theta) = I_0\cos^2\theta$, the beam

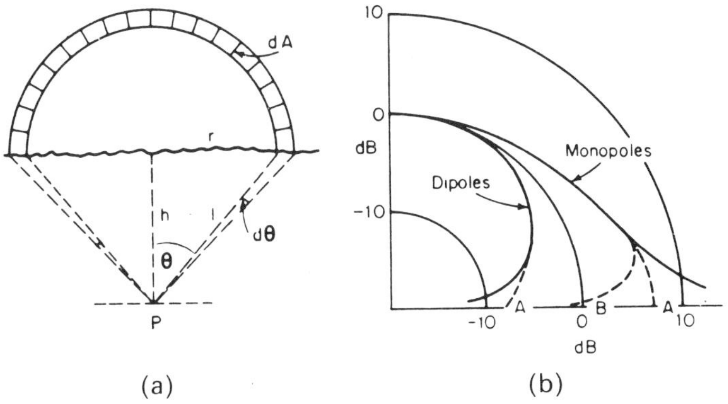

Fig. 7.1 Simple model for the vertical directionality of ambient noise: (a) geometry with straight-line propagation paths; (b) directional patterns for surface distribution of monopoles and dipoles. The dashed segments near the horizontal show the effect of attenuation and refraction near the sea floor (A) and near the deep sound channel axis (B) (Urick, 1983; *Principles of Underwater Sound*, 3rd edn; reproduced with permission of McGraw-Hill Publishing Company).

pattern of the noise received at a depth below the surface is

$$N(\theta) = I_0 \cos\theta \tag{7.4b}$$

Equations (7.4a) and (7.4b) are plotted in Fig. 7.1(b) in polar coordinates. At angles near the horizontal (where $\theta = 90°$) the effects of attenuation, refraction and boundary multipaths prevent the curves from going to either zero or infinity. Using this simple model, a receiver located within the deep sound channel would not be expected to receive noise arriving along paths near the horizontal since range-independent ray tracing would show that such ray paths do not exist. The beam pattern of the noise from monopole sources would therefore have a maximum at an oblique angle above and below the horizontal (typically $\pm 10°$ to $\pm 15°$ off the horizontal axis in temperate, deep-sea regions), with a minimum at $\theta = 90°$. This simple conceptual picture is not always valid in realistic environments; specifically, more sophisticated models would consider the effects of range-dependent refraction, bottom reflection and multipath arrivals.

For high-frequency noise sources (>1000 Hz), many investigators (e.g. Anderson, 1958; Becken, 1961; Von Winkle, 1963) have suggested a function of the form $I(\theta) = I_0 \cos^m \theta$, where I_0 is the intensity radiated by a small area of sea surface in the downward direction ($\theta = 0°$) and m is an integer. Values of $m = 1$, 2 or 3 have been obtained, depending upon conditions and methods of measurement. Most measurements center roughly about $m = 2$, a value consistent with the hypothesis of a dipole source formed by the actual source and its image in the sea surface.

For low-frequency noise sources (< 500 Hz), the agreement between available observations (see Fig. 6.7) and the findings of this simple model (Fig. 7.1) further suggests that a distribution of monopoles ($m = 0$; thus, $I(\theta) = I_0$) adequately models the distant shipping noise.

7.4 THE RANDI MODEL – A SPECIFIC EXAMPLE

The Research Ambient Noise Directionality (RANDI) model is one example of an ambient noise model. RANDI calculates the vertical and horizontal directionality of low-frequency (10–500 Hz) ambient noise for a selected ocean environment (Wagstaff, 1973).

7.4.1 Transmission loss

RANDI utilizes one of three sources of transmission loss inputs:

1. self-contained linear ray-trace routine
2. input data consisting of transmission loss versus range and arrival angle arrays
3. deep sound channel propagation algorithm.

7.4.2 Noise sources and spectra

RANDI considers six sources of isotropic and anisotropic surface and volumetric noise: (1) shipping, (2) wind–wave, (3) biological, (4) sea state zero, (5) rain and (6) distant sources. A seventh source, that of a target, is also included as an option (Fig. 7.2). Figure 7.3 illustrates the modeled noise field. The surface noise is generated by an infinite number of point sources distributed along a horizontal noise source plane located at a depth of about 6 m, corresponding to the nominal drafts of surface vessels. Empirical algorithms are used to calculate squared pressure spectrum levels for each of the seven noise generation mechanisms. All of the noise (and target) squared pressure spectrum levels are then integrated over a user-specified bandwidth using an input frequency–response function. Noise level in this context refers to the mean squared spectrum level.

7.4.3 Directionality

To determine the horizontal directionality of the noise field, the ocean is divided into n wedge-shaped regions (called sectors) with the receiver at the center. The vertical thickness of the sectors is equal to the ocean depth at the receiver location. Noise calculations are performed independently in each sector defined by the user. The total squared noise pressure, as measured by an omnidirectional hydrophone, is obtained by summing the

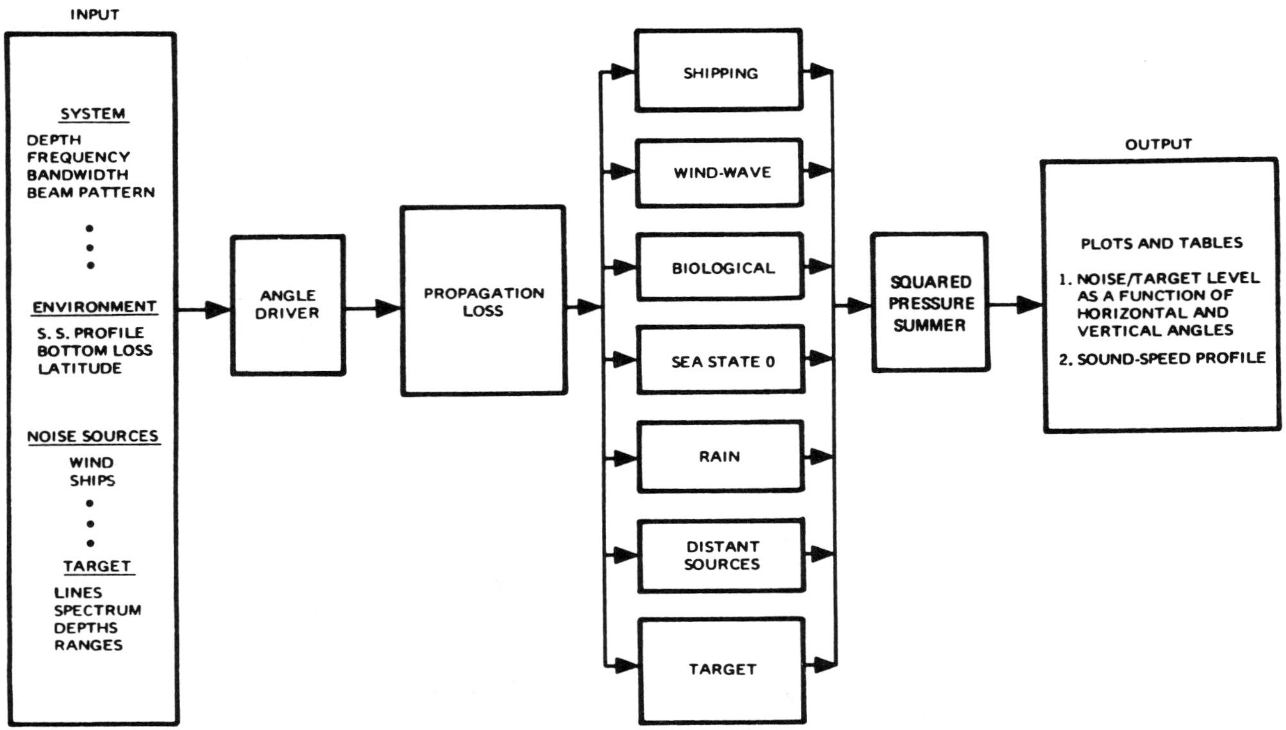

Fig. 7.2 Block diagram of the RANDI noise model (Wagstaff, 1973).

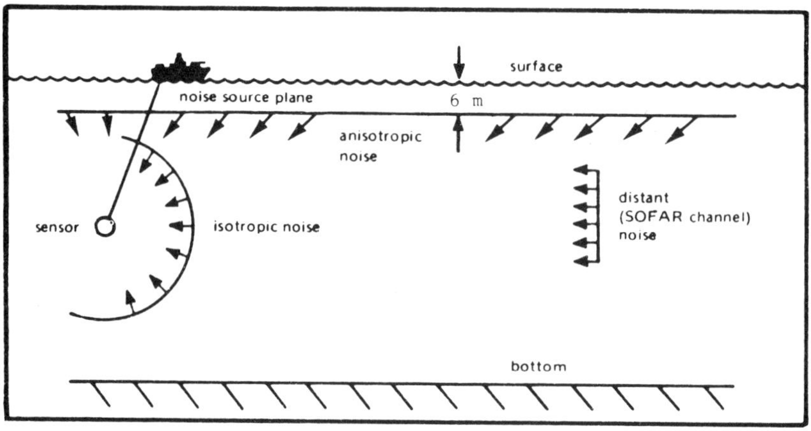

Fig. 7.3 The noise field used in the RANDI model (Wagstaff, 1973).

squared noise pressures for the *n* sectors. An example of horizontal directionality of the noise field is presented in Fig. 7.4 (note that the units for noise level are referenced to degrees).

The vertical directionality of the noise field is determined for a differential vertical angle by first calculating the area defined by the intersections of the corresponding ray bundles with the anisotropic noise field plane. Next, these areal extents are multiplied by the local effective squared noise pressure levels. The resulting levels are reduced by the appropriate transmission loss and then summed. At this point, the contribution of the isotropic noise sources is considered. These new levels are then convolved with the vertical response of the receiver array. This whole process is then repeated for each new differential angle. An example of vertical directionality of the noise field is presented in Fig. 7.5 (note that the units for noise level are referenced to steradians).

7.4.4 Recent developments

The latest version of this model (RANDI III) has been developed for application to shallow-water and coastal areas in the low- to mid-frequency range (~ 10–300 Hz). The receiver can be either a horizontal or a vertical line array (Breeding, 1993). The dominant sources of noise are assumed to be shipping, distant storms and local winds. The ships are treated as discrete sources, and the shipping noise is propagated over great-circle routes to the receiver elements using a wide-angle, Finite Element Parabolic Equation (FEPE) model. The arriving contributions from all ships are added coherently at the receiver, and the array response is determined by beamforming with various shading schemes. Environ-

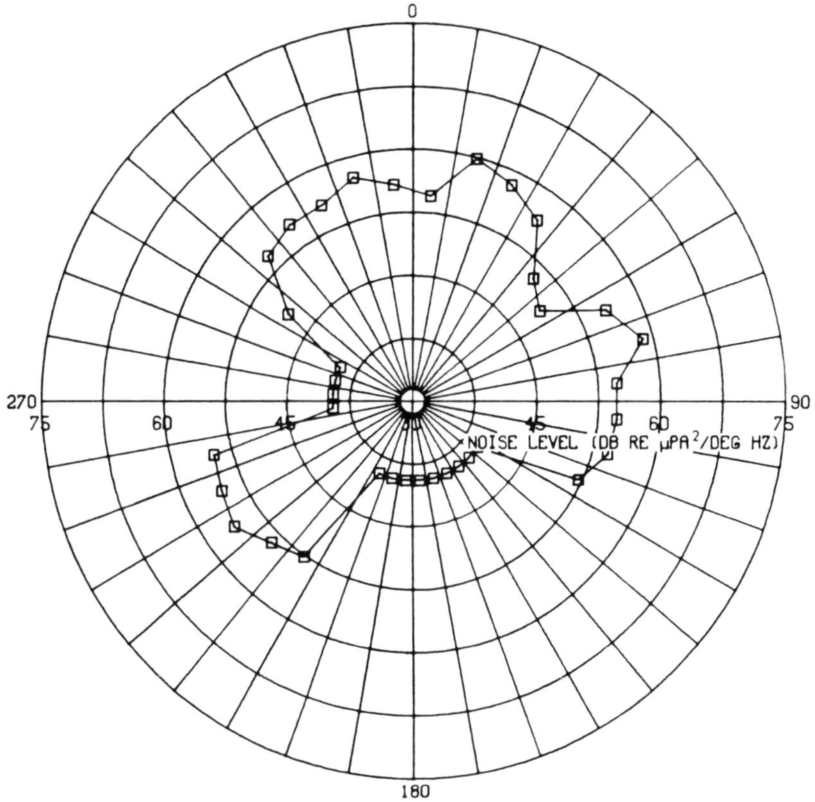

Fig. 7.4 Per-degree horizontal directionality of ambient noise in 10° sectors generated by the RANDI noise model for a frequency of 100 Hz at a depth of 91 m in the North Pacific Ocean (Wagstaff, 1973).

mental and shipping information is automatically extracted from US Navy standard databases: HITS for determination of shipping densities; DBDB for bathymetry; GDEM for sound speed profiles; and LFBL for bottom interaction parameters (refer to Table 10.5).

7.5 THE NOISE NOTCH

The use or ray-tracing techniques in noise models can sometimes introduce artifacts into the computed noise fields. The so-called noise notch is one feature that is frequently evidenced in the vertical directionality of low-frequency noise fields.

Figure 7.6 presents a vertical noise directionality diagram produced by the Fast Ambient Noise Model (FANM). This particular example is for

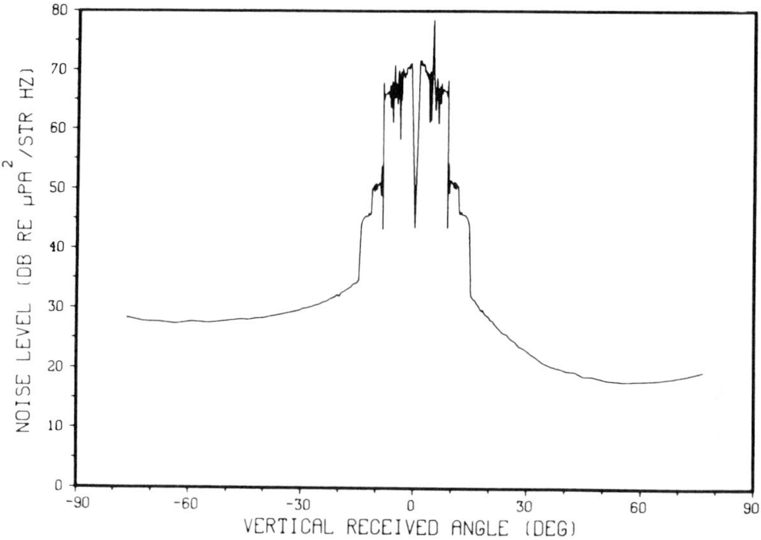

Fig. 7.5 Ambient noise vertical directionality for the 10° horizontal sector centered at 15° in Fig. 7.4. Rays with positive angles arrive at the receiver from below the horizontal plane; a ray with an angle of − 90° arrives from the surface directly over the receiver (Wagstaff, 1973).

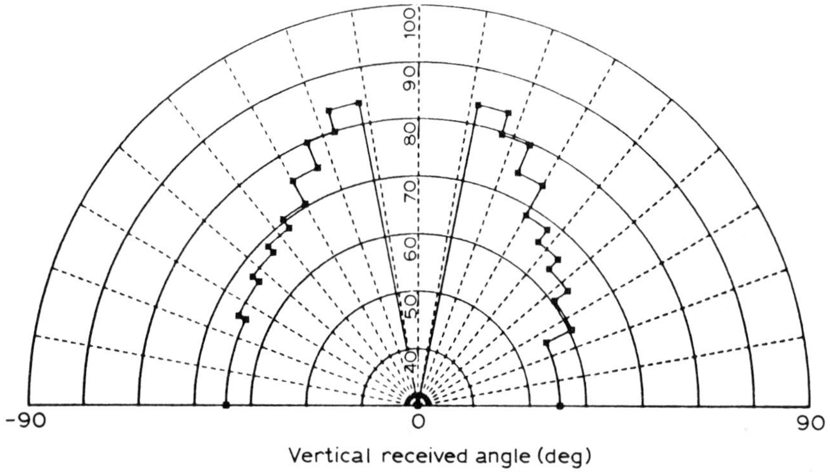

Fig. 7.6 Vertical noise directionality (dB re $\mu Pa^2/Hz$/steradian) computed by the FANM noise model at a frequency of 50 Hz. Rays with positive angles arrive at the receiver from below the horizontal plane.

a low-frequency (50 Hz) case under the assumption of a range-independent ocean environment. Consistent with the conceptual picture discussed in Chapter 6, most of the low-frequency noise arrives at near-horizontal angles. Furthermore, slightly higher levels are observed in the direction of the sea surface than in the direction of the sea floor. This result is also expected in that the bottom-reflected rays are more highly attenuated than are the corresponding surface-interacting rays; Figs 7.5 and 7.6 both show this effect. The feature of real interest, however, is the horizontal notch (or null) which is predicted by range-independent ray theory. The notch feature is not always observed in actual measurements at sea. This situation then presents a paradox. The generation of such features is not limited to ray-tracing techniques. Carey *et al.* (1987), for example, were able to produce similar features using a parabolic equation (PE) model under similar range-independent environmental assumptions.

Anderson (1979) made measurements of the vertical directionality of noise in the North Pacific Ocean in September 1973. He did not observe the horizontal null in the directional spectrum of the noise that is predicted by ray theory for a horizontally homogeneous (range-independent) ocean.

This aspect can be further explored by first examining basic ray-theoretical considerations in typical deep-ocean, range-independent environments. Specifically, this notch (or null) is the result of ray arrivals being excluded from a region, centered on the horizontal axis of the receiver, limited in width to $\pm\theta_R$, where θ_R defines the limiting ray as

$$\theta_R = \cos^{-1}\frac{c_R}{c_S} \qquad (7.5)$$

where θ_R = limiting ray angle at the receiver
c_R = sound speed at the receiver position
c_S = sound speed at the source position

This corresponds to those rays leaving the source horizontally ($\theta_s = 0$); equation (7.5) then follows directly from Snell's law. Figure 7.7 illustrates the geometry for defining the limiting ray. The sound speed profile is based on observational data presented by Anderson (1979). Assuming that the noise source plane is near the surface, and that the receiver is located at 1200 m (just below the deep sound channel axis), then equation (7.5) yields

$$\theta_R = \cos^{-1}\left[\frac{1485\,\text{m/s}}{1530\,\text{m/s}}\right] = \cos^{-1}(0.971) \approx 14^\circ$$

Thus, ray-theoretical considerations indicate that noise from the surface cannot arrive within about 14° of the horizontal axis of the receiver, at least under the environmental and geometrical conditions assumed here. These conditions, however, are typical of many deep-ocean environments which support deep sound channel propagation.

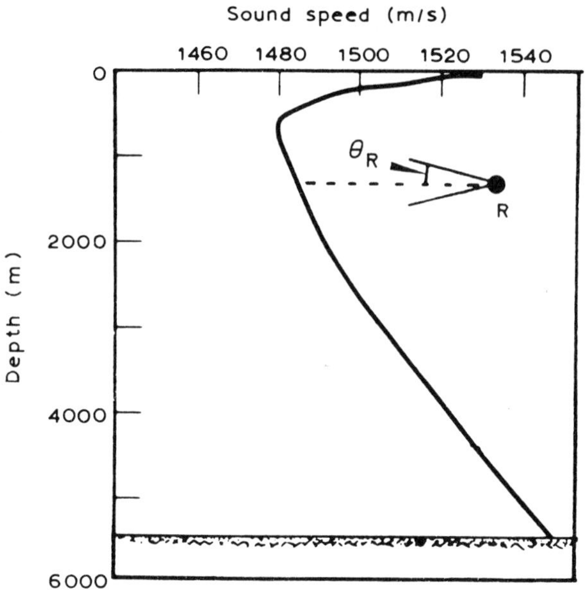

Fig. 7.7 Geometry for calculation of limiting rays. Noise source plane is located near the sea surface; receiver (R) is located at a depth of 1200 m. In this example, limiting ray angles (θ_R) are $\pm 14°$ off the horizontal axis of the receiver. (Sound speed profile is from Anderson, 1979).

The apparent paradox created by the discrepancy between ray theory and observations can be explained in terms of range-dependent propagation effects. In particular, two effects can contribute to filling the noise notch through conversion of near-surface shipping and weather noise into shallow-angle propagation paths near the sound channel axis: slope conversion and horizontal sound speed variations (e.g. Anderson, 1979). These mechanisms were addressed in Chapter 6. Slope conversion considers the reflection and scattering of sound into the deep sound channel from ships transiting the continental shelves and slopes of the ocean basins. The variation of sound speed as a function of latitude results in the axis of the sound channel shoaling at high latitudes, thus allowing far northern (Kibblewhite *et al.*, 1976) or far southern (Bannister, 1986) shipping and weather noise to enter the deep sound channel and to propagate for long distances.

Carey and Wagstaff (1986) reviewed low-frequency (< 500 Hz) physical noise models and measurements. They confirmed that coherent signals from surface ships were a dominant characteristic of the horizontal ambient noise field. The vertical directionality of the noise showed a broad angular distribution centered about the horizontal axis at lower frequencies (<200 Hz), and a dual-peaked ($\pm 10^\circ$ to $\pm 15^\circ$ off the horizontal) distribution at frequencies between 300 and 500 Hz. In the frequency range 20–200 Hz, the data exhibited a smooth variation with frequency. However, the tonal nature of surface ship radiated noise spectra would suggest a spiky spectral variation. Smooth spectral variations in the vertical directionality of the noise field are also characteristic of low-shipping areas. Carey and Wagstaff thus suggested that environmental noise sources such as wind-driven noise, in addition to shipping, may be required to explain the broad angular and frequency characteristics. Carey *et al.* (1990) conducted additional simulations using a PE model. These simulations confirmed the role of downslope conversion as a low-pass filter in determining the vertical noise field at mid-basin.

The primary mechanism for introducing energy into the sound channel is the downslope conversion process; however, diffuson may also be a candidate. Dashen and Munk (1984) have addressed other aspects of this problem.

Hodgkiss and Fisher (1990) made a series of direct measurements that confirmed the contribution of the downslope conversion mechanism to the near-horizontal noise distribution. They noted that the effects of absorption appeared to diminish the near-horizontal energy with increasing distance from the coast, and that these effects were more pronounced at higher frequencies.

7.6 BEAM NOISE STATISTICS MODELS

Models of beam noise statistics use a statistical approach to model the low-frequency ambient noise field in the ocean. To be of practical use to large-aperture, narrow-beam passive sonar systems, these models must include sonar-specific beam pattern characteristics.

The probability measures of beam noise depend on array configuration, orientation, location and season. For detection predictions, the measure of interest is the total power in selected frequency bands. For prediction of false alarm rates, the desired measure is a characterization of the narrowband components of shipping noise (Moll *et al.*, 1979). Only detection predictions will be addressed here.

Using the formalism developed by Moll *et al.* (1979), an expression for the total noise power in a specified band at the beamformer output can be presented. As a consequence of the principle of superposition of the instantaneous pressures of sound from multiple point sources, the averaged noise power at the beamformer output (Y) can be expressed as

$$Y = \sum_{i=1}^{m} \sum_{j=1}^{n} \sum_{k=1}^{A_{ij}} S_{ijk} Z_{ijk} B_{ijk} \qquad (7.6)$$

where m = number of routes in the basin
n = number of ship types
A_{ij} = number of ships of type j on route i; a random variable
S_{ijk} = source intensity of the kth ship of type j on route i; a random variable that is statistically independent of the source intensity of any other ship
Z_{ijk} = intensity transmission ratio from ship ijk to the receiving point
B_{ijk} = gain for a plane wave arriving at the array from ship ijk

The probability density function for Y can then be obtained from its characteristic function.

7.7 DATA SUPPORT REQUIREMENTS

The utilization of noise models requires a specialized database containing information on shipping routes, shipping density by merchant vessel type, and radiated noise levels by vessel type. Vessel types are usually differentiated by freighters, tankers and fishing craft. Further distinctions can be made according to gross tonnage. A ship-count database that may be automatically accessed by ambient noise models is the HITS (Historical Temporal Shipping) database which is incorporated in AUTO SHIPS (see Chapter 10). The resolution of HITS is 1° squares by ship type; spatial coverage is essentially world wide (Estalote *et al.*, 1986).

The complexity of the databasing task can be better appreciated by examining Fig. 7.8, which shows generalized shipping routes (from Solomon *et al.*, 1977). The routes are generalized in the sense that the lines connecting the various ports are not necessarily indicative of the exact paths followed by the merchant traffic. Factors such as seasonal climates and severe weather patterns can alter the exact routes followed by the ships at any particular time.

Mapping transformations between rectangular (for the ship positions and routes) and polar (for noise model sector geometry applications) further complicate model implementation. Figure 7.9 illustrates one particular implementation which relates polar sector geometries to Cartesian latitude and longitude divisions. Here, a sector is defined in terms of sector angle (SECANG), sector range (SECRNG), angle increment (ANGINC) and range increment (RNGINC).

An example of the spatial variability of ambient noise levels generated by a noise model is presented in Fig. 7.10 for the Mediterranean Sea. At low frequencies (<100 Hz), most noise is due to shipping traffic. The regions of high noise levels (denoted by H) coincide with established shipping routes (compare with Fig. 7.8). Noise variations of as much as 20 dB can occur over relatively short distances.

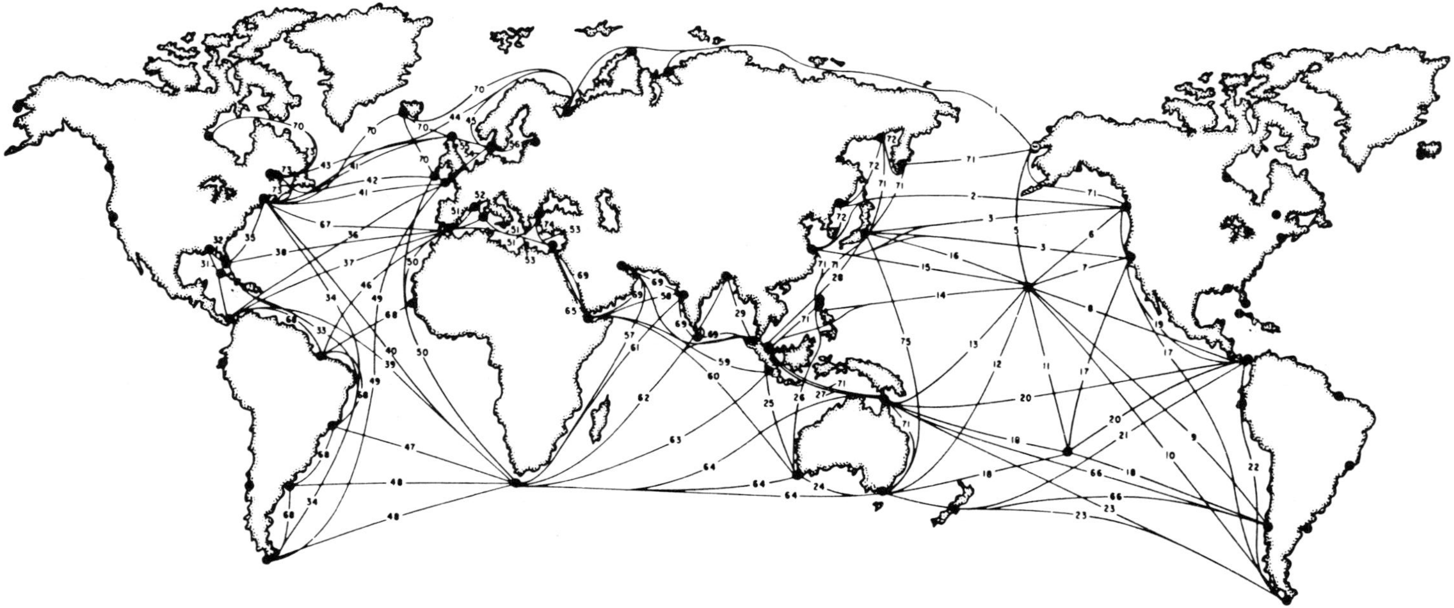

Fig. 7.8 Generalized shipping routes (Solomon *et al.*, 1977).

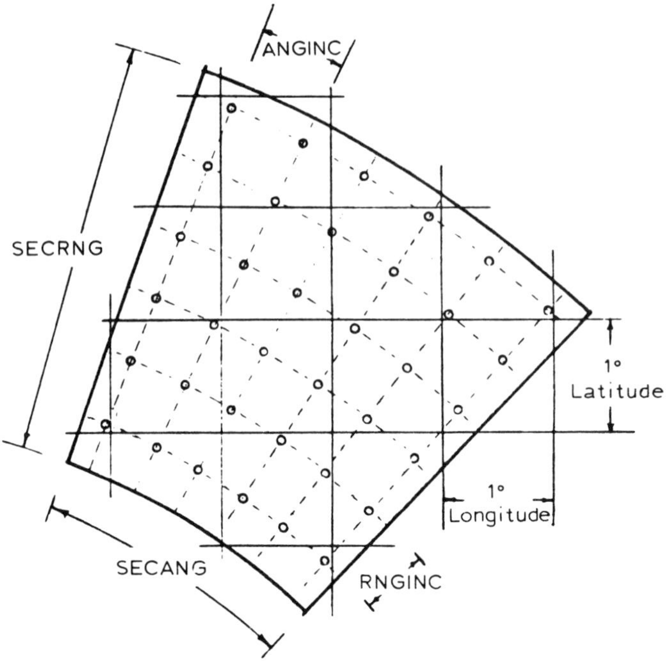

Fig. 7.9 Mapping transformation from a rectangular to a polar coordinate system. Shipping density data are usually represented in a Cartesian coordinate system while noise models utilize a polar reference frame. This particular example is from the FANM moise model (Baker, 1976).

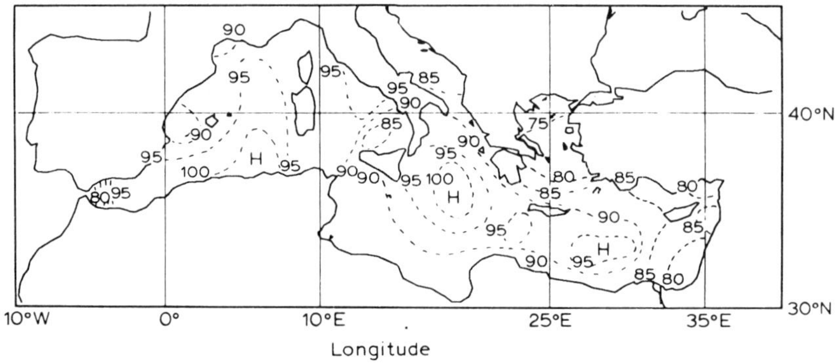

Fig. 7.10 Ambient noise level (spectrum level, dB re 1 μPa) variations in the Mediterranean Sea as simulated by a noise model. Simulation is valid for low frequency (< 100 Hz) noise for a shallow receiver depth during the winter season. The contour interval is 5 dB; H signifies areas of high noise levels.

7.8 NUMERICAL MODEL SUMMARIES

A summary of available noise models is presented in Table 7.1. This summary segregates the models according to the categories of ambient noise and beam noise statistics. Models in the latter category are further segregated according to analytic and simulation approaches. Numbers within parentheses following the model names refer to appropriate documentation. When direct references are not available, secondary sources of information are cited. Model documentation can range from informal programming commentaries to journal articles to detailed technical reports containing a listing of the actual computer code. Abbreviations and acronyms are defined in Appendix A. This summary does not claim to be exhaustive.

Table 7.1. Summary of underwater acoustic noise models

Ambient noise	*Beam noise statistics*
ANDES [1]	*Analytic*
AMBENT [2]	BBN Shipping Noise [9]
CNOISE [3]	BTL [10]
DANES [4]	USI Array Noise [11]
DUNES [5]	Sonobuoy Noise [12]
FANM [6]	
Normal Mode Ambient Noise [7]	*Simulation*
RANDI I/II/III [8]	BEAMPL [13]
	DSBN [14]
	NABTAM [15]

References to noise models: [1] Renner (1986a, 1986b, 1988); [2] McConnell (1983), Robinson and McConnell (1983); [3] Estalote (1984) – secondary reference; [4] Osborne (1979), Lukas *et al.* (1980b); [5] Bannister *et al.* (1989); [6] Cavanagh (1974a, 1974b), Lasky and Colilla (1974), Baker (1976), Long (1979); [7] Kuperman and Ingenito (1980); [8] Wagstaff (1973), Hamson and Wagstaff (1983), Schreiner (1990), Breeding (1993); [9] Mahler *et al.* (1975), Moll *et al.* (1977, 1979); [10] Goldman (1974); [11] Jennette *et al.* (1978); [12] McCabe (1976); [13] Etter *et al.* (1984) – secondary reference; [14] Cavanagh (1978a, 1978b); [15] Etter *et al.* (1984) – secondary reference.

8

Reverberation I. Observations and physical models

8.1 BACKGROUND

Reverberation is defined as that portion of the sound received at a hydrophone which is scattered by the ocean boundaries or by volumetric inhomogeneities. Accordingly, reverberation-producing scatterers in the sea can be grouped into three classes: sea surface, sea floor, and ocean volume. Surface and bottom reverberation both involve a two-dimensional distribution of scatterers and therefore can be considered jointly as boundary reverberation. Volume reverberation is produced by the marine life and inanimate matter distributed in the sea, and also by fine-scale features of the ocean itself. Useful reviews of oceanic scattering and reverberation have been provided by Andersen and Zahuranec (1977), Farquhar (1970), Ellis *et al.* (1993) and Pierce and Thurston (1993).

Reverberation has several features which distinguish it from noise (Bartberger, 1965; Moritz, 1982). Principal among these is the fact that reverberation is produced by the sonar itself; thus, the spectral characteristics of reverberation are essentially the same as the transmitted signal. The intensity of reverberation varies with the range of the scatterers and also with the intensity of the transmitted signal.

Based on its characteristic temporal and spatial correlation properties, oceanic reverberation can be segregated according to diffuse and facet components (Gerstoft and Schmidt, 1991). Diffuse reverberation results from scattering by the small-scale, stochastic structure of the oceanic waveguide (e. g. surface and bottom roughness and bottom inhomogeneity). Facet reverberation results from abrupt changes in the bathymetric and subbottom features of the ocean (e.g. seamounts and faults). Due to its stochastic nature, diffuse reverberation is characterized by a relatively low correlation; alternatively, the deterministic nature of oceanic facets gives rise to 'signal-like' reverberation.

The fundamental ratio upon which reverberation depends is called the scattering (or backscattering) strength (Urick, 1983, Chapter 8). It is the ratio (in dB) of the intensity of sound scattered by a unit area (for boundary reverberation) or volume (for volume reverberation), referred to a distance of 1 m, to the incident plane-wave intensity

$$S_{b,v} = 10 \log \frac{I_s}{I_i} \tag{8.1}$$

where $S_{b,v}$ = scattering strength for boundary (b) or volume (v) reverberation
I_s = intensity of sound scattered by a unit area or unit volume of water
I_i = intensity of incident plane wave

The computation of volume and boundary reverberation levels based on scattering strengths is discussed in Chapter 9.

8.2 VOLUME REVERBERATION

The major source of volume reverberation in the sea has been established as biological (e.g. Johnson *et al.*, 1956). Different marine organisms affect different bands of the active sonar spectrum. At frequencies in excess of 20 kHz, the scatterers are zooplankton. At frequencies between 2 and 10 kHz, the dominant scatterers are the various types of fish that possess a swim bladder (an air-filled sac that enables fish to maintain and adjust their buoyancy). Acoustically, the bladder amounts to an internal air bubble that becomes resonant at a frequency that depends upon the size and the depth of the fish. Love (1978) has reviewed the current understanding and modeling of swim bladder resonant acoustic scattering .

A volume scattering strength model has been developed by Saenger (1984) which is valid over the frequency range 1–15 kHz. The depth variability is a function of seasonal bioacoustic constants (organized by ocean province), the mean water density profile and its gradient, and the acoustic frequency. The absolute level of the volume scattering strength depends upon the seasonal standing crop of scatterers, which may exhibit interannual variability. Love (1975) also developed a volume reverberation model based on fish distribution data.

The use of fishery data to predict volume reverberation was explored further by Love (1993). In 1988 and 1989, volume reverberation measurements were made in the Norwegian Sea and North Atlantic Ocean in the frequency range 800–5000 Hz. Below 5 kHz, volume reverberation is usually caused by scattering from the swimbladders of relatively large fish. The scattering strength (S_L) of a layer of dispersed, nonacoustically interacting fish is

$$S_L = 10\log_{10}\sum_{i=1}^{n} \sigma_i(f) \times 10^{-4}$$

where n = number of fish in the layer
σ = acoustic cross-section of an individual fish (cm^2) at any given frequency (f)

The parameter σ represents a swimbladder-bearing fish as a spherical shell enclosing an air cavity in water. Love (1993) concluded that a possible pitfall of relying solely on fishery data to predict low-frequency volume reverberation is that fishery research concentrates on species of present or potential commercial value. Species of no commercial value could be ignored in such data, despite their potential contribution to low-frequency volume reverberation.

8.2.1 Deep scattering layer

Measurements of the depth variation of volume scattering strengths show an overall decrease with increasing depth. This is consistent with the general distribution of biological organisms within the sea. There is often a well-marked increase at certain depths. The depth of increased volume scattering is called the deep scattering layer (DSL).

The DSL generally exhibits a diurnal migration in depth, being at a greater depth by day than by night, and with a rapid depth change near sunrise and sunset. The depth of the DSL can be expected to lie between 180 and 900 m by day in mid-latitudes, and to be shallower by night. In the Arctic, the DSL lies just below the ice cover.

Greene and Wiebe (1988) have reported on the use of high-frequency (420 kHz), dual-beam acoustics to measure target strength distributions of zooplankton and micronekton. Their results are consistent with the present picture of the diel vertical migration of sound-scattering layers in the ocean. At this frequency, the most abundant targets (off the northeast coast of the United States) corresponded to animals the size of mature krill and had target strengths between −62 and −71 dB. The use of acoustic techniques to estimate biomass has also been explored (e.g. Penrose *et al.*, 1993).

8.2.2 Column or integrated scattering strength

Some kinds of scatterers in the sea, such as the deep scattering layer (DSL) or a layer of air bubbles just below the sea surface, lie in layers of finite thickness instead of being diffusely distributed throughout an irregular volume. Such layered reverberation is more easily considered as a form of

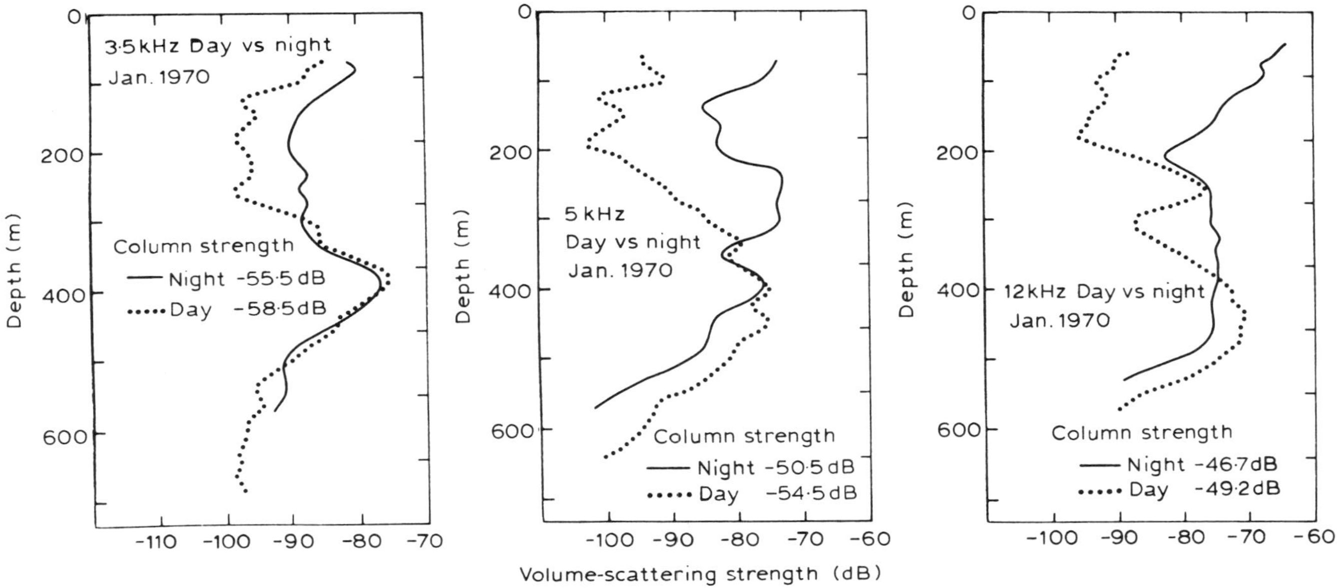

Fig. 8.1 Comparison of day and night profiles of volume scattering strength versus depth during January 1970 in the Eastern Pacific Ocean: (a) 3.5 kHz; (b) 5 kHz; and (c) 12 kHz. Column strengths are calculated by integrating the volume scattering strengths over the entire profile (Vent, 1972).

boundary reverberation. When the layer is made infinitely great, the quantity S_b (equation (8.1)) becomes the scattering strength of the entire water column and is called the column, or integrated, scattering strength. It can be readily measured in the field by means of explosive sound signals and sonobuoys, and it is a convenient single number to characterize the total amount of volume reverberation existing at the time and location the data were obtained.

Measurements of volume reverberation have been obtained, for example, by Vent (1972) and by Gold and Renshaw (1978). Figure 8.1 compares day and night profiles of volume scattering strength versus depth during January 1970 in the Eastern Pacific Ocean at three frequencies: 3.5 kHz, 5 kHz and 12 kHz. The corresponding column strengths are also indicated. Note the large differences in scattering strengths between night and day.

8.2.3 Vertical scattering plumes

Pelagic organisms tend to favor horizontal orientations, thus forming what is termed scattering layers. In certain areas of the oceans, the seepage of gas or liquids from the seabed produces vertical scattering plumes which are detectable over the frequency range 10–1000 kHz (Hovland, 1988). A special case of particular interest to underwater acousticians concerns the plumes formed over some seamounts. A scattering plume over the Hancock Seamount in the Hawaiian chain was investigated by echo sounder (38 kHz). This plume was detectable over a vertical extent of approximately 300 m directly above the seamount. These seepages appear to attract various marine organisms which might further complicate the scattering dynamics in the water column overlying seamounts.

8.3 BOUNDARY REVERBERATION

8.3.1 Sea-surface reverberation

The roughness of the sea surface and the presence of trapped air bubbles makes the sea surface an effective but complex scatterer of sound. Scattering can occur out-of-plane as well as within the vertical plane containing the source and receiver.

Sea-surface scattering strengths have typically been measured using nondirectional (mostly explosive) sources and receivers as well as directional sonars in which a sound beam is formed so as to intercept the sea surface at a desired angle. The scattering strength of the sea surface has been found to vary with angle, frequency and the roughness of the surface. Sea-surface roughness is usually characterized either by the near surface wind speed or by wave heights. The measured scattering strengths show

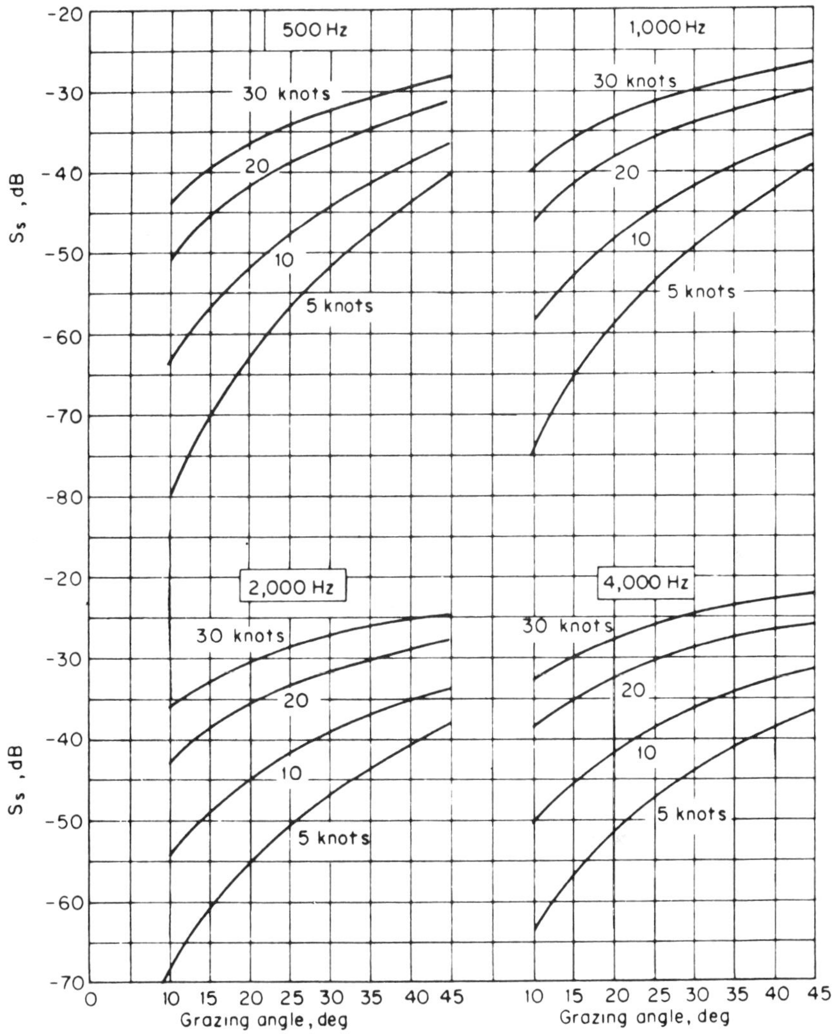

Fig. 8.2 Low-frequency scattering strengths of the sea surface (S_s) computed from the empirical expressions given by Chapman and Harris (1962) (Urick, 1983; *Principles of Underwater Sound*, 3rd edn; reproduced with permission of McGraw-Hill Publishing Company).

a strong variation with frequency at low frequencies and low grazing angles, and little variation with frequency at high frequencies and high grazing angles.

A series of measurements conducted by Chapman and Harris (1962) were analyzed in octave bands between 0.4 and 6.4 kHz, and the results were fitted by the empirical expressions:

$$S_s = 3.3\beta \log\frac{\theta}{30} - 42.4 \log\beta + 2.6 \tag{8.2}$$

$$\beta = 158[v f^{1/3}]^{-0.58}$$

where S_s = surface scattering strength (dB)
θ = grazing angle (deg)
v = wind speed (knots)
f = frequency (Hz)

Chapman and Scott (1964) later validated these results over the frequency range 0.1–6.4 kHz for grazing angles below 80°. Figure 8.2 presents curves computed from this formula for several frequencies with wind speed as a parameter.

McDaniel (1993) reviewed recent advances in the physical modeling of monostatic sea-surface reverberation in the frequency range 200 Hz–60 kHz (see also the earlier review by Fortuin (1970)). Three sources of surface reverberation were considered: rough surface scattering, scattering from resonant bubbles, and scattering from bubble clouds (or plumes).

In the absence of subsurface bubbles, rough surface scattering is adequately explained by the composite roughness model (e.g. Thorsos, 1990). This model partitions treatment of surface scattering into two regimes according to the wind–wave spectrum: large-scale waves (using the Kirchhoff approximation) and small-scale waves (using a modified Rayleigh approximation). McDaniel summarized recent advances in surface reverberation modeling according to acoustic frequency:

High-frequency reverberation (3–25 kHz). Backscattering at grazing angles above 30° is in agreement with rough surface scattering theories; at lower grazing angles, anomalously higher backscattering strengths are believed due to scatter from resonant microbubbles. Backscattering in coastal waters is higher (by an order of magnitude) than in the open ocean under similar wind conditions; this effect is attributed, in part, to the greater generation of microbubbles in coastal waters. McDaniel (1993) also observed that many recent backscattering strength measurements are lower than would be predicted using the Chapman–Harris (1962) empirical model.

Low-frequency reverberation (< 1 kHz). Anomalous scatter similar to that observed at higher frequencies is evident, apparently due to entrained air; however, the nature of the physical processes governing the scattering at these lower frequencies is not evident from an examination of the available data.

Extensive measurements of low-frequency (70–950 Hz) sea-surface backscattering strengths were made during the critical sea test (CST) experiments for grazing angles ranging from 5 to 30°, and for wind speeds ranging from 1.5 to 13.5 m/s (Ogden and Erskine, 1994). Analysis of these measurements revealed several regimes in the frequency versus wind speed (f–U) domain corresponding to at least two different scattering mechanisms. Perturbation theory (Thorsos, 1990) was found to provide adequate descriptions of the data at high frequencies for calm seas, and at lower frequencies for all wind speeds, where air–water interface scattering is the dominant mechanism. The Chapman–Harris empirical relationship adequately described surface backscattering for rougher seas at higher frequencies where scattering from bubble clouds is presumed to dominate the scattering process. In the transition region where these two effects are competing, the scattering strengths depended upon the details of the surface and wind characteristics.

Ogden and Erskine (1994) proposed a formula for computing the total scattering strength at the sea surface (S_{total}) as a combination of perturbation theory (S_{pert}) and the Chapman–Harris empirical relationship (S_{CH}):

$$S_{\text{total}} = \alpha S_{\text{CH}} + (1 - \alpha) S_{\text{pert}} \tag{8.3}$$

where

$$S_{\text{pert}} = 10 \log\left[1.61 \times 10^{-4} \tan^4\theta \exp\left(-\frac{1.01 \times 10^6}{f^2\, U^4 \cos^2\theta}\right)\right] \tag{8.4}$$

S_{CH} = Chapman–Harris empirical formula (equation (8.2))

$$\alpha = \frac{U - U_{\text{pert}}}{U_{\text{CH}} - U_{\text{pert}}} \tag{8.5}$$

This formula is valid for grazing angles (θ) less than 40°, for wind speeds (U) less than 20 m/s (measured at a height of 19.5 m above the sea surface) over the frequency range (f) 50–1000 Hz.

For practical applications, the algorithm uses a minimum wind speed of 2.5 m/s since, at lower wind speeds, swell is likely to dominate the scattering process. Also, an arbitrary cutoff of 1° has been specified for the grazing angle. In the hatched area of the f–U domain of Fig. 8.3, **only** the Chapman–Harris formula (S_{CH}) should be used (i.e. $\alpha = 1$); in the blank area of the $f - U$ domain, **only** the perturbation theory formula (S_{pert}) should be used (i.e. $\alpha = 0$). In the stippled area corresponding to the transition region, the full equation (S_{total}) should be used where α is evaluated at the input frequency (f) and input wind speed (U). Figure 8.3 can be used to determine the wind speeds at the perturbation theory (lower) boundary (U_{pert}) and at the Chapman–Harris (upper) boundary (U_{CH}). For convenience, Ogden and Erskine (1994) put these boundaries

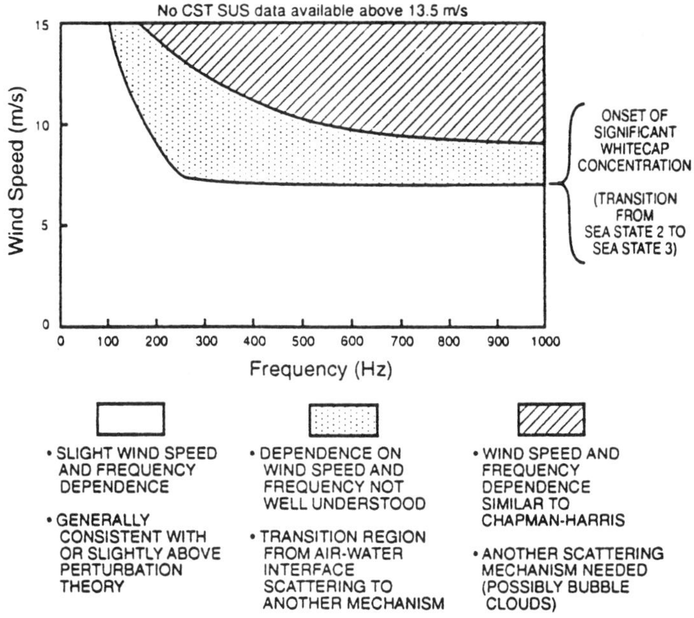

Fig. 8.3 Frequency–wind speed (f–U) domain for sea-surface scattering strengths (Ogden and Erskine, 1994).

into analytical form. The lower boundary was approximated by two line segments: from 240 to 1000 Hz the segment was defined by $y = 7.22$, while from 50 to 240 Hz the segment was given by $y = 21.5 - 0.0595x$. The upper boundary was approximated by the cubic equation:

$$y = 20.14 - 0.0340x + 3.64 \times 10^{-5}x^2 - 1.330 \times 10^{-8}x^3.$$

8.3.2 Under-ice reverberation

The detection, localization and classification of targets with active sonars under an ice canopy is limited by reverberation from the rough under-ice surface and by false targets presented by large ice features such as keels.

Sea ice is the dominant cause of reverberation in Arctic regions. For undeformed first-year ice, reverberation levels at frequencies above about 3 kHz are approximately equivalent to that expected from an ice-free sea surface with a 30 knot wind.

Because of the variation in the under-ice surface, scattering strength measurements as a function of grazing angle reveal different characteristics. When the under-ice surface is relatively smooth, the scattering strength increases with grazing angle as in the open ocean. When ridge keels are present, low-grazing-angle sound waves strike them at a near-normal incidence and substantial reflection occurs. Measurements of the scattering strength of the ice-covered sea for two Arctic locations at different times of the year have been summarized by Brown (1964) and by Milne (1964). Both sets of data show an increase of scattering strength with increasing frequency and grazing angle.

Reverberation under the ice can be minimized by using highly directional sources and receivers to exploit optimum ray paths. Such paths could provide time discrimination between target echoes and reverberation for those targets spatially separated from the sea surface (Hodgkiss and Alexandrou, 1985).

8.3.3 Sea-floor reverberation

The sea floor, like the sea surface, is an effective reflector and scatterer of sound. Scattering can occur out-of-plane as well as within the vertical plane containing the source and the receiver.

A correlation of scattering strength with the size of the particles in a sedimentary bottom has consistently been observed (e.g. McKinney and Anderson, 1964). The sea floor can then be classified according to sediment composition (sand, clay, silt) and correlated with the scattering strength. For example, mud bottoms tend to be smooth and have a low impedance contrast to water, while coarse sand bottoms tend to be rough, with a high impedance contrast. There can be large spreads in the measured data for apparently the same bottom type. This may be due in part to the refraction and reflection of sound within the subbottom sediment layers.

A Lambert's law relationship (Urick, 1983, Chapter 8) between scattering strength and angle appears to provide a good approximation to the observed data for many deep-water bottoms at grazing angles below about 45°.

Lambert's law refers to a type of angular variation which many rough surfaces appear to satisfy for both the scattering of sound and light. According to Lambert's law, the scattering strength varies as the square of the sine of the grazing angle.

Mackenzie (1961) analyzed limited reverberation measurements at two frequencies (530 and 1030 Hz) in deep water. The scattered sound responsible for reverberation was assumed to consist of nonspecular reflections obeying Lambert's law:

$$S_B = 10 \log \mu + 10 \log \sin^2 \theta \tag{8.6}$$

where S_B = bottom scattering strength (dB)
μ = bottom scattering constant
θ = grazing angle (deg)

The term $10 \log \mu$ was found to be constant at -27 dB for both frequencies.

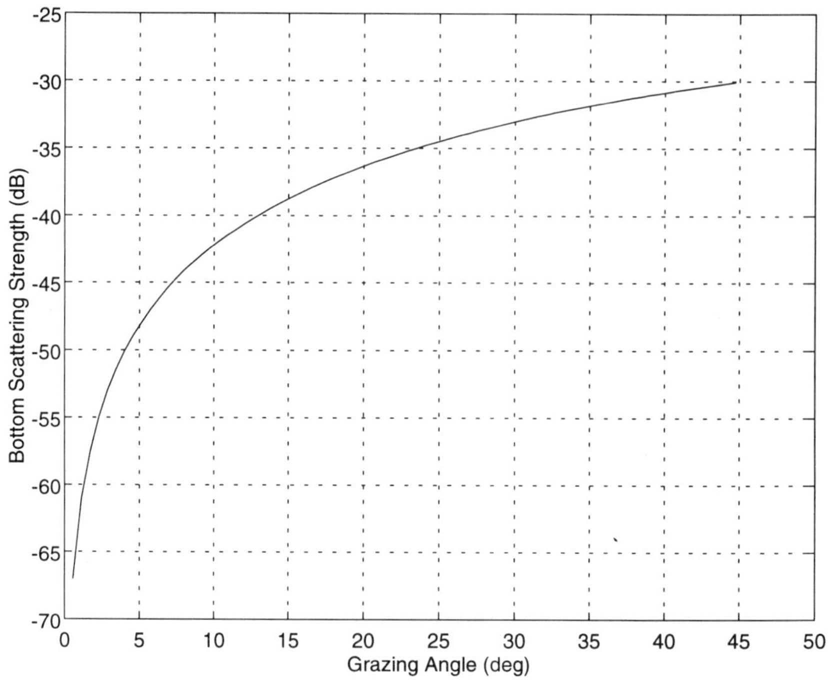

Fig. 8.4 Bottom scattering strengths (S_B) calculated according to equation (8.6) using a value of -27 dB for the term $10 \log \mu$.

This value has been largely substantiated by other measurements over a broad range of frequencies. The bottom scattering strength (S_B) has been graphed in Fig. 8.4 according to equation (8.6), using a value of -27 dB for $10 \log \mu$.

As part of the US Navy's Acoustic Reverberation Special Research Project (ARSRP), sponsored by the Office of Naval Research (ONR), an area on the northern Mid-Atlantic Ridge has been designated as a natural laboratory. This area is a natural reverberation site because its steep rock inclines, deep subsurface structures and sediment structure provide opportunities to study many aspects of bottom reverberation. The resulting data have permitted detailed comparisons with numerical codes of bistatic reverberation (e.g. Smith *et al.*, 1993).

In long-range acoustic propagation problems in the deep ocean, interactions with the sea floor generally occur at subcritical grazing angles. Under conditions of smooth basaltic surfaces and low grazing angles, most waterborne energy should be internally reflected; furthermore, the scattering functions should be monotonic with respect to grazing angle, as would be predicted by Lambert's law. However, in ocean areas characterized by a flat, basaltic sea floor, the observed scattering functions are not always monotonic with respect to angle; rather, peaks in the scattering functions sometimes occur at angles corresponding to compressional and shear head (or interface) wave propagation. In order to investigate this seemingly anomalous behavior, Swift and Stephen (1994) generated scattering functions using a model that also included volumetric heterogeneities below a flat, basaltic sea floor. They executed the model over a wide range of length scales for the imbedded scatterers when ensonified by a Gaussian pulse beam at a grazing angle of 15°. At this angle, a truly homogeneous sea floor composed of basaltic rocks would produce total internal reflection; however, they found that bottom models containing 10% velocity perturbations (representing the volumetric heterogeneities) produced significant levels of upward-scattered energy. In reality, energy leaks below the sea floor as an evanescent phase existing only when the incident energy is subcritical, and this energy only penetrates to depths of a few wavelengths. Interactions between this energy and those volumetric heterogeneities just below the surface excite interface waves (as well as compressional and shear waves in the bottom), in agreement with the observed scattering functions. Swift and Stephen (1994) also found that the scattering functions were influenced by the presence of velocity gradients (both compressional and shear) below the sea floor. Specifically, in the absence of velocity gradients, bottom-propagated energy is not refracted upward; thus, it interacts with yet deeper velocity anomalies and produces greater scattering.

The utility of the reverberation field as an inverse-sensing technique is analogous to that of the ambient noise field (section 6.6). Makris (1993), for example, inverted the reverberation field to image the sea floor. This was accomplished by using simultaneous inversion of multiple reverberation measurements made at different locations.

9

Reverberation II. Mathematical models

9.1 BACKGROUND

Mathematical models of reverberation generate predictions of boundary and volumetric reverberation using the physical models of boundary and volumetric scattering strengths developed in Chapter. 8.

The development of reverberation models has proven to be formidable for two reasons (e.g. Moritz, 1982; Goddard, 1993): there are theoretical difficulties in solving complex boundary value problems for which analytical tools are poorly developed, and there are practical difficulties in identifying and measuring all the parameters affecting the reverberation process.

9.2 THEORETICAL BASIS FOR REVERBERATION MODELING

9.2.1 Basic approaches

The scattering of sound by volumetric inhomogeneities in the ocean can be described either by Rayleigh's law or by geometrical acoustic scattering (Albers, 1965, p. 121). Rayleigh's law applies when the size (d) of the scattering particle is much smaller than the wavelength (λ) of the incident sound. Geometrical acoustic scattering is valid when d is much larger than λ. An intermediate condition applies when d is approximately equal to λ. Specifically, the following relationships hold:

(1) If $d \ll \lambda$, the pressure of the scattered sound is proportional to f^2 (where f is the acoustic frequency) and to the volume of the scatterer, regardless of its shape (Rayleigh's law).
(2) If $d \approx \lambda$, the pressure is a complicated function of frequency and also varies with the acoustic properties of the scatterer together with the characteristics of the ocean medium or boundaries.

(3) If $d \gg \lambda$, the scattering is independent of frequency and depends only on the acoustic properties of the scatterer and its cross-section (geometrical acoustic scattering).

This situation has, in part, encouraged the development of two different approaches to the modeling of reverberation in the ocean: cell-scattering models and point-scattering models.

Cell-scattering models assume that the scatterers are uniformly distributed throughout the ocean. Thus, the ocean can be divided into cells, each containing a large number of scatterers. Summing the contribution of each cell yields the total average reverberation level as a function of time after transmission. A scattering strength is used per unit area or volume, as appropriate. This approach is the most commonly used in sonar modeling.

Point-scattering models are based on a statistical approach in which the scatters are assumed to be randomly distributed throughout the ocean. The reverberation level is then computed by summing the echoes from each individual scatterer. This approach is particularly well suited, for example, to under-ice reverberation modeling.

Reverberation is a time-domain problem. Reverberation levels can be represented as a function of range from the receiver by identifying the reverberation level corresponding to the time after transmission that the leading edge of the backscattered signal arrives at the receiver.

Reverberation models are typically combined with environmental, propagation, noise and signal-processing models to form a new class of models referred to as active sonar models (see Chapter 10). The transmission loss component of many reverberation models is based on ray-theoretical considerations. Models based on wave theoretical considerations are usually not well suited to calculations at the high frequencies associated with traditional sonar applications because of the required computational time (section 4.9). However, for modern low-frequency applications, wave-theoretical techniques have been used with some success. For example, normal mode techniques have been used to model reverberation in oceanic waveguides (Bucker and Morris, 1968; Zhang and Jin, 1987; Ellis, 1993), and a two-way parabolic equation model that computes backscattered energy (Collins and Evans, 1992; Orris and Collins, 1994) has been used in reverberation simulations (e.g. Schneider, 1993).

In the analysis of sonar performance, reverberation and ambient noise are jointly considered as the background masking level against which a signal must be detected. In many situations, it is useful to understand which contributor (noise or reverberation) is most responsible for creating the interfering background. Such diagnostic information can be used to improve new sonar designs or to optimize the performance of existing sonars.

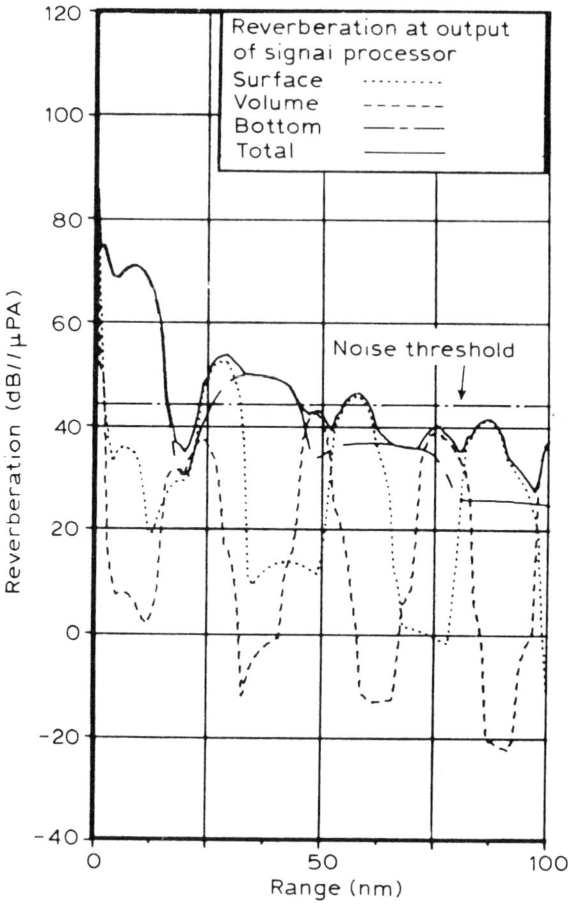

Fig. 9.1 Averaged reverberation (surface, bottom, volume and total) levels versus range as predicted by a reverberation model; the noise threshold is also plotted for comparison (adapted from Hoffman, 1979).

Figure 9.1 illustrates a graphic product generated by a long-range reverberation model based on the cell-scattering approach. The reverberation level corresponds to the time the leading edge of the signal arrives at the receiving array after bouncing off the target at the specified range. The plotted reverberation level is averaged over pulse length and corrected for processing gain. This corresponds to reverberation in the receiver bandwidth at the output of the signal processor before the thresholding device. Each of the bottom, surface and volume components of reverberation is the incoherent sum of the respective intensities derived from ray tracing. The noise threshold displayed on the reverberation plot is the combined ambient and self-noise in the receiver bandwidth at the output of the signal processor. In this example, the noise threshold is calculated by correcting spectrum level noise for array gain, bandwidth and processing gain. Thus, the range (time) at which the background masking level changes from reverberation-limited to noise-limited can be determined.

9.2.2 Advanced developments

A numerical method known as transmission-line modeling (TLM) has been used to model the scattering of acoustic waves off underwater targets. This method models the target and propagation space using meshes, rendering the problem discrete in both space and time. The mesh method models target features in considerable detail, operates readily in both two and three dimensions, and is capable of dealing with complicated scattering processes (Orme *et al.*, 1988). The method has already found use in volume scattering modeling, and in modeling time-varying propagation phenomena in multipath channels; it is now also being applied to transducer structure modeling (Coates *et al.*, 1990).

A combination of the boundary element method (BEM) and wavenumber integration (WI), referred to as the hybrid BEM–WI approach, has been used extensively in the seismic community to model elastic scattering problems. Wavenumber integration reduces the 2D Helmohltz equation to a 1D ordinary differential equation in a range-independent environment. BEM combines an integral representation of the wave field within a volume with a point representation of stresses and displacements on the boundary between fluid and elastic media (Schmidt, 1991).

Keiffer and Novarini (1990) demonstrated that the wedge assemblage (or facet ensemble) method offers a unique description of sea-surface-scattering processes in the time domain by allowing for separation of the reflected and scattered components.

Jackson *et al.* (1986) applied the composite roughness model (see section 8.3.1) to high-frequency bottom backscattering. Lyons *et al.* (1994) enhanced this model by incorporating the effects of volume scattering and scattering from subbottom interfaces.

9.3 CELL-SCATTERING MODELS

Simplifying assumptions are often necessary in order to make reverberation modeling feasible. These assumptions may seem to restrict the results to purely idealized situations; however, the resulting expressions for reverberation have been found to be practical for sonar design and prediction purposes. The assumptions necessary are the following (Urick, 1983, Chapter 8):

(1) Straight-line propagation paths, with all sources of attenuation other than spherical spreading neglected. The effect of absorption on the reverberation level can be accommodated.
(2) A homogeneous distribution of scatterers throughout the area or volume producing reverberation at any given time.
(3) A sufficiently high density of scatterers to ensure that a large number of scatterers occurs in an elemental volume (dV) or area (dA).
(4) A pulse length short enough for propagation effects over the range extension of the elemental volume or area to be neglected.
(5) An absence of multiple scattering (i.e. the reverberation produced by reverberation is negligible).

In the following sections, basic theoretical relationships will be developed for volume and boundary reverberation, as adapted from Urick (1983, Chapter 8).

9.3.1 Volume reverberation theory

Let a directional projector in an ideal medium (as implied by the above assumptions) insonify a volume of water containing a large number of uniformly distributed scatterers. The beam pattern of the projector can be denoted by $b(\theta, \phi)$, and the axial intensity at unit distance will be I_0; the source level SL $= 10 \log I_0$. The intensity at 1 m in the (θ, ϕ) direction is $I_0 b(\theta, \phi)$ by the definition of the beam pattern function. Let there be a small volume dV of volume scatterers at range r. The incident intensity at dV will be $I_0 b(\theta, \phi)/r^2$. The intensity of the sound scattered by dV at a point P distant 1 m back toward the source will be $(I_0 b(\theta, \phi)/r^2) s_v \, dV$, where s_v is the ratio of the intensity of the scattering produced by a unit volume, at a distance of 1 m from the volume, to the intensity of the incident sound wave. The quantity $10 \log s_v$ is the scattering strength for volume reverberation and is denoted by the symbol S_v. In the region near the source, the reverberation contributed by dV will have the intensity $(I_0/r^4) b(\theta, \phi) s_v \, dV$ and will produce a mean-squared voltage output of

$$R^2 (I_0/r^4) b(\theta, \phi) b'(\theta, \phi) s_v \, dV$$

at the terminals of a hydrophone having a receiving beam pattern $b'(\theta, \phi)$ and voltage response R. By assumption, the elemental volumes dV can be made so small that their total contribution can be summed up by integration, and the coefficient s_v is then a constant. Furthermore, the equivalent plane-wave reverberation level (RL) can be defined as the intensity (in dB) of an axially incident plane wave producing the same hydrophone output as the observed reverberation, in which case the following relationship is obtained:

$$\mathrm{RL} = 10 \log \left[\frac{I_0}{r^4} s_v \int_V b(\theta, \phi) b'(\theta, \phi) \, dV \right] \tag{9.1}$$

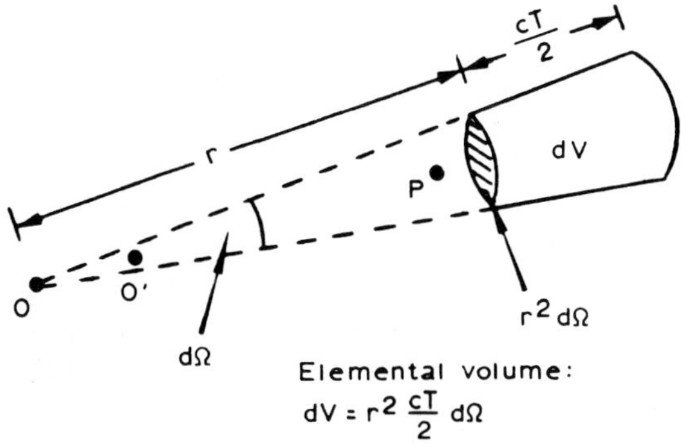

Fig. 9.2 Geometry for volume scattering.

In order to proceed further, the elemental volumes dV must be examined more closely. As shown in Fig. 9.2, take dV to be an infinitesimal cylinder of finite length with ends normal to the incident direction. The area of the end face of this cylinder dV may be written r^2 dΩ, where dΩ is the elemental solid angle subtended by dV at the source. Using a pulsed sonar, the extension in range of dV is such that the scattering produced by all portions of dV arrive back near the source at the same instant of time. Another way of viewing this is that the scattering of the front end of the pulse by the rear scatterers in dV will arrive back at the source at the same instant as the scattering of the rear of the pulse by the front scatterers in dV. The extension in range is thus $c\tau/2$, where τ is the pulse length and c is the speed of sound (Urick, 1983, pp. 241–2). The elemental volume can then be expressed as

$$\mathrm{d}V = r^2 \frac{c\tau}{2} \mathrm{d}\Omega \tag{9.2}$$

and the reverberation level (RL) is

$$\begin{aligned} \mathrm{RL} &= 10 \log \left[\frac{I_0}{r^4} r^2 \frac{c\tau}{2} s_{\mathrm{v}} \int b(\theta, \phi) b'(\theta, \phi) \, \mathrm{d}\Omega \right] \\ &= \mathrm{SL} - 40 \log r + S_{\mathrm{v}} + 10 \log \left[r^2 \frac{c\tau}{2} \int b(\theta, \phi) b'(\theta, \phi) \, \mathrm{d}\Omega \right] \end{aligned} \tag{9.3}$$

where the range extension $c\tau/2$ is small compared to the range r.

9.3.2 Boundary reverberation theory

Boundary reverberation pertains to the reverberation produced by scatterers distributed over the sea surface and the sea floor.

The derivation of an expression for the equivalent plane-wave level of boundary reverberation proceeds in a manner similar to that for volume reverberation. The result is

$$\mathrm{RL_b} = 10\log\left[\frac{I_0}{r^4}s_\mathrm{b}\int b(\theta,\,\phi)\;b'(\theta,\,\phi)\mathrm{d}A\right] \tag{9.4}$$

where $\mathrm{d}A$ is an elemental area of the scattering surface and $10\log s_\mathrm{b}$ is the scattering strength (S_b) for boundary reverberation. If $\mathrm{d}A$ is taken as a portion of a circular annulus in the plane of the scatterers with the center directly above or below the transducer, then

$$\mathrm{d}A = \frac{c\tau}{2} r\,\mathrm{d}\phi \tag{9.5}$$

where $\mathrm{d}\phi$ is the subtended plane angle of $\mathrm{d}A$ at the center of the annulus. The reverberation level then becomes

$$\begin{aligned}\mathrm{RL_b} &= 10\log\frac{I_0}{r^4}s_\mathrm{b}\frac{c\tau}{2}r\int_0^{2\pi} b(\theta,\,\phi)b'(\theta,\,\phi)\,\mathrm{d}\phi \\ &= \mathrm{SL} - 40\log r + S_\mathrm{b} + 10\log\left[\frac{c\tau}{2}r\int_0^{2\pi} b(\theta,\,\phi)b'(\theta,\,\phi)\,\mathrm{d}\phi\right]\end{aligned} \tag{9.6}$$

9.4 THE REVMOD MODEL – A SPECIFIC EXAMPLE

REVMOD (Reverberation Spectrum Model), originally developed by C.L. Ackerman and R.L. Kesser at the Applied Research Laboratory, The Pennsylvania State University, has been expanded and further documented by Hodgkiss (1980, 1984). REVMOD is a set of computer programs which models surface, bottom and volume reverberation. It is based on the cell-scattering approach.

REVMOD considers the effects of: motion of the sonar platform; transmit signal windowing; transmit and receive beam patterns; scatterer velocity distributions; and sound absorption. Model constraints include specification of a constant sound speed profile and nonreflection of sound at the surface and bottom boundaries.

Geometrically, REVMOD divides the illuminated volume of the ocean into cells and evaluates scatterer motion relative to the sonar platform for a measure of the spectral shifting and spreading due to the environment.

Acoustically, REVMOD determines the backscattering level for each cell. The contributions from each cell are then summed to compute the total reverberation power spectra received at the sonar.

REVMOD is composed of three major software modules:

(1) RVMDS – computes the scattering function resulting from the combined effects of the environment, vehicle dynamics, and transmit and receive beam patterns;
(2) RVMDT – convolves the scattering function with the transmit signal energy spectrum to yield the reverberation power spectrum;
(3) RVMDR – convolves the reverberation power spectrum with the receiver impulse response energy spectrum.

Module RVMDS computes the scattering function for reverberation and describes the effects of the ocean medium on a transmit signal with carrier frequency ω_c.

Module RVMDT completes the reverberation model by combining a detailed description of the transmit signal (pulse length, envelope shape and source level) with the environmental characterization provided by module RVMDS.

Module RVMDR creates a model of a receiver operating in a reverberant ocean environment. A matched filter envelope detector uses the reverberation spectra generated by module RVMDT.

Optional software can also be included to model a matched filter envelope detection receiver appropriate for post-processing the reverberation spectra. Thus, the effects of utilizing a receiver impulse response window, which differs from that of the transmitter waveform, can be investigated.

A mathematical description of the scattering function (RVMDS) will be provided here. Calculations of the reverberation power spectrum (RVMDT) and subsequent convolution with the receiver impulse response energy spectrum (RVMDR) are discussed by Hodgkiss (1980, 1984).

REVMOD generates a scattering function description of the ocean medium which does not require a detailed description of the transmit signal. Thus, the model is appropriate for investigations of the influence of the medium on various transmit-signal types.

Acoustic backscatter can be modeled as the passage of a transmit waveform through a linear, time-varying filter:

$$\tilde{s}_{\mathrm{r}} = \sqrt{E_{\mathrm{t}}} \int_{-\infty}^{\infty} \tilde{f}(t-\lambda)\tilde{b}\left(t-\frac{\lambda}{2},\ \lambda\right)\mathrm{d}\lambda \tag{9.7}$$

where $\tilde{s}_r(t)$ = received backscattered signal
E_t = transmit signal energy
$\tilde{f}(t-\lambda)$ = transmit waveform
t = time
λ = time delay
$\tilde{b}(t-\lambda/2, \lambda)$ = time-varying impulse function of the filter

The basic geometry of REVMOD consists of a spherical shell representing that portion of the ocean medium illuminated by the signal wavefront

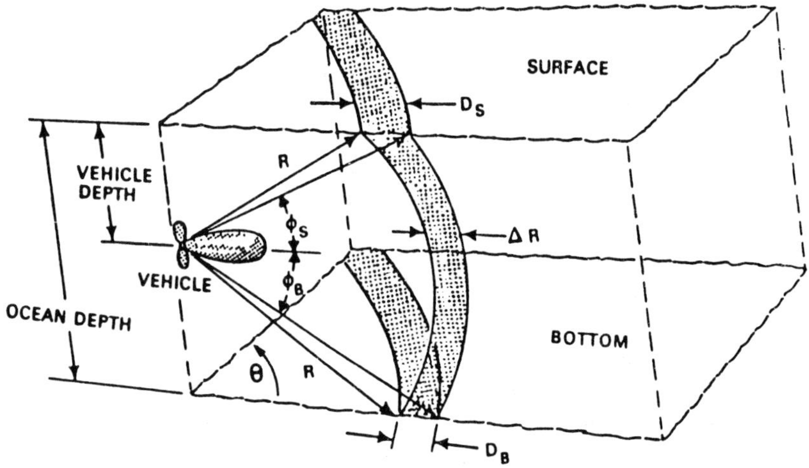

Fig. 9.3 REVMOD reverberation model geometry (Hodgkiss, 1984; *IEEE J. Oceanic Engr.*, **10**, 285–9; copyright by IEEE).

corresponding to a range (R) after transmit (refer to Fig. 9.3). The shell thickness is governed by the variables D_S, D_B and ΔR for surface, bottom and volume scattering, respectively. The spherical shell is further subdivided into a grid of cells, each of which contributes to surface, bottom or volume backscatter. The location of each cell is specified in terms of an azimuth–elevation angle pair (θ_j, ϕ_i). Using this geometry, REVMOD calculates the scattering functions corresponding to the surface, bottom and volume backscatter of a transmit waveform at specified ranges of interest. Specifically, the normalized attenuation ($\hat{A}_{i,j}$) of the transmitted signal (including propagation and beam pattern effects) is calculated for each surface, bottom and volume grid cell as follows.

Surface scattering

$$\begin{aligned} 10\log(\hat{A}_{i,j}) = {} & 10\log\{D_S(R\Delta\theta)\} + S_S - 40\log(R) - (2\alpha R) \\ & + 10\log\{\hat{P}_T(\theta_j, \phi_S)\hat{P}_R(\theta_j, \phi_S)\} + 10\log\{c/(2\Delta R)\} \end{aligned} \quad (9.8)$$

Volume scattering

$$\begin{aligned} 10\log(\hat{A}_{i,j}) = {} & 10\log\{\Delta R(R\Delta\theta)(R\Delta\phi)\} + S_V - 40\log(R) - (2\alpha R) \\ & + 10\log\{\hat{P}_T(\theta_j, \phi_i)\hat{P}_R(\theta_j, \phi_i)\} + 10\log\{c/(2\Delta R)\} \end{aligned} \quad (9.9)$$

Bottom scattering

$$\begin{aligned} 10\log(\hat{A}_{i,j}) = {} & 10\log\{D_B(R\Delta\theta)\} + S_B - 40\log(R) - (2\alpha R) \\ & + 10\log\{\hat{P}_T(\theta_j, \phi_B)\hat{P}_R(\theta_j, \phi_B)\} + 10\log\{c/(2\Delta R)\} \end{aligned} \quad (9.10)$$

where $\hat{A}_{i,j}$ = scattering level (normalized to 1 s in range) from the (i,j)th grid cell (s^{-1})

D_S = lateral dimension of surface-scattering patch (m)

D_B = lateral dimension of bottom-scattering patch (m)

R = range (m)

ΔR = increment in range over which scattering is averaged (m)

$\Delta\theta$ = angular grid cell width in azimuth (rad)

$\Delta\phi$ = angular grid cell height in elevation (rad)

S_S = surface-scattering coefficient (dB/m^2)

S_V = volume-scattering coefficient (dB/m^3)

S_B = bottom-scattering coefficient (dB/m^2)

α = sound absorption coefficient (dB/m)

$\hat{P}_T(\theta_j, \phi_i)$ = transmit beam pattern (normalized to 1 at $\theta_j = \theta_T$ and $\phi_i = \phi_T$)

$\hat{P}_R(\theta_j, \phi_i)$ = receive beam pattern (normalized to 1 at $\theta_j = \theta_R$ and $\phi_i = \phi_R$)

θ_j = azimuth to center of jth grid column (rad)

ϕ_i = elevation to center of ith grid row (rad)

(θ_T, ϕ_T) = transmit beam vector

(θ_R, ϕ_R) = receive beam vector

ϕ_S = elevation angle to surface (rad)

ϕ_B = elevation angle to bottom (rad)

c = sound speed (m/s)

Hodgkiss (1984) provided sample results from REVMOD for the case of a bottom-mounted transducer with a steerable, axially symmetric beam pattern (Fig. 9.4). Wind-driven surface waves and a surface current both contribute to spectral broadening and Doppler shifting of the transmit spectrum upon backscattering from the sea surface. The volume-backscattering component contributes only to spectral broadening of the transmit spectrum since the water column is assumed motionless, on the average. In this particular example, bottom backscattering is not a factor owing to the bottom-mounted transducer geometry.

For the purposes of discussion here, only a downwind transducer orientation is considered, although Hodgkiss also addressed upwind and crosswind orientations. This downwind case considers only the reverberation received at a range of 400 m. A quantized Hanning transmit envelope (Fig. 9.5) and a Hanning receive envelope (Fig. 9.6) are used in conjunction with the beam pattern shown previously in Fig. 9.4. The transmit beam vector is horizontal, and the receive beam vector is inclined towards the surface 25° off the horizontal. The transmit and receive envelope durations were 102 and 86 ms, respectively. See Ziomek (1985, Chapters 3 and 4), for example, for a discussion of complex apertures and directivity functions.

The Doppler of the surface component of the scattering function reflects a complex interaction with this moving boundary (negative Doppler

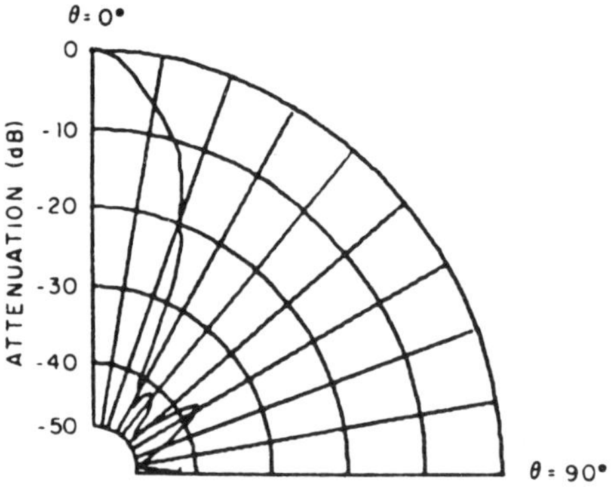

Fig. 9.4 Transmit and receive beam pattern used in an example from the REVMOD reverberation model (Hodgkiss, 1984; *IEEE J. Oceanic Engr.*, **10**, 285–9; copyright by IEEE).

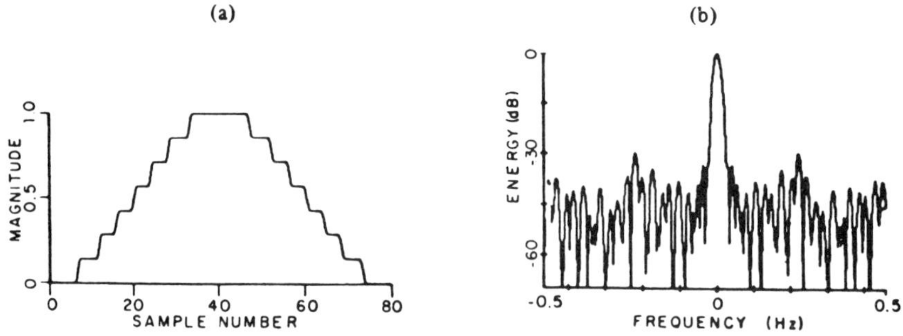

Fig. 9.5 Sampled envelope function (a) and energy spectra (b) for quantized Hanning transmit envelope used in an example from the REVMOD reverberation model (Hodgkiss, 1984; *IEEE J. Oceanic Engr.*, **10**, 285–9; copyright by IEEE).

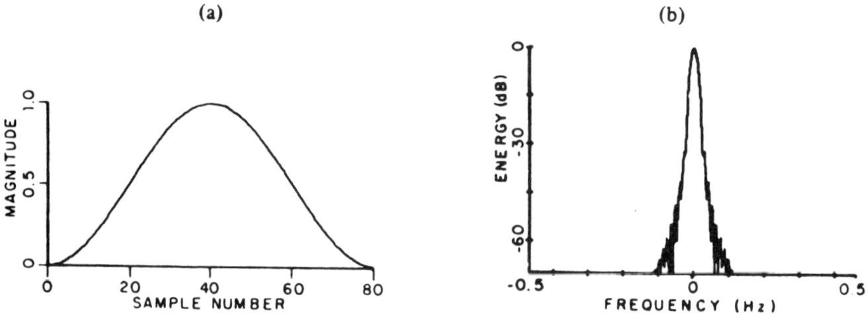

Fig. 9.6 Sampled envelope function (a) and energy spectra (b) for Hanning receive envelope used in an example from the REVMOD reverberation model (Hodgkiss, 1984; *IEEE J. Oceanic Engr.*, **10**, 285–9; copyright by IEEE).

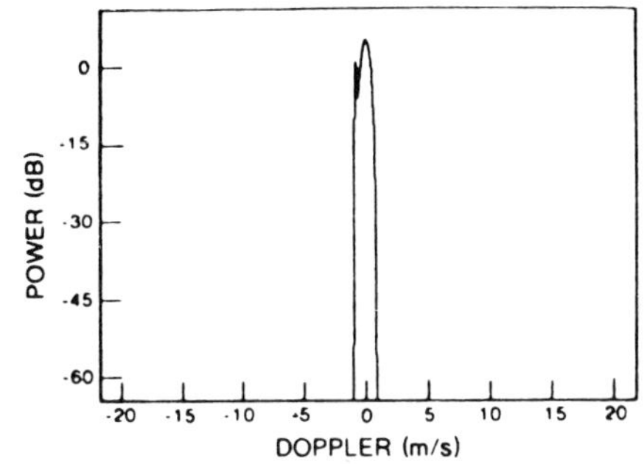

Fig. 9.7 Example from the REVMOD reverberation model showing the surface-scattering function for a downwind case (Hodgkiss, 1984; *IEEE J. Oceanic Engr.*, **10**, 285–9; copyright by IEEE).

ridge), as shown in Fig. 9.7. This ridge corresponds to the Doppler-shifting effects of the surface wave and current velocity vectors.

9.5 BISTATIC REVERBERATION

9.5.1 Computational considerations

Most active sonars are configured in what is termed a monostatic geometry, meaning that the source and receiver are at the same position. In some sonar systems, however, the source and receiver are separated in range or depth, or both, in what is termed a bistatic configuration. Bistatic geometries are characterized by a triangle of source, target and receiver positions, and by their respective velocities (Cox, 1989). Such geometries are commonly employed in sonobuoy applications (Bartberger, 1985) and also in active surveillance applications (Franchi *et al.*, 1984). Geometries involving multiple sources and receivers are termed multistatic.

A bistatic sonar must detect the target signal against the background noise level that is the intensity sum of: the ambient noise; the surface, bottom and volume reverberation; and the energy propagating directly from the source to the receiver over a set of one-way paths referred to as direct arrivals (Fig. 9.8). These direct arrivals represent a significant difference between bistatic and monostatic sonars. Since the direct (one-way) paths are very intense, they must be carefully considered in predicting the performance of bistatic sonars (Bartberger, 1985). Each different propagation path may require separate predictions of transmission loss. Bistatic reverberation models are commonly confronted with substantial bookkeeping requirements associated with sorting the various reverberation and noise contributions. In the monostatic situation, reverbera-

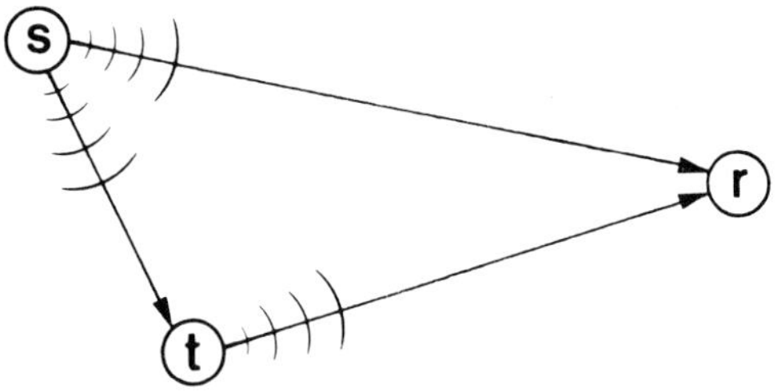

Fig. 9.8 Geometry for bistatic reverberation showing relative positions of the source (s), target (t) and receiver (r) in either the vertical or the horizontal plane.

tion is due to backscatter. In bistatics, forward and out-of-plane scattering are important (Cox, 1989).

The bistatic problem is further complicated by the presence of certain bathymetric features, particularly seamounts, which can introduce a discrimination problem between acoustic returns from these geologic features and from targets of interest. Eller and Haines (1987) have described seamount statistics as well as the characteristics of acoustic reverberation from seamounts.

A real-time bathymetric reflection and scattering prediction system has been developed to support at-sea acoustical surveys (Haines *et al.*, 1988). The system utilizes gridded databases (to generate 3D environmental scenarios) in conjunction with the ASTRAL propagation model and the ANDES noise model. In its present configuration, the system operates on a HP-9020 desk-top computer and provides color graphic displays of predicted ocean bottom reverberation as well as echoes from seamounts.

A computer model called ORBS (Ocean Refraction and Bathymetric Scattering) has been developed to calculate the directional distribution of bottom-scattered acoustic energy received at an array (Baer *et al.*, 1985; Wright *et al.*, 1988). ORBS incorporates the refractive effects of a varying sound speed structure in the ocean volume as well as the effects of out-of-plane scattering caused by rough bathymetry. A seamount may be considered where it makes sense to consider only a single encounter with the sea floor. The numerical techniques used include the split-step solution to the parabolic equation away from the boundary, propagation of the coherence function near the boundary, and a modified Kirchhoff formulation to incorporate scattering.

9.5.2 The bistatic acoustic model – a specific example

The Bistatic Acoustic Model (BAM) predicts the performance of a sonar system consisting of a sound source and a receiver separated in both range and depth (Bartberger, 1991a; Vendetti *et al.*, 1993a). BAM computes the echo-to-background ratios for targets located at a set of bipolar grid locations in a horizontal plane containing a specific target. The bipolar grid points are defined by the intersections of circles of constant range about the source and receiver. There are three separate versions of BAM corresponding to three different sonar system configurations: vertical line array, horizontal line array, or volumetric array. BAM consists of four computer programs:

(1) BISON – computes the ray arrival structure, the surface, bottom and volume reverberation time histories, and the vertical distribution of ambient noise;
(2) BISTAT – computes the echo-to-background ratios on a bipolar grid within a specified circle about the receiver;
(3) CTEST – generates the time series of the direct arrivals, the surface, bottom and volume reverberation, and the combined interference of all of these components;
(4) PLOSS – calculates the source-to-target transmission loss and the target-to-receiver transmission loss as a function of range.

A sample output product from BAM is illustrated in Fig. 9.9, which shows computed echo-to-background levels. These levels are numerically coded to facilitate printer output. The digital pattern in Fig. 9.9 indicates the potential for target detection under the hypothetical problem geometry and acoustic conditions:

- frequency = 2 kHz
- source depth = 300 m (4-element omni source array)
- receiver depth = 500 m (6-element omni receiver array)
- target depth = 350 m (no Doppler)
- source–receiver range = 10 km
- maximum target–receiver range = 50 km
- pulse length = 5 s

The problem area is delimited by a circle centered about the receiver with a radius equal to the maximum target–receiver range. Also shown is the cumulative area coverage (in units of square kilometers) contained within specified contours of echo-to-background level. In this example, the maximum area coverage possible would be $\pi \times (50\,\text{km})^2 \approx 7854\,\text{km}^2$. Thus, for echo-to-background levels $\geqslant -10$ dB, only 63% ($4942\,\text{km}^2/7854\,\text{km}^2$) of the specified search area would be covered effectively. Clearly, this information is useful in assessing the effectiveness of candidate sensors and search tactics. For example, optimum search tactics could be developed by varying selected parameters and then assessing the resulting changes in area coverage.

The elliptically shaped region of poor detection (< -10 dB), located immediately between the source and receiver, is a consequence of the

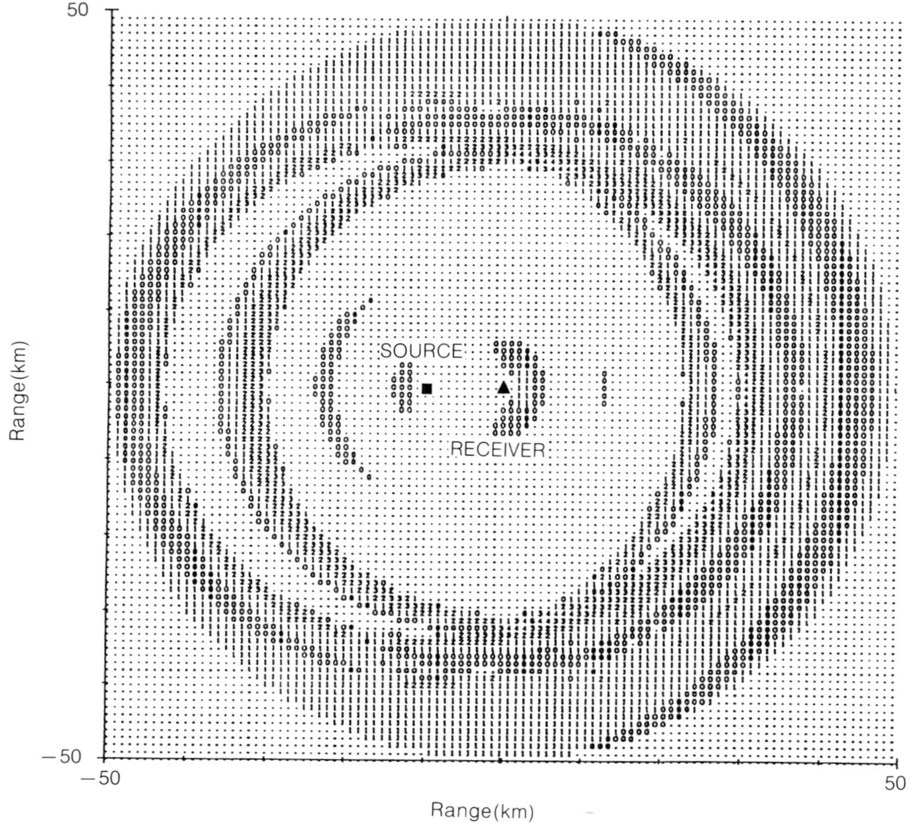

Symbol	Echo-to-background level
·	< −10 dB
0	−10 to −6
1	−6 to −2
2	−2 to 2
3	2 to 6
4	6 to 10
5	10 to 14
6	14 to 18
7	18 to 22
8	> 22

Controur level	Cumulative area
≥ −10 dB	4942 km^2
≥ −6	3427
≥ −2	871
≥ 2	283
≥ 6	45
≥ 10	2
≥ 14	0
≥ 18	0
≥ 22	0

Fig. 9.9 Sample output from the Bistatic Acoustic Model (BAM) showing echo-to-background levels. Also shown is the cumulative area coverage contained within specified contours of echo-to-background level.

direct arrivals, which represent the transmission of energy directly from the source to the receiver. These direct arrivals, also referred to as the direct blast, consist of a number of multipath arrivals extending over a time interval that is determined by geometry, environment, and system characteristics. The direct blast often masks target echoes of interest. The extent of this masking region is approximately determined by the inequality (Cox, 1989):

$$\tau_1 + \tau_2 < \tau_3 + \tau_b$$

where τ_1 = travel time, source to target
τ_2 = travel time, target to receiver
τ_3 = travel time, source to receiver
τ_b = effective masking time of direct blast

The effects of the direct blast can be mitigated to some extent by: controlling the vertical beam pattern of the receiver or source, or both, to minimize the contribution of steep-angle (boundary-interacting) arrivals; nulling the receiver in the direction of the source; or using Doppler to discriminate between direct-blast energy and target echoes.

9.6 POINT-SCATTERING MODELS

9.6.1 Computational considerations

Point-scattering models are based on a statistical approach which assumes that the scatterers are randomly distributed throughout the ocean. The echoes from each individual scatterer are then summed to compute the reverberation level.

This approach is not as widely used as the cell-scattering technique. When dealing with scatterers whose dimensions are comparable to the acoustic wavelength under consideration, the point-scattering approach may be the preferred alternative. This situation arises in two common problems: high-frequency sonars and under-ice scattering.

A point-scattering model suitable for high-frequency applications has been described by Princehouse (1977). A model appropriate for under-ice scattering is briefly described below.

9.6.2 The under-ice reverberation simulation model – a specific example

Bishop (1987, 1989a, 1989b) and Bishop *et al.* (1986, 1987) developed a bistatic, high-frequency ($\geqslant$2 kHz), under-ice acoustic-scattering model to evaluate the scatter produced by a pulse, originating from an arbitrarily located source, as detected by an arbitrarily located receiver. This model is referred to as Under-Ice Reverberation Simulation.

This model uses measured two-dimensional under-ice acoustic profile data and several empirical results that relate geometric parameters of the large-scale under-ice relief features such as ice keels to construct a three-dimensional, bimodal under-ice surface consisting of first-year ice keels and sloping flat ice regions. A first-year keel is modeled as an ensemble of randomly oriented ice blocks on a planar surface inclined at some slope angle with respect to a horizontal plane at sea level. The keel is characterized by length, draft, width, ice thickness and aspect angle. A region of flat ice is modeled as a smooth planar surface whose slope angle is less than some critical angle that serves to distinguish a flat ice feature from an ice keel. The Kirchhoff approximation is used to evaluate the target strength of a facet of an ice block. This approximation assumes that reflection coefficients appropriate for an infinite plane wave at an infinite plane interface can be used in the local scattering geometry of a rough surface (in some texts, this approximation is referred to as the tangent plane method). The target strength of a keel is calculated in range increments as the coherent sum of the backscatter from all scattering facets contained within one-half the pulse length projected onto the keel. The model has been used to show the effects of various ice and acoustic parameters on reverberation and target strength frequency distributions.

The Under-Ice Reverberation Simulation model was used to calculate monostatic backscattering and reverberation for one realization at three

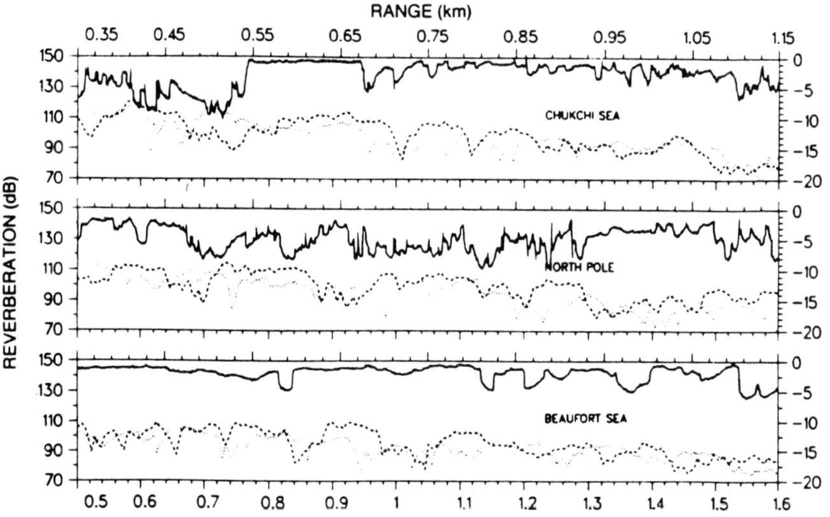

Fig. 9.10 Under-ice profiles (solid line), and measured (dashed line) and modeled (dotted line) under-ice monostatic reverberation time series for three Arctic sites (Bishop, 1989b).

Arctic sites: a site near the North Pole; a site in the Chukchi Sea; and a site in the Beaufort Sea (Bishop, 1989a, 1989b).

To parametrize the large-scale surface roughness at these sites, the keel frequency and mean keel relief were calculated. The Chukchi Sea site was the roughest of the three sites with a keel frequency of 16.0 keels/km and an average keel relief of 4.1 m. The North Pole site had a keel frequency of 13.7 keels/km and an average keel relief of 4.0 m. The Beaufort Sea site was the least rough of the sites with a keel frequency of 4.5 keels/km and an average keel relief of 3.4 m.

Figure 9.10 illustrates the results obtained for these three sites. The under-ice profiles for 1.15 km transects at each site are indicated by a heavy solid line. The measured and modeled under-ice monostatic reverberation levels are indicated by the dashed and dotted lines, respectively. The time scale (in seconds) is indicated at the bottom of the figure.

These results support the intuitive notion that relatively high reverberation levels are associated with areas of high ice roughness. More specifically, average keel draft determines the average scattering facet surface area (which determines the average amplitude of a specular return) and the keel frequency determines the total number of scattering facets (which determines the probability of scatter occurrence).

Differences between measured and modeled results are thought to be due to significant scatterers present in the under-ice surface but not present in either the measured or modeled data.

Table 9.1. Summary of underwater acoustic reverberation models

Cell scattering		*Point scattering*
BAM[a] [1]	PEREV [7]	REVGEN [13]
BISAPP[a] [2]	RASP[a] [8]	Under-Ice Reverberation Simulation[a,b] [14]
BISSM[a] [3]	REVMOD [9]	
DOP [4]	REVSIM [10]	
EIGEN/REVERB [5]	RUMBLE[a] [11]	
MAM [6]	TENAR [12]	

[a] Bistatic geometries.
[b] Under-ice.

References to reverberation models: [1] Bartberger (1985, 1991a), Vendetti *et al.* (1993a); [2] Pomerenk and Novick (1987); [3] Caruthers *et al.* (1990), Caruthers and Novarini (1993), Caruthers and Yoerger (1993); [4] Marsh (1976); [5] Sienkiewicz *et al.* (1975); [6] Bartberger (1991b), Vendetti *et al.* (1993b); [7] Smith *et al.* (1993); [8] Franchi *et al.* (1984), Palmer and Fromm (1992); [9] Ackerman and Kesser (1973), Hodgkiss (1980, 1984); [10] Chamberlain and Galli (1983); [11] Bucker (1986), Kewley and Bucker (1987); [12] Luby and Lytle (1987); [13] Princehouse (1977); [14] Bishop *et al.* (1986, 1987), Bishop (1987).

9.7 NUMERICAL MODEL SUMMARIES

A summary of stand-alone underwater acoustic reverberation models is presented in Table 9.1. The models are segregated according to cell-scattering or point-scattering approaches. Also indicated are those models capable of handling bistatic geometries and those applicable to under-ice environments. Numbers within parentheses following the model names refer to appropriate documentation. Model documentation can range from informal programming commentaries to journal articles to detailed technical reports containing a listing of the actual computer code. Abbreviations and acronyms are defined in Appendix A. This summary does not claim to be exhaustive.

10

Sonar performance models

10.1 BACKGROUND

The ultimate purpose of sonar performance modeling is twofold: first, advanced sonar concepts can be optimally designed to exploit the ocean environment of interest; and, second, existing sonars can be optimized for operation in any given ocean environment.

In the case of existing sonars, performance prediction products can be tailored to individual sonar systems by providing the sonar operators with on-scene equipment mode selection guidance. When combined with current tactical doctrine, this information product is commonly referred to as a tactical decision aid (TDA). These decision aids are used by the force commanders to optimize the employment of ASW resources in any particular tactical environment at sea. Performance prediction products help both the force commanders and sonar operators to better understand and thus exploit the ocean environment from a tactical standpoint. Sonar performance models also support various naval underwater acoustic surveillance activities.

Beyond this dimension of modeling lies the realm of engagement modeling. In essence, engagement models use sonar detection performance data (either measured or predicted by sonar performance models) to simulate overall ASW combat system performance. These simulations are usually conducted in the context of an integrated naval force, set in opposition to a hypothetical threat force. The output data typically include exchange ratios, which are useful in determining force level requirements and in developing new tactics. The subject of engagement modeling quickly becomes classified and thus will not be discussed further in this book.

Sonar performance models use active and passive sonar equations to generate performance predictions. The input variables required for solution of these equations are generated by mathematical models of propagation, noise and reverberation. This hierarchy of models was previously illustrated in Fig. 1.1.

Sonar performance models can logically be segregated into active sonar models and passive sonar models, as would be suggested by the distinction between active and passive sonar equations.

The complexity of the sonar performance modeling problem is a natural consequence of the naval operations that these models support. In modern naval battle group operations, for example, transmission loss must be calculated along a multiplicity of paths connecting widely separated sources and receivers. Noise interference from nearby consort vessels must be factored into sonar performance calculations together with distant merchant shipping and local weather noises. Furthermore, scattering and reverberation from bathymetric features and volumetric false targets must be efficiently and realistically modeled. The demands placed on computational efficiency and database management are enormous. Underwater acoustic modelers face challenges that severely tax existing mathematical methods. Moreover, with the advent of multistatic scenarios involving multiple sets of separated sources and receivers, true three-dimensional (3D) modeling is no longer a theoretical luxury but rather a practical necessity.

This situation has become even more complex with the heightened interest in shallow-water sonar operations. Acoustic interactions with highly variable sea-floor topographies and compositions further compound already intensive scattering and reverberation calculations. In addition, the sound speed field in shallow-water areas is often characterized by high spatial and temporal variability. As a result, statistical approaches have been explored in order to obtain meaningful predictions of active sonar performance in shallow-water environments. In the MOCASSIN model, for example, Schneider (1990) provided the option to specify a stochastic sound speed field in order to approximate the horizontal variability typical of coastal regions.

10.2 SONAR EQUATIONS

The sonar equations are simple algebraic expressions used to quantify various aspects of sonar performance. These equations vary according to active or passive sonars, and within active sonars they further differentiate according to noise or reverberation limited conditions.

The various system and environmental parameters that make up the sonar equations are listed in Table 10.1 together with reference locations and brief descriptions. A condensed statement of the equations is as follows (Urick, 1983, Chapter 2):

Table 10.1 Sonar parameter definitions and reference locations

Parameter	*Symbol*	*Reference*	*Definition*
Source level	SL	1 m from source on its acoustic axis	$10 \log \frac{\text{intensity of source}}{\text{reference intensity}^{a}}$
Transmission loss	TL	1 m from source and at target or receiver	$10 \log \frac{\text{signal intensity at 1 m}}{\text{signal intensity at target or receiver}}$
Target strength	TS	1 m from acoustic center of target	$10 \log \frac{\text{echo intensity at 1 m from target}}{\text{incident intensity}}$
Noise level	NL	At hydrophone location	$10 \log \frac{\text{noise intensity}}{\text{reference intensity}^{a}}$
Receiving directivity index	DI	At hydrophone terminals	$10 \log \frac{\text{noise power generated by an equivalent nondirectional hydrophone}}{\text{noise power generated by actual hydrophone}}$
Reverberation level	RL	At hydrophone terminals	$10 \log \frac{\text{reverberation power at hydrophone terminals}}{\text{power generated by signal of reference intensity}^{a}}$
Recognition differential	RD	At hydrophone terminals	$10 \log \frac{\text{signal power to just perform a certain function}}{\text{noise power at hydrophone terminals}}$

[a] The reference intensity is that of a plane wave of rms pressure 1 μPa.
(Urick, 1983; *Principles of Underwater Sound*, 3rd edn; reproduced with permission of McGraw-Hill Publishing Company.)

Active sonars (monostatic)

Noise background

$$SL - 2TL + TS = NL - DI + RD_N \tag{10.1}$$

Reverberation background

$$SL - 2TL + TS = RL + RD_R \tag{10.2}$$

Passive sonars

$$SL - TL = NL - DI + RD \tag{10.3}$$

where RD has been subscripted to make it clear that this parameter is quantitatively different for noise (N) and for reverberation (R). In this application, RD has the same meaning as detection threshold (DT) as used by Urick (1983, Chapter 12). The sonar equations are valid for the condition in which the signal excess is zero (see Table 10.2). The apparent simplicity of the individual components of the sonar equations can be misleading. The complexity of these parameters will be better appreciated through the discussions presented in sections 10.3 and 10.4.

Table 10.2. Terminology of various combinations of the sonar parameters

Name	*Parameters*	*Remarks*
Echo signal level	$SL - 2TL + TS$	The intensity of the echo as measured in the water at the hydrophone
Noise masking level	$NL - DI + RD_N$	Another name for these two combinations is minimum detectable echo level
Reverberation masking level	$RL + RD_R$	
Signal excess	$SL - 2TL + TS - (NL - DI) - RD$	Detection just occurs, under the probability conditions implied in the term RD, when the signal excess is zero
Performance figure	$SL - (NL - DI)$	Difference between the source level and the noise level measured at the hydrophone terminals
Figure of merit (FOM)	$SL - (NL - DI) - RD$	Equals the maximum allowable one-way transmission loss in passive sonars, or the maximum allowable two-way loss for TS = 0 dB in active sonars

(Urick, 1983; *Principles of Underwater Sound*, 3rd edn; reproduced with permission of McGraw-Hill Publishing Company.)

These equations are valid for sound levels in a 1 Hz wide frequency band (referred to as spectrum level). For frequency bands greater than 1 Hz, the sound levels in equations (10.1)–(10.3) must be corrected for bandwidth (e.g. Kinsler *et al.*, 1982, pp. 409–17).

It is convenient in practical work to assign separate names to different combinations of terms in the sonar equations. A summary of commonly used terms is presented in Table 10.2. The combination of terms labeled figure of merit (FOM) is the most useful because it combines environmental, sonar and target parameters into one convenient quantitative measure of sonar performance. Since the FOM quantitatively equals the transmission loss under the conditions specified in the sonar equation, the FOM can give an immediate indication of the range at which an active or passive sonar can detect its target when corresponding transmission loss curves are already available. In the case of active sonars that are reverberation (versus noise) limited, the FOM is not constant but varies with range (time) and thus fails to be a useful indicator of active sonar performance.

Bistatic geometries are characterized by a triangle of source, target and receiver positions. Bistatic sonar equations, which differ from their monostatic counterparts, are required (Cox, 1989).

The parameters listed in Table 10.1 can be related to the functions of the basic acoustic models: propagation models generate estimates of transmission loss (TL); noise models predict noise levels (NL); and reverberation models provide estimates of the reverberation level (RL).

For passive sonars, the sonar equipment will define DI and RD; the target will define SL. TS is not considered in purely passive sonar models. For active sonars, the sonar equipment will difine SL, DI and RD; the target will define TS.

Useful overviews of sonar signal processing have been assembled by Knight *et al.* (1981) and Nielsen (1991).

10.3 THE NISSM MODEL – A SPECIFIC EXAMPLE

The Navy Interim Surface Ship Model (NISSM) II, developed by Weinberg (1973), is a computer program designed to predict the performance of active sonar systems using ray-theoretical techniques. The measure of performance is expressed in terms of probability of detection versus target range for a given false-alarm rate. Selected intermediate results can also be displayed, including ray traces, transmission loss, and boundary and volume reverberation. This model is applicable to range-independent ocean environments.

NISSM uses a cell-scattering model of reverberation. The most time-consuming phase in the execution of NISSM is the volume reverberation computation. The volume scattering strength is expressed in terms of a column (or integrated) scattering strength, and a careful integration of the volume reverberation integral is performed which considers multipath structure.

Because the NISSM II model encompasses all the modeling categories addressed thus far, and because it is widely used in the sonar-modeling community, it is considered to be representative of the more general class of active sonar models. Examples of selected NISSM outputs are presented at the end of this section.

10.3.1 Propagation

The depth-dependent sound speed $c(z)$ and the inclination angle (θ) at a point on a ray in the x–z plane can be related by Snell's law (see Fig. 10.1) as

$$\frac{c(z)}{\cos\theta} = c_v$$

where $c(z)$ is the sound speed as a function of depth, θ is the inclination angle, and c_v is the vertex sound speed. Snell's law implies that the range (x) and travel time (T) along the ray can be calculated as

$$x = x_a + \int_{z_a}^{z} \frac{c(z)|dz|}{\sqrt{c_v^2 - c(z)^2}} \tag{10.4}$$

$$T = T_a + \int_{z_a}^{z} \frac{c_v|dz|}{c(z)\sqrt{c_v^2 - c(z)^2}} \tag{10.5}$$

Officer (1958, Chapter 2) provides a rigorous development of these equations.

The range derivative ($\partial x/\partial c_v$) is given by

$$\left.\frac{\partial x}{\partial c_v}\right|_{Z=\text{const}} = \left.\frac{\partial x_a}{\partial c_v}\right|_{Z=Z_a} - \int_{z_a}^{z} \frac{c(z)c_v|dz|}{[c_v^2 - c(z)^2]^{3/2}} \tag{10.6}$$

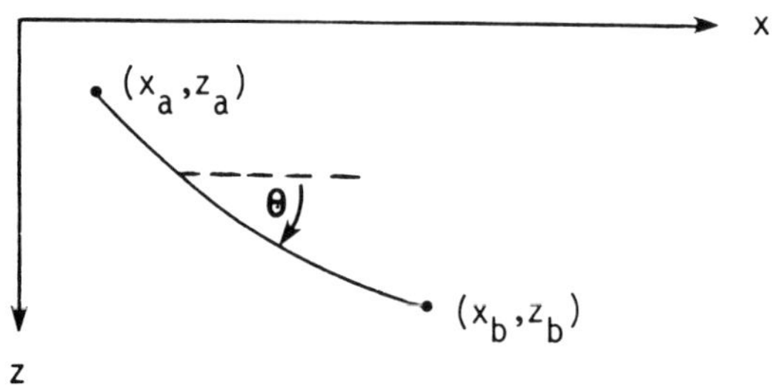

Fig. 10.1 Ray geometry showing inclination angle along a ray, as used in the NISSM active sonar model (Weinberg, 1973).

(which equals zero at caustics) and is used in computing the geometrical spreading loss factor (η_{sp}):

$$\eta_{sp} = |p| = \left| c_v \tan\theta \tan\theta_0 \, x \frac{\partial x}{\partial c_v} \right|^{-1/2} \tag{10.7}$$

where θ_0 is the inclination angle of the ray at a point source of unit magnitude situated at $x_a = 0$, $z_a = z_0$.

The relative geometrical acoustic pressure at field point (x,y) at time (t) is

$$p = \eta_{sp} \, e^{-i\omega(t-T)} \tag{10.8}$$

where $\omega = 2\pi f$ and $f =$ frequency.

Since the range, travel time and range-derivative integrals are symmetric with respect to the initial and final depths, if follows that the pressure satisfies the law of reciprocity. This law states that the acoustic pressure will remain unchanged if the source and receiver positions are interchanged.

The sound speed profile is approximated by a continuous function of depth (Weinberg, 1971) using Leroy's (1969) sound speed formula. Earth curvature corrections are also applied.

The relative acoustic pressure along a ray is found by multiplying the relative geometrical acoustic pressure (p) by the surface, bottom and absorption loss factors:

$$p = \eta_{sp} (\eta_{sur})^{n_{sur}} (\eta_{bot})^{n_{bot}} \, e^{-i\omega t + i\omega T + i\Phi + \alpha S} \tag{10.9}$$

and the transmission loss (TL) is given by

$$\mathrm{TL} = -20 \log |p| \tag{10.10}$$

where $N_{sp} = -20 \log \eta_{sp}$ is the geometrical spreading loss (dB)
$N_{sur} = -20 \log \eta_{sur}$ is the surface loss per bounce (dB/bounce), n_{sur} is the number of bounces
$N_{bot} = -20 \log \eta_{bot}$ is the bottom loss per bounce (dB/bounce), n_{bot} is the number of bounces
$A = -20 \log \alpha / \log_e (10)$ is the absorption coefficient (dB/km); α is expressed in nepers per unit distance
$t =$ elapsed time (s)
$T =$ travel time (s)
$\Phi =$ accumulated phase shift (rad)
$S =$ arc length (km)
$\omega =$ frequency (rad/s)

The surface loss may be input as a table of surface loss per bounce versus angle of incidence. The bottom loss may either be input as a table of bottom loss per bounce versus angle of incidence, or internally computed either by Marine Geophysical Survey (MGS) data or by an empirical equation.

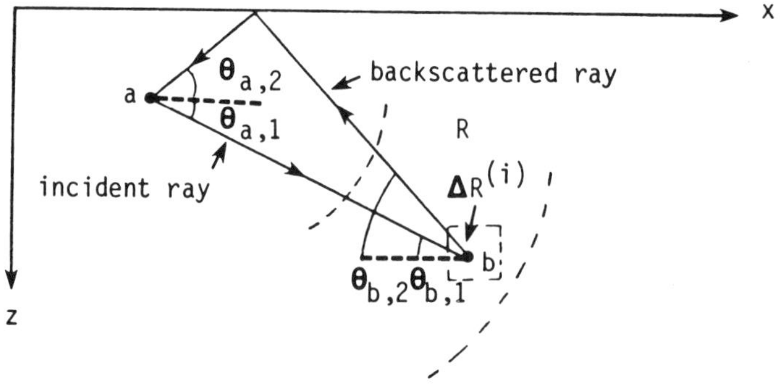

Fig. 10.2 Closed ray path used in reverberation calculations in the NISSM active sonar model (Weinberg, 1973).

The absorption coefficient can be represented in one of three ways: (1) input as a table of absorption per unit distance, versus frequency; (2) calculated by Thorp's (1967) equation; or (3) calculated by the equation attributed to H.R. Hall and W.H. Watson (1967, unpublished manuscript).

Eigenrays are found by tracing a preselected fan of test rays to the target depth. When two adjacent rays of the same type bracket a target range, a cubic spline interpolation is performed to determine the eigenray. The principal ray types have been defined previously in Fig. 4.2.

Shadow zone propagation is characterized according to Pekeris (1946) as

$$|p^2| = |p_b^2| \frac{x_b}{x} \exp[-\alpha(x - x_b)] \qquad \alpha = \frac{5.93}{c_{sur}} g_{sur}^{2/3} f^{1/3} \tag{10.11}$$

where p = pressure in the shadow zone at (x,z)
p_b = pressure on the shadow boundary at (x_b,z)
c_{sur} = surface sound speed (km/s)
g_{sur} = surface sound speed gradient (s^{-1})
f = frequency (Hz)

A uniform asymptotic expansion at a caustic is used to calculate the pressure in the vicinity of a caustic (Ludwig, 1966).

Phase shifts are accumulated as follows: π radians decrease in Φ at the surface; $\pi/2$ radians decrease in Φ at caustics; and zero phase shift at the bottom.

The computation of surface-duct propagation by ray theory does not allow for leakage from the surface layer into the underlying medium. Therefore, the AMOS equations are used (Marsh and Schulkin, 1955).

10.3.2 Reverberation

The basic theory for reverberation modeling was described in Chapter 9. Referring to Fig. 10.2, the acoustic pressure at the end of the closed path is determined in NISSM as

$$|p_{\text{rev}}|^2 = \int_R |d_{\text{tra}}\,\eta_{\text{tra}}(\theta_{\text{a},1})\,p_1 p_2\,\eta_{\text{rec}}(\theta_{\text{a},2})|^2 \mu \mathrm{d}R \qquad (10.12)$$

where $D_{\text{tra}} = 20 \log d_{\text{tra}}$ is the reference level of the transmitting array
$P_1 = -20 \log |p_1|$ is the transmission loss of the incident ray
$P_2 = -20 \log |p_2|$ is the transmission loss of the backscattered ray
$N_{\text{tra}} = -20 \log \eta_{\text{tra}}$ is the transmitter response (dB)
$N_{\text{rec}} = -20 \log \eta_{\text{rec}}$ is the receiver response (dB)
R = region containing the scatterers corresponding to rays with travel time (t): $T - \tau \leq t \leq T$
τ = pulse length (s)
$T = t_0 + t_1 + t_2;\ 0 \leq t_0 \leq \tau$
t_1 = travel time of incident ray
t_2 = travel time of backscattered ray
μ = backscattering strength (volume, surface or bottom)

NISSM usually computes eight paths to each representative scatterer [$\Delta R^{(i)}$]; thus, there may be as many as 56 (i.e. $8^2 - 8$) differently oriented closed paths per scatterer.

For volume reverberation, the expression for p_{rev} is replaced by the double summation (refer to Fig. 10.3)

$$|p_{\text{vol}}|^2 = \sum_{i,j} |d_{\text{tra}}\eta_{\text{tra}}(\theta_{\text{a},1}^{(i,j)})p_1^{(i,j)}p_2^{(i,j)}\eta_{\text{rec}}(\theta_{\text{a},2}^{(i,j)})|^2 \mu_{\text{vol}}^{(j)} x^{(i)} \Delta x^{(i,j)} \Delta z^{(j)} \Delta\phi \qquad (10.13)$$

where $\Delta\phi$ = horizontal beamwidth (rad)
$x^{(i)}$ = range of the (i,j)th scatterer (km)
$z^{(j)}$ = depth of the (i,j)th scatterer (km)
$U_{\text{vol}}^{(j)} = 10 \log \mu_{\text{vol}}^{(j)}$ is the volume scattering strength per unit volume of ocean

The surface reverberation is calculated using the expression for p_{rev} (equation (10.12)), except now

$U_{\text{sur}} = 10 \log \mu_{\text{sur}}$ is the surface scattering constant per unit area of surface (dB) (Chapman and Harris, 1962)

Bottom reverberation is calculated in a manner analogous to that for surface reverberation, where

$U_{\text{bot}} = 10 \log \mu_{\text{bot}}$ is the bottom scattering constant per unit area of sea floor (dB) (Mackenzie, 1961)

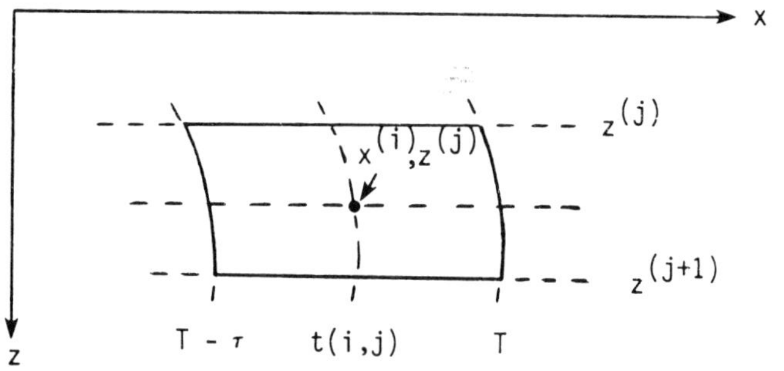

Fig. 10.3 Ensonified region used in reverberation calculations in the NISSM active sonar model (Weinberg, 1973).

The total reverberation is assumed to be the random-phase addition of surface and bottom echoes (also referred to as fathometer returns), and volume, surface and bottom reverberation.

10.3.3 Target echo

Target echo (or echo signal level) at a particular time is determined by

$$|p_{\text{echo}}|^2 = \sum_i |d_{\text{tra}}\,\eta_{\text{tra}}(\theta_{\text{a},1}^{(i)})\,p_1^{(i)}\,p_2^{(i)}\,\eta_{\text{rec}}(\theta_{\text{a},2}^{(i)})|^2 \mu_{\text{tar}}^{(i)} \tag{10.14}$$

where $U_{\text{tar}} = 10 \log \mu_{\text{tar}}$ is the target strength. The summation is taken over all closed paths with round-trip travel times between $T - \tau$ and T.

10.3.4 Noise

Noise is separated into two components: self-noise and ambient noise. Assuming isotropic noise fields:

$$P_{\text{noise}} = 10 \log\,[\eta_{\text{DI}}(p_{\text{self}}^2 + p_{\text{amb}}^2)] \tag{10.15}$$

where $\text{DI} = -10 \log \eta_{\text{DI}}$ is the directivity index (dB)
$P_{\text{self}} = 20 \log p_{\text{self}}$ is the self-noise spectrum density
$P_{\text{amb}} = 20 \log p_{\text{amb}}$ is the ambient noise spectrum density

10.3.5 Signal-to-noise ratio

Target echo to masking background, for a narrowband process, may be approximated by the signal-to-noise ratio (s/n):

$$\text{s/n} = \frac{\Delta f_{\text{echo}}\,p_{\text{echo}}^2}{\Delta f_{\text{noise}}\,p_{\text{noise}}^2 + \Delta f_{\text{rev}}\,p_{\text{rev}}^2} \tag{10.16}$$

where $P_{echo} = 20 \log p_{echo}$ is the target spectrum density at time T (dB)
$P_{rev} = 20 \log p_{rev}$ is the reverberation spectrum density at time T (dB)
$P_{noise} = 20 \log p_{noise}$ is the noise spectrum density at the beamformer output (dB)
Δf_{echo} = equivalent bandwidth for the received echo (Hz)
Δf_{rev} = equivalent bandwidth for the received reverberation (Hz)
Δf_{noise} = equivalent bandwidth for the received noise (Hz)

Assuming that the target echo is Gaussian distributed and centered about the receiving(Doppler shifted) frequency, with a standard deviation of $\tau/2$, then

$$\Delta f_{echo} = \phi(\tau \Delta f) - \phi(-\tau \Delta f) \tag{10.17}$$

where Δf is the receiving bandwidth (Hz), τ is the pulse length (s), and where the function $\phi(x)$ is defined as

$$\phi(x) = \frac{1}{\sqrt{2\pi}} \int_{-\infty}^{x} e^{-y^2/2} dy \tag{10.18}$$

is the normal probability function (e.g. Burdic, 1991; Hassab, 1989).

Reverberation is also assumed to be Gaussian with a standard deviation of $\tau/2$, but is centered about the transmitting frequency. Therefore, reverberation energy falling outside the echo band is $\Delta f_{rev} p_{rev}^2$, where

$$\Delta f_{rev} = \phi(2\tau \Delta f_{Dop} + \tau \Delta f) - \phi(2\tau \Delta f_{Dop} - \tau \Delta f) \tag{10.19}$$

and where

$$\Delta f_{Dop} = 2f \frac{V_{cl}}{c_s} \tag{10.20}$$

where Δf_{Dop} = Doppler shift (Hz)
f = frequency (Hz)
V_{cl} = closing speed (km/s)
c_s = sound speed (km/s)

If the noise-spectrum density can be considered constant over the receiving band, then

$$\Delta f_{noise} = \Delta f \tag{10.21}$$

10.3.6 Probability of detection

Probability of detection is modeled using a narrowband, square-law envelope detector (Fig. 10.4).

The input $x(t)$ is assumed to be either a stationary zero-mean Gaussian signal, $s(t)$, and noise $n(t)$, or noise alone. It is assumed that the signal

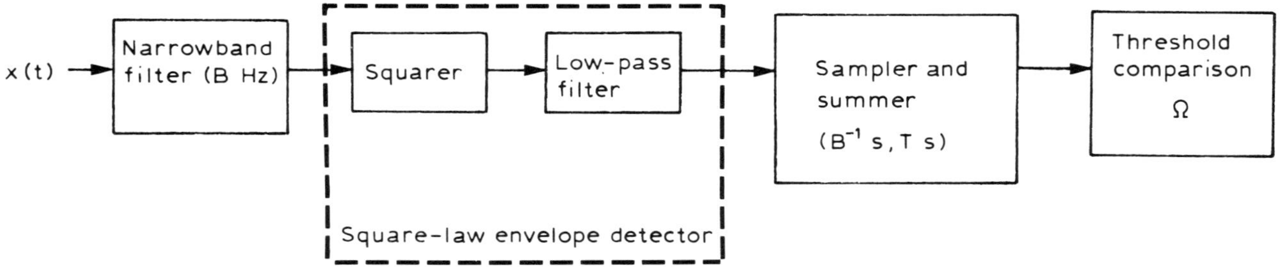

Fig. 10.4 Narrowband detector employed in the NISSM active sonar model (Weinberg, 1973).

spectrum has the same bandwidth as the narrowband filter and is centered on it, and that the noise is flat over the frequency interval. Samples of the squared-envelope, which are taken every $1/B$ seconds, are accumulated for an observation time of T seconds. The threshold (Ω) is fixed.

If P_F is a given probability of a false alarm, a quantity A is determined by inverting the following expression:

$$P_F = e^{-A} \sum_{m=0}^{M-1} \frac{1}{m!} A^m \tag{10.22}$$

where $M = BT + 1$. Once A is known, the probability of detection (P_D) is given by

$$P_D = e^{-A/(1+S/N)} \sum_{m=0}^{M-1} \frac{1}{m!} \left(\frac{A}{1 + S/N} \right)^m \tag{10.23}$$

where S/N is the maximum value with respect to time of s/n at a particular target.

10.3.7 Model outputs

NISSM II can generate data appropriate for graphical presentation of ray diagrams, transmission loss versus range, reverberation (surface, bottom, volume and total) versus time, signal-to-noise ratio versus range, and probability of detection versus range.

Sample outputs are illustrated for reverberation (Fig. 10.5), signal-to-noise ratio (Fig. 10.6) and probability of detection (Fig. 10.7). Representative illustrations of ray diagrams and transmission loss have already been presented elsewhere in this book. These examples are presented for illustratives purpose only and are not intended to portray a complete model run.

The reverberation-versus-time plots illustrated in Fig. 10.5 are valid for a deep-water environment with a shallow sonic layer depth, a critical depth and sufficient depth excess to support convergence zone propagation, as well as surface duct and bottom bounce paths. At about 11 s, a strong bottom bounce return is evident in the bottom reverberation curve. At above 63 s, a strong convergence zone (CZ) return is evident in both the volumetric and surface reverberation curves (but not in the bottom reverberation curve).

The signal-to-noise ratio plot in Fig. 10.6 indicates a high (positive) ratio at ranges corresponding to direct path (0–2 km) and convergence zone (near 50 km) detections. These sonar detection opportunities are vividly demonstrated in Fig. 10.7; here, high probabilities of detection are associated with direct path and CZ ranges.

It is sometimes useful to plot different combinations of active sonar parameters on the same graph. From Table 10.2, for example, echo signal

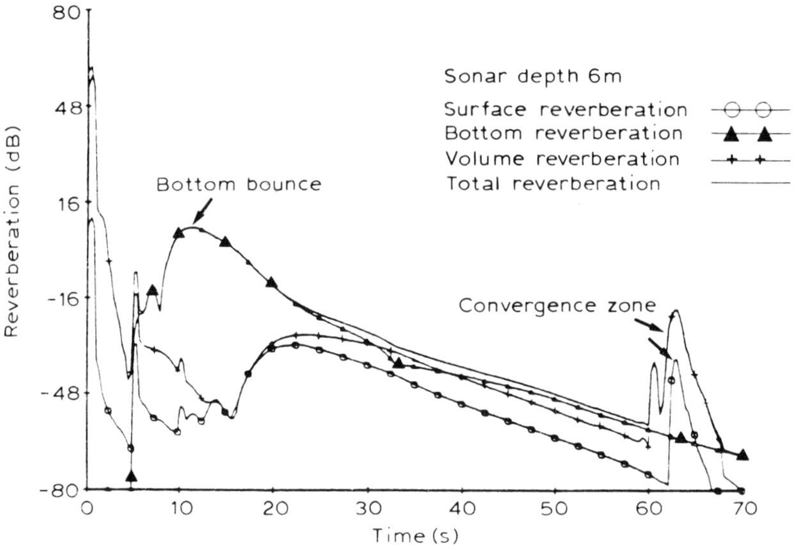

Fig. 10.5 Sample reverberation versus time plot generated by the NISSM active sonar model (Weinberg, 1973). The bottom bounce contribution is readily visible in the bottom reverberation curve; the convergence zone contribution is evident in both the surface and volume reverberation curves.

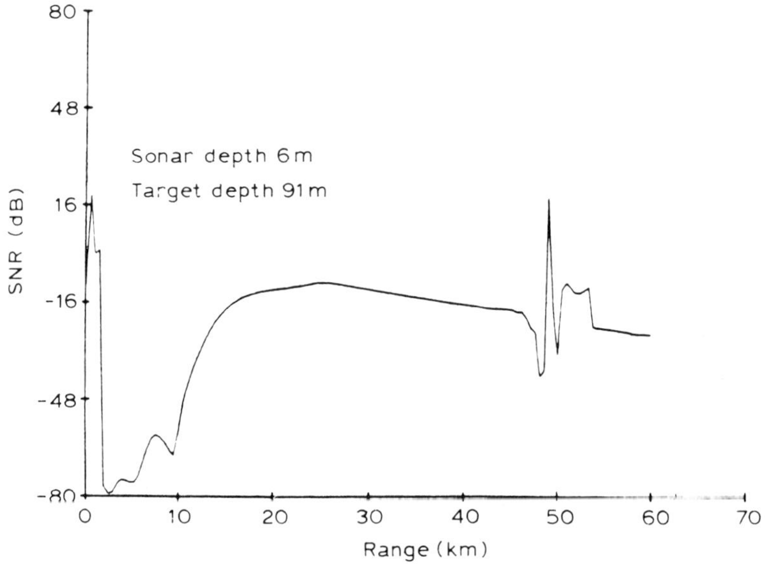

Fig. 10.6 Sample signal-to-noise ratio (SNR) plot generated by the NISSM active sonar model based in part on the data presented in Fig. 10.5 (Weinberg, 1973).

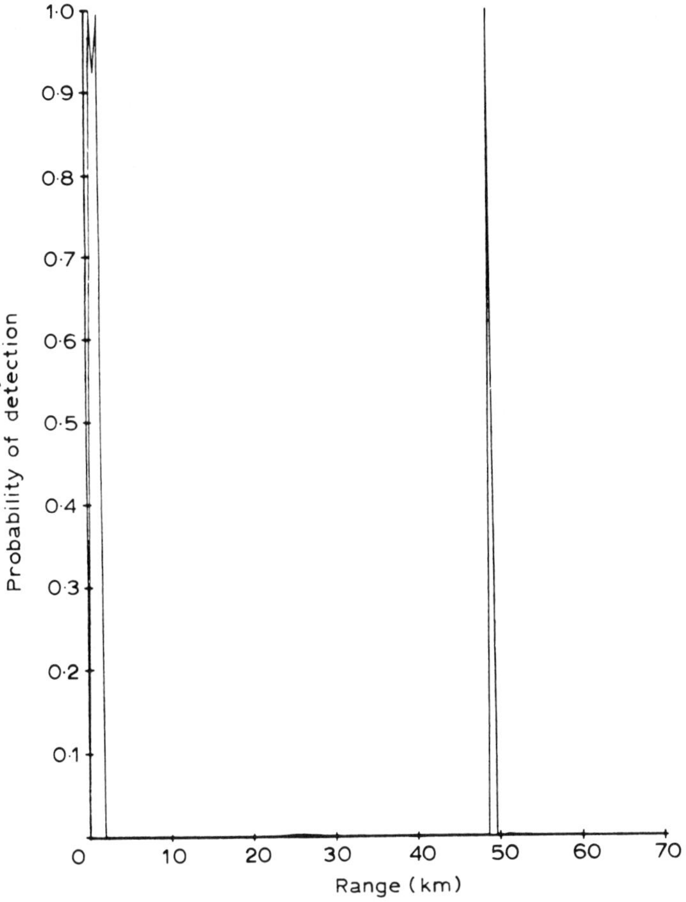

Fig. 10.7 Sample probability of detection plot generated by the NISSM active sonar model based in part on the data presented in Fig. 10.6 (Weinberg, 1973).

level can be plotted together with the noise and reverberation masking levels, as in Fig. 10.8. With such information, signal excess as a function of range can be computed, as in Fig. 10.9, which provides a direct indication of sonar detection performance. Specifically, all areas under the signal excess curve, but above the horizontal zero signal excess line, are associated with detection opportunities. That is, detection is accomplished when the signal excess is greater than, or equal to, zero. Thus, NISSM and similar models can provide graphic assessments of active sonar performance under the stated environmental and system-specific conditions. The intermediate quantities further provide diagnostic tools for analyzing particular aspects of active sonar design and operation problems.

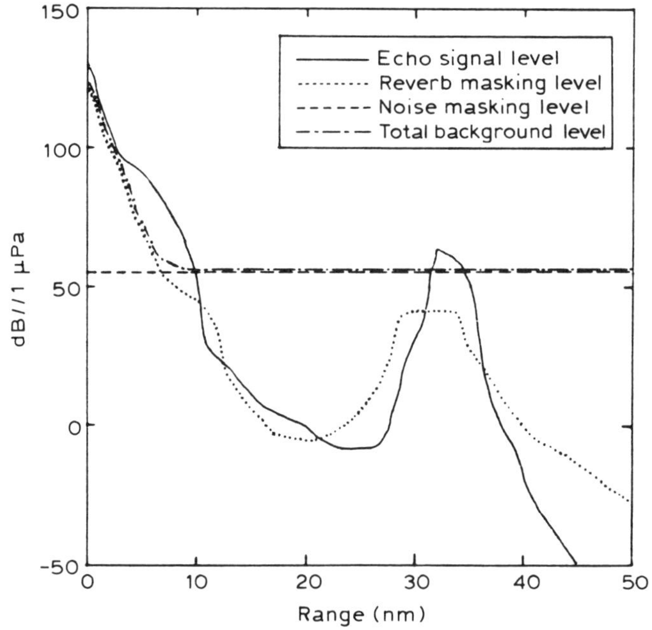

Fig. 10.8 Sample graphical combination of selected sonar parameters. See Table 10.2 for an explanation of terms.

10.4 MODEL-OPERATING SYSTEMS

The trend toward modular designs in the programming of underwater acoustic models is evidenced by recent interest in Model-Operating Systems (MOS) (e.g. Eller, 1990; Holmes *et al.*, 1990). These systems provide a framework for the direct linkage of data management software with computer-implemented codes of acoustic models. Such systems relieve model operators of much of the tedium associated with data entry operations.

Model-operating systems can facilitate comparative model evaluations (see Chapter 11) by standardizing the hardware and software configurations of different modeling techniques. The resulting uniformity also encourages a higher degree of configuration management and thus assists in the process of model evaluation. Sonar trainers now commonly utilize sophisticated range-dependent acoustic models to generate more realistic training environments for the operators of sonar equipment (Miller, 1982, 1983). Requirements for near-real time program execution are frequently satisfied using modular approaches similar to those employed in the construction of model-operating systems.

In general, the development of a sonar model-operating system is based on several considerations (Locklin and Webster, 1980): (1) it is assumed that the complex sonar performance modeling problem can be

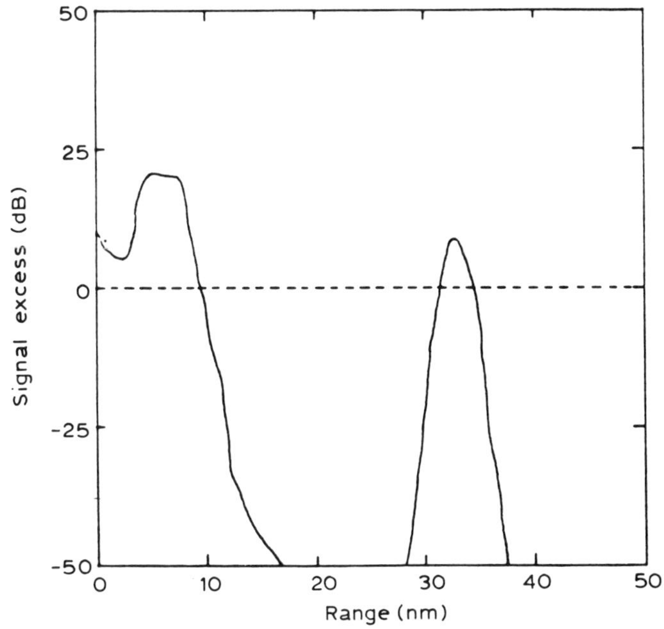

Fig. 10.9 Signal excess as a function of range based on the data presented in Fig.10.8. Active sonar detection ranges are indicated by regions of positive signal excess.

decomposed into a set of generic functions, each of which solves a well-defined and bounded portion of the overall problem (decomposition is a powerful technique to break down large problems into manageable units); (2) it is assumed that the problem can be solved on a computer through the use of an automated data system and that the development of this system can be accomplished such that its computational structure bears a precise relationship to the problem's decomposed functions; and (3) it is assumed that the functions implemented on a computer have the means to communicate with each other in a prescribed manner.

10.4.1 System architecture

Model-operating systems fall into two general classes of software architecture: bundled and executive. The principal distinction between these two architectures concerns the packaging of the software and databases to meet specific modeling requirements. Bundled systems are tailored to a narrow range of modeling applications and, in general, have only one method implemented to perform a generic modeling function. Executive systems are tailored to respond to a broader range of modeling applications by allowing the user to interactively control the hierarchical selection of functions. Furthermore, executive systems allow for several methods of performing a single generic function, thus providing a dynamic formulation of a modeling application.

Two different modeling environments can be addressed by the MOS: research and production. These two environments are not different with respect to what computing functions are performed, but rather how the computing functions are used by the analyst. This difference in usage should be reflected in the MOS architecture in order to provide the analyst with an optimized capability to perform both types of activities, if necessary. These environments are briefly discussed below to provide potential users with a perspective and a choice of the appropriate style of computing suitable to their specific needs.

The research environment entails activities such as experimentation, iterative problem solution, computation of intermediate model results and model input sensitivity studies, among others. Practical applications include sonar system design and environmental-acoustics research. This implies that the MOS must be highly interactive with respect to the interface between the analyst and the MOS, and usually suggests an executive architecture.

The production environment is typified by operations which are definable, structurally organized and ordered. To a large extent, these activities, once defined, have specific end products, solve a complete problem and are performed repetitively using the same (or slightly modified) functions. Practical applications include routine sonar performance predictions and sonar operational trainers. This implies that the computing tools can be organized and structured beforehand, and then used repetitively thereafter. It further indicates that a more optimal composition of functions may be needed, with less user interactive capability, and less flexibility with respect to the computing function composition. This usually suggests a bundled architecture.

The functional organization of the MOS features a uniform way of viewing the system structure for solution of both the passive and active sonar equations. While these two problems are distinctly different, this need not imply that the active MOS and passive MOS must be constructed as two independent systems. Rather, the redundancy in system control software and in certain lower level functions common to both the active and passive sonar problems makes separate approaches less attractive.

For illustrative purposes, a hypothetical functional decomposition of both the passive and active sonar equations will be described below.

10.4.2 Sonar modeling functions

A representative decomposition of the sonar equations into generic modeling functions is shown in the system data flows in Fig. 10.10 for the

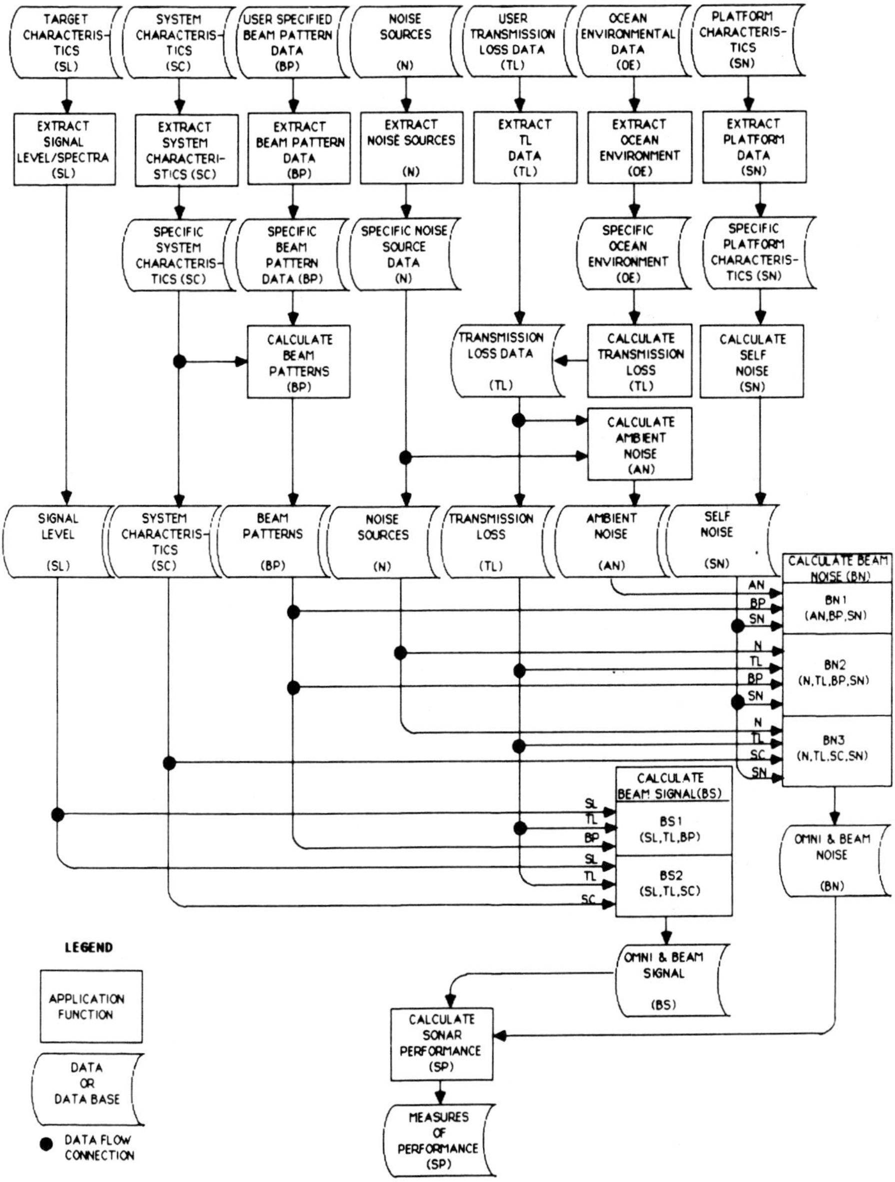

Fig. 10.10 Sample acoustic modeling data flow concept showing model operating system application functions for the passive sonar equation (Locklin and Webster, 1980).

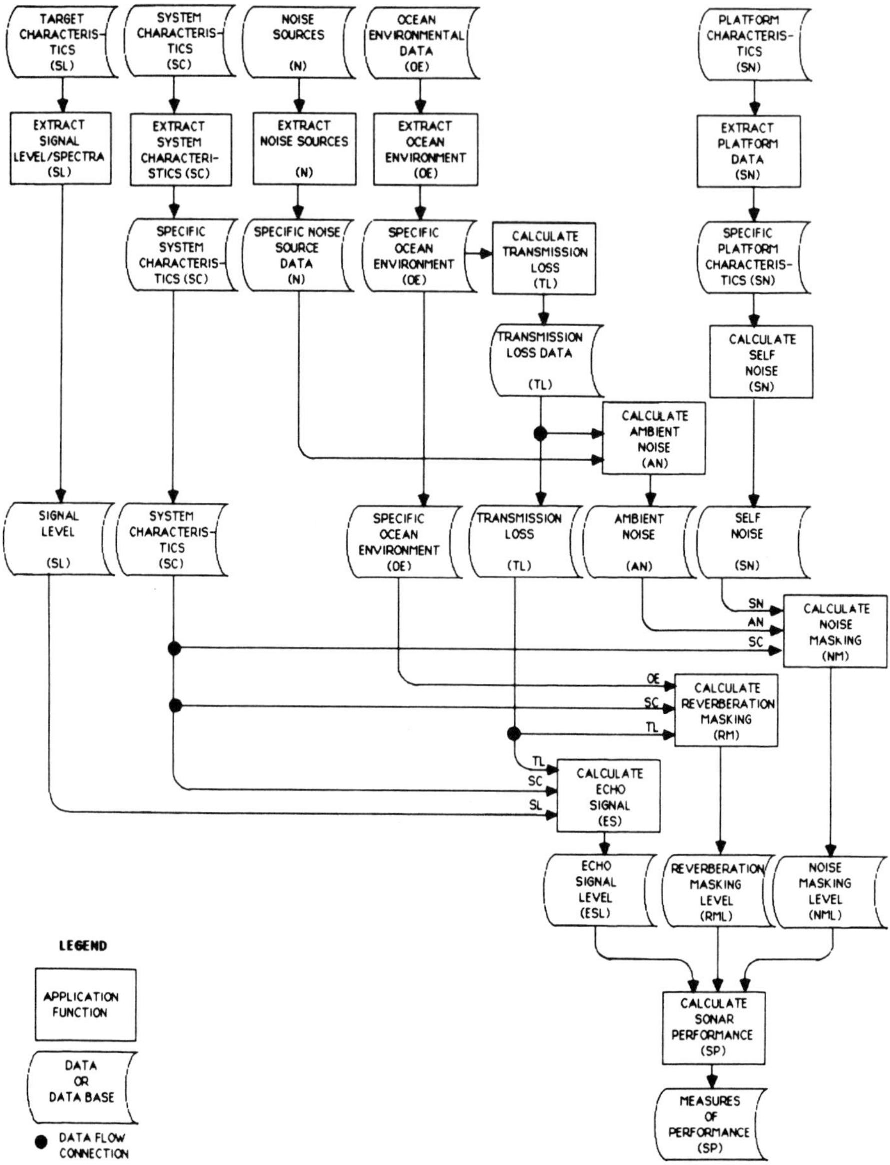

Fig. 10.11 Sample acoustic modeling data flow concept showing model operating system application functions for the active sonar equation (Locklin and Etter, 1988).

passive sonar equation (Locklin and Webster, 1980) and in Fig. 10.11 for the active sonar equation (Locklin and Etter, 1988). These particular functional decompositions are appropriate for incorporation into an executive architecture.

Although no decisionary logic options or looping controls are shown in these figures, a prerequisite processing sequence is nonetheless established. This sequence starts with an extraction process operating on databases which results in the generation of data sets specific to the analyst's application. In the case of passive sonars (Fig. 10.10), subsequent calculations use the extracted data to develop the data sets required by either the beam noise (BN) or beam signal (BS) functions which contribute to the final sonar performance (SP) function. This particular conceptual picture identifies three methods for generating BN, and two methods for BS, although a greater or lesser number of methods could be accommodated. This reflects the various levels of complexity possible in noise modeling and passive sonar system parameterization, respectively. In the case of active sonars (Fig. 10.11), the three key processes involve calculation of the echo signal level (ESL), the reverberation-masking level (RML), and the noise-masking level (NML) for a specific active sonar configuration. The signal level/spectra (SL) function for target characteristics describes the nature of the signal reflected from the target and considers the effects of target strength (TS), as specified in the active sonar equation. Furthermore, the beam pattern (BP) function is implicit in the system characteristics (SC) function. The final sonar performance (SP) function would then contain the algorithms necessary to generate signal excess and probability of detection outputs.

An important consideration for Figs 10.10 and 10.11 involves the usage of 'EXTRACT' and 'CALCULATE', and the symbols denoting application functions and data. 'EXTRACT' implies such operations as database access, retrieval, reformatting and output in a standard form, and includes the computational algorithms necessary to prepare the required data. 'CALCULATE' implies that the primary emphasis is placed on some computational model or algorithm, with all input and output handled by standardized file interface utilities. The application functions shown in Figs 10.10 and 10.11 are the highest level of modularization envisioned in these particular representations. A further decomposition may be possible depending upon the particular requirements of the MOS. Each identified function would be associated with families of software application programs and associated data storage facilities.

10.4.3 System usage

Model-operating systems, by design, facilitate and thus encourage the generation of multiple solutions for any given sonar performance problem.

Divergent answers obtained from different models can alert the user to potential problems in either the problem specification or in the underlying physics (i.e. violation of domains of applicability). Unexplained divergences can at least be honestly represented by 'performance envelopes' which likely bracket a realistic answer.

Many criteria drive the design of any particular MOS. Design decisions will likely entail tradeoffs between cost and ease of use, and between accuracy and understanding. Some degree of compromise will be involved. There is always the danger of 'creeping elegance' in which product embellishments not contained in the original design specifications add to the life-cycle costs and delay the implementation schedule. There is also a common desire expressed by potential MOS users to obtain 'all the information' generated by models. While noble in principle, in practice what is usually more important is how information is presented rather than how much information is presented. Creative visualization techniques which logically combine essential information on a very few charts are highly desirable. The generation of high data volumes in 3D modeling, for example, necessitates the development of data fusion and animation techniques which permit the rapid assimilation and comprehension of complex acoustic interactions with dynamically changing ocean environments. Finally, documentation of MOS design and operation is essential for system maintenance and operator training.

10.4.4 The generic sonar model – a specific example

The Generic Sonar Model (GSM) was designed to provide sonar system developers and technologists with a comprehensive modeling capability for evaluating the performance of sonar systems and for investigating the ocean environment in which these systems operate (Weinberg, 1981, 1985b).

The development approach used a modular design, adhered to a strict programming standard, and implemented off-the-shelf software when practical.

GSM is presently restricted to an ocean environment that is independent of range and time. In particular, the oceanic sound speed profile is layered, the surface and bottom are assumed to be horizontal reflecting boundaries, and sonar beam patterns are assumed constant over all azimuthal angles.

From a sonar system design standpoint, the significant capabilities of GSM include the modeling of passive signal excess versus range or frequency, active signal excess versus range, range and bearing errors versus range, LOFAR (Low-Frequency Analysis and Recording) diagrams, and autocorrelation and crosscorrelation functions.

The next lower level of investigation involves computation of the signal, noise, reverberation, target echo and detection threshold. These, in turn, require inputs of ambient noise, self-noise, scattering strengths, scattering spectra, sonar system responses, source level, directivity index, filter equalizers, and eigenrays.

A range-dependent (3D) version of GSM is being developed by Dr Henry Weinberg of the Naval Undersea Warfare Center in New London, Connecticut (USA). This new system is designated CASS (Comprehensive ASW Sonar Simulation). The propagation component of CASS is GRAB (Gaussian Ray Bundles), which models acoustic propagation in a range-dependent ocean environment.

10.5 DATA SOURCES AND AVAILABILITY

Databases with which to support sonar performance model development, evaluation and operation are available in a variety of formats. Many have automated data retrieval features that make them attractive for application to model-operating systems.

Available automated oceanographic and acoustic databases have been identified and summarized by Etter *et al.* (1984). Two categories of data banks were described: primary data banks and modified data banks.

Primary data banks contain original or modestly processed data (e.g. National Oceanographic Data Center, 1992). Modified data banks are distinguished by the fact that they modify or extrapolate data derived from primary data banks in order to satisfy operational requirements for compactness or for ease of handling (e.g. Naval Oceanographic Office, 1993).

Parameter summaries by data bank are provided in Table 10.3 for representative primary data banks, and in Table 10.4 for selected modified data banks. Not all of these databases are fully documented, and some may not be available for use outside the custodian's facilities. This summary does not claim to be exhaustive.

A subset of these collective databases has been designated by the US Navy as standards for use in sonar modeling. These databases are maintained in the Oceanographic and Atmospheric Master Library (OAML), which is chartered to provide fleet users with standard models and databases (Willis, 1992; Naval Oceanographic Office, 1993). The designated OAML databases are summarized in Table 10.5 together with brief descriptions of their contents, temporal resolution, and spatial coverage. Dissemination of these databases is strictly controlled by the US Navy.

The accessibility of many of these databases has been facilitated by the US Navy's technology transfer program. One such example is NEONS (Naval Environmental Operational Nowcasting System) developed by the Naval Research Laboratory (NRL). NEONS manages three types of

Table 10.3 Summary of primary data banks. Abbreviations and acronyms are defined in Appendix A

Data bank	*Custodian*	*Database parameters*			
FNMOC Data Files	FNMOC	*Marine meteorological data*			
		• Clouds	• Marine ship file	• Precipitation	• Sensible and evaporative heat flux
		• Ship surface observations	• Solar radiation	• Surface air temperature	• Surface marine winds
		• Surface pressure	• Total heat flux		
		Oceanographic data			
		• Combined sea height direction and period	• Insolation, reflected radiation	• MBT, XBT, STD, hydrocasts	• Operational bathythermograph observations
		• Potential mixed layer depth	• Primary layer depth	• Primary wave period and direction	• Sea-surface temperature
		• Sea-surface temperature anomaly	• Secondary wave period and direction	• Sensible and evaporative heat flux; total heat flux	• Significant wave height
		• Swell direction, period and height	• Temperature at 200, 400, 600, 800, 1000 and 1200 ft	• Temperature 100 and 200 ft below primary layer depth	• Temperature differences over 200 ft thick layers between the surface and 1200 ft
		• Temperature gradient 100 ft above and below primary layer depth	• Temperature at the top of the thermocline	• Thermocline gradient	• Wave direction, period and height
		• White caps			
Acoustic Data Files	SPAWAR	• Marine environmental/acoustic data			
AEAS Environmental-Acoustic Data Bank	NRL	• Ambient noise • Hydrocast data • Wave height	• Bathymetry • Shipping density	• Beam data • Sound speed profiles	• Current profiles • Transmission loss

DBDB (Digital Bathymetric Data Base)	NAVOCEANO	• Bathymetry			
DoD Bathymetric Data Library	DMA	• Deep scattering layer (DSL) depth	• Soundings		
NAWC Bottom Loss Data Files	NAWC	• Bottom loss data			
NAVDAB	NCCOSC	• Transmission loss	• Ambient noise	• Scattering and reverberation	
NCDC Raw Spectral Ocean Wave Model Data Files	NCDC	• Wave spectra			
NAVOCEANO Data Files	NAVOCEANO	• Ambient noise	• Bathymetry	• Bioluminescence	• Boring and fouling organisms
		• Bottom loss, transmission loss	• Climatology	• Computed sound speed, sigma-t, specific volume	• Conductivty
		• Core samples	• Dangerous marine animals	• Geomagnetics	• Gravity
		• Plankton	• Sediment samples	• Seismic profile	• Subsurface currents
		• Surface currents	• Temperature, salinity, oxygen versus depth	• Volume reverberation	• Ice type/thickness
NCDC Marine Climatic Data Files	NOAA/NCDC	• Air temperature	• Dew point temperature	• Low clouds	• Present weather
		• Pressure	• Sea-surface temperature	• Total clouds	• Visibility
		• Waves	• Wind		

(*Continued*)

Table 10.3 (*Continued*)

Data bank	*Custodian*	*Database parameters*			
NGDC Marine Geology and Geophysics Data Files	NOAA/NGDC	• Airborne magnetic survey (elements D, I, F)	• Digital hydrographic survey	• Marine bathymetry	• Marine gravity
		• Marine magnetic survey (total intensity F only)	• Megascopic core description, marine geological sample index, grain size analysis	• Seismic profiles	• Summary bathymetric and topographic files
NODC Data Files	NOAA/NODC	• Computed sound speed • Surface currents	• Marine chemical parameters • Temperature	• Marine meteorological parameters	• Salinity
Scripps Sediment Data Bank	SIO	• Manganese nodule analyses files • Source reference files	• Sea floor photograph file • Stations file	• Sediment analyses file	• Sediment description file
Volume Scattering Strength Data Bank	NRL	• Integrated volume scattering strength			

Table 10.4 Summary of modified data banks. Abbreviations and acronyms are defined in Appendix A

Data bank	*Custodian*	*Database parameters*		
ASEPS V4 Ocean Evironmental Data Set	NRL	• Bathymetry • Surface shipping density	• Bottom types • Watermass provinces	• Sound speed profiles
Auto Ocean/Auto Ships/HITS	NRL	• Bathymetry • Wave heights	• Sound speed profiles	• Shipping density
GDEM Data Files	NAVOCEANO	• Salinity model	• Sound speed profile model	• Temperature model
ICAPS Data Files	NAVOCEANO	• Acoustic bottom type • Temperature	• Bathymetry	• Salinity
OMS Data Files SIMAS Data Files	NUWC	• Biological scattering coefficients • Salinity • Wind speed	• Bottom loss provinces • Sound speed profiles	• Bottom scattering coefficients • Water depth
SYNBAPS	NRL	• Bathymetric soundings and contours		

Table 10.5 US Navy Standard (OAML) databases. Abbreviations and acronyms are defined in Appendix 1

Database	*Contents*	*Temporal resolution*	*Spatial coverage*
High-Frequency Bottom Loss (HFBL)	Analysis of MGS data ($f > 1$ kHz)	Fixed	Global
Low-Frequency Bottom Loss (LFBL)	Reflective and refractive characteristics of ocean bottom (derived from BLUG) ($f < 1$ kHz)	Fixed	Global
Generalized Digital Environmental Model (GDEM)	Synthetic temperature and salinity profiles based on analysis of ocean station data, XBT files and MOODS	Seasonal	Global
Historical Wind Speed (HWS)	Analysis of historical data: ocean surface wind speed and direction statistics	Monthly	Global
Ocean Floor Depth (DBDB)	Uncorrected echo sounder depth measurements	Fixed	Global
Volume Scattering Strength (VSS)	Integrated volume scattering strength and scattering strength profiles at selected frequencies	Seasonal	Global
Shipping Noise (SN)			
• Low Resolution	Omnidirectional and horizontally directional shipping noise and spectra	Seasonal	Northern hemisphere

(*Countinued*)

Table 10.5 *(Continued)*

Database	*Contents*	*Temporal resolution*	*Spatial coverage*
• High Resolution	Omnidirectional shipping noise and spectra	Seasonal	Northern hemisphere
• Historical Ice Edge (HIE)	MIZ noise	Monthly	Arctic
• Directional Shipping Noise (SN)	Estimated horizontal directionality of shipping noise	Seasonal	Northern hemisphere
Wind and Residual Noise (WRN)	Wind-generated noise spectra and regional (residual) noise sources	Monthly	Northern hemisphere
Historical Temporal Shipping (HITS)	Shipping densities (fishing vessels, merchants, tankers, large tankers, and super tankers)	Monthly	Global
ICECAP	Under-ice roughness and ridge frequency	Seasonal	Arctic

environmental data: observations, images and gridded data (Schramm, 1993). NRL designed NEONS for compatibility with computer-industry and international data-exchange standards. NEONS is being utilized to support civilian distribution of FNMOC data products via the Naval/ NOAA Oceanographic Data Distribution System (NODDS).

10.6 NUMERICAL MODEL SUMMARIES

Available active sonar performance models are summarized in Table 10.6. Also indicated three representative model-operating systems. Numbers in parentheses refer to appropriate documentation. Model documentation can vary from informal programming commentaries to journal articles to detailed technical reports containing a listing of the actual computer code. Abbreviations and acronyms are defined in Appendix A. This summary does not claim to be exhaustive.

Table 10.6 Summary of active sonar performance models and model-operating systems

Active sonar	*Model-operating systems*
Active RAYMODE[a] [1]	CASS [12]
ASPM [2]	GENERIC Sonar Model (GSM) [13]
CASTAR [3]	PRISM [14]
CONGRATS [4]	
LIRA [5]	
LORA [6]	
MOCASSIN [7]	
NISSM II [8]	
SEARAY [9]	
SONAR [10]	
SST [11]	

[a] US Navy standard model.

References to active sonar models and model-operating systems: [1] Smith (1982), Medeiros (1982b, 1985b), Naval Oceanographic Office (1991b, 1991d); [2] Berger *et al.* (1994); [3] Naval Oceanographic Office (1993) – secondary reference; [4] Weinberg (1969), Cohen and Einstein (1970), Cohen and Weinberg (1971); [5] Hoffman (1979); [6] Hoffman (1976); [7] Schneider (1990); [8] Weinberg (1973); [9] Tuovila (1989); [10] Marsh and Poynter (1969), Bertuccelli (1975); [11] Goddard (1989, 1994); [12] Weinberg (unpublished manuscript); [13] Weinberg (1982, 1985b); [14] Chaika *et al.* (1979).

The majority of the active sonar models listed in Table 10.6 are intended for use in ASW scenarios, although two of these models (CASTAR and SEARAY) were designed for use in mine-hunting scenarios.

Since sonar model operating systems normally utilize existing underwater acoustic models and standard oceanographic databases, these systems are unique only in the sense of the number and types of models and databases included, and the particular architecture employed.

11

Model evaluation

11.1 BACKGROUND

Model evaluation is defined as the systematic gathering and promulgation of information about models in order to determine model limitations and domains of applicability. Model evaluation should be viewed as a process rather than as a specific result.

The intent of this chapter is not to furnish a compendium of evaluation results but rather to describe the evaluation process itself. Such a description can provide useful insights into the benefits and shortcomings of model evaluation. The fundamental steps involved in the evaluation process will be described and only a select number of examples will be used to further illustrate the process.

At first glance, the process of model evaluation might appear to be very straightforward. When examined more closely, however, a number of factors are seen to complicate and frustrate the process.

To be credible, the evaluations must be based on statistically comprehensive comparisons with benchmark data. Such benchmark data can include field measurements, outputs from other models or closed-form analytical solutions. Each of these data sources has limitations. Field measurements, as we have seen, are quite limited with regard to temporal, spatial and spectral coverage; furthermore, these data sets are often classified on the basis of national security concerns. Outputs from other models are useful only when those models themselves have been properly evaluated. Analytical solutions can provide very accurate albeit idealistic comparisons for a relatively small number of useful problems.

Model development is an evolving process. Any particular version of a model quickly becomes outdated and the attendant evaluation results become obsolete. Therefore, comparisons of model accuracy are valid only for the particular model version tested. To ignore progress in model developments gained through previous evaluation experiences is to do an injustice to those models. Accordingly, only the most recent and authoritative evaluation results should be considered when assessing models.

Rivalries among different governmental, academic and industrial laboratories frequently add a political aspect to the model evaluation process. Many of the well-known models have gained their notoriety, at least in part, through the zealous advocacy of their developers and sponsors.

Clearly what is needed then is an autonomous clearing house to provide for both the unbiased evaluation of models and the timely promulgation of the results. An intermediate step in this direction has been taken by the naval sonar modeling community through the establishment of configuration management procedures. These procedures provide a disciplined method for controlling model changes and for distributing the models (and selected test cases) to qualified users. Configuration management comprises four major activities: (1) configuration identification and use of product baseline; (2) configuration change control; (3) status accounting and documentation of all product changes; and (4) reviews, audits and inspections to promote access to information and decision making throughout the software life cycle.

The US Chief of Naval Operations established the Oceanographic and Atmospheric Master Library (OAML) in 1984. OAML is chartered to provide fleet users with standard models and databases while ensuring consistent, commonly based environmental service products (Willis, 1992).

11.2 PAST EVALUATION EFFORTS

Comprehensive model evaluation efforts in the past have been very limited. Notable efforts have been sponsored by the US Navy but have only addressed propagation models. None of these efforts is presently active and their results are therefore only of limited academic interest. All of these efforts were successful in addressing the immediate concerns at the time, but were unsuccessful in the longer term due to the evolutionary nature of model developments and the lack of widely accepted evaluation criteria.

The Model Evaluation Program (MEP) was administered by the (now defunct) Acoustic Environmental Support Detachment (AESD) under ONR (Office of Naval Research) sponsorship. Emphasis was placed on model-to-model comparisons and the evaluation results were not widely disseminated.

A methodology for comparing acoustic propagation models against both other models and measured data was developed by the Panel on Sonar System Models (POSSM). POSSM was administered by the Naval Underwater Systems Center (NUSC) and was sponsored by the US Naval Sea Systems Command (Lauer and Sussman, 1976, 1979). Some of these results will be discussed later in this chapter. Among the many observations made by POSSM was the lack of documentation standards for acoustic models (Lauer, 1979).

In an effort to promote standardization, the US Navy established the Acoustic Model Evaluation Committee (AMEC) in 1978. The specific charter of AMEC was to develop a management structure and administrative procedures for the evaluation of acoustic models of propagation, noise and reverberation, although initial attention was restricted to propagation models. Specific evaluation factors included model accuracy, running time, core storage, ease of effecting slight program alterations and available ancillary information. The activities of AMEC culminated in evaluation reports for two propagation models: FACT and RAYMODE. As a result of AMEC's activities, RAYMODE was designated the (interim) Navy standard model for predicting acoustic propagation in the ocean. Subsequently, AMEC was disbanded.

More recent publications have proposed various theoretical and practical procedures for model evaluation (McGirr, 1979, 1980; Hawker, 1980; Pedersen and McGirr, 1982). Additional evaluation studies have been described for both propagation models (Hanna, 1976; Eller and Venne, 1981; Hanna and Rost, 1981) and reverberation models (Eller *et al.*, 1982). Practical test cases suitable for all types of propagation models have been reported by Spofford (1973b), Davis *et al.* (1982) and Chin–Bing *et al.* (1993b). In May, 1994, a Reverberation Workshop was conducted in Gulfport, Mississippi (USA), by the Office of Naval Research. The purpose of this workshop was to assess the fidelity of low-frequency (30–400 Hz) scattering and reverberation models. Six test cases comprising either field-measurement data sets or analytic problems with known reference solutions were used to establish the accuracy and state of current reverberation model development; uniform plot standards facilitated intercomparison of test case results. The results of this workshop are to be published in a proceedings volume.

11.3 ANALYTICAL BENCHMARK SOLUTIONS

In the absence of comprehensive experimental data, underwater acousticians have explored the use of analytical benchmark solutions to assess the quality of numerical models (Felsen, 1986; Jensen, 1986; Jensen and Ferla, 1988; Robertson, 1989). Such benchmarks emphasize idealized, but exactly solvable, problems. In the case of propagation models, two benchmark problems have been investigated in detail: upslope propagation in a wedge-shaped channel and propagation in a plane parallel waveguide with range-dependent sound speed profile (Felsen, 1987). These two problems are illustrated in Figs 11.1 andd 11.2, respectively (Jensen and Ferla, 1988, 1990).

BENCHMARK PROBLEMS ASSOCIATED WITH A WEDGE-SHAPED WAVEGUIDE

Parameters common to all three cases

wedge angle: $\theta_0 = \arctan 0.05 \simeq 2.86°$
frequency: $f = 25$ Hz
isovelocity sound speed in water column: $c_1 = 1500$ m/s
source depth: 100 m
source range from the wedge apex: 4 km
water depth at source position: 200 m
pressure-release surface

Case 1: *Pressure-release bottom.* This problem should be done for a line source parallel to the apex, i.e. 2-D geometry.

Case 2: *Penetrable bottom with zero loss.*
sound speed in the bottom: $c_2 = 1700$ m/s
density ratio: $\rho_2/\rho_1 = 1.5$
bottom attenuation: 0 dB/λ
This problem should be done for a point source in cylindrical geometry.

Case 3: *Penetrable lossy bottom.* As in Case 2 except with a bottom loss of 0.5 dB/λ.

Receiver depths

Case 1: 30 m
Cases 2 and 3: 30 m and 150 m

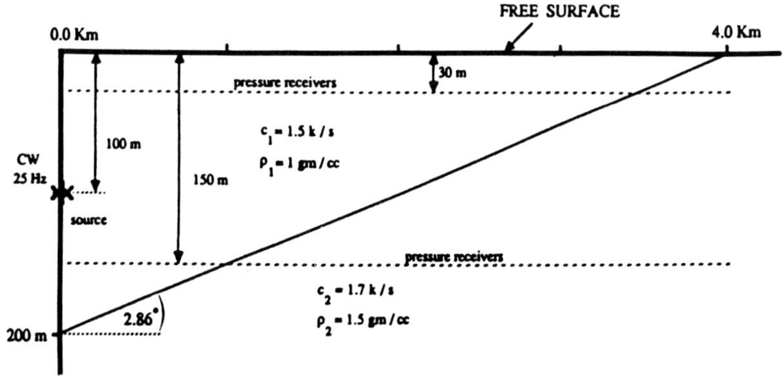

Fig. 11.1 Analytical benchmark problem: wedge geometry for cases 1, 2 and 3 (Jensen and Ferla, 1988).

BENCHMARK PROBLEMS ASSOCIATED WITH A PLANE-PARALLEL WAVEGUIDE

$$\frac{c(r,z)}{c_0} = \left\{1 + \left(\frac{\pi l_1}{L}\right)^2 e^{-2\pi r/L} + \left(\frac{2\pi l_2}{L}\right)^2 e^{-4\pi r/L} - \frac{2\pi l_1}{L}\left[1 - \left(\frac{2\pi l_2}{L}\right) e^{-2\pi r/L}\right] \cos\left(\frac{\pi z}{L}\right) e^{-\pi r/L} - \left(\frac{4\pi l_2}{L}\right) \cos\left(\frac{2\pi z}{L}\right) e^{-2\pi r/L}\right\}^{-1/2},$$

where $c_0 = 1500$ m/s is a reference sound speed, z is the depth below the pressure-release surface, L is the channel depth, and r is range from the source; l_1 and l_2 are parameters with the values $l_1 = 0.032L$ and $l_2 = 0.016L$.

Consider two cases:

Case 4: *Low-frequency, shallow water.*

$f = 25$ Hz	(frequency),
$L = 500$ m	(channel depth),
$R = 4$ km	(range coverage).

Case 5: *High-frequency, deep water.*

$f = 100$ Hz	(frequency),
$L = 3$ km	(channel depth),
$R = 20$ km	(range coverage).

Note: The field should be computed for both source and receiver at depth $z = \frac{1}{2}L$.

Fig. 11.2 Analytical benchmark problem: plane-parallel waveguide for cases 4 and 5 (Jensen and Ferla, 1988).

A special series of papers coordinated by the Acoustical Society of America described the results of a concerted effort to apply standard benchmark problems (with closed form solutions) to the evaluation of propagation models. The propagation models selected for evaluation were based on different formulations of the wave equation (Felsen, 1990; Jensen and Ferla, 1990; Buckingham and Tolstoy, 1990; Thomson, 1990; Thomson *et al.*, 1990; Stephen, 1990; Collins, 1990b; Westwood, 1990). The results of this evaluation provided several important insights (Jensen and Ferla, 1990):

1. one-way (versus two-way) wave equation solutions do not provide accurate results for propagation over sloping bottoms;
2. the COUPLE model provides a full-spectrum, two-way solution of the elliptic-wave equation based on stepwise-coupled normal modes; this code is ideally suited for providing benchmark results in general range-dependent ocean environments; however, the solution technique is computationally intensive and is therefore impractical at higher frequencies;
3. IFDPE provides a limited spectrum, one-way solution of the parabolic approximation to the full wave equation; the implicit finite-difference solution technique is computationally efficient, and accurate one-way results are provided for energy propagating within $\pm 40^\circ$ of the horizontal axis;
4. PAREQ provides a narrow-angle, one-way solution of the parabolic approximation to the full wave equation; the split-step Fourier solution technique is computationally efficient, and accurate one-way results are provided for energy propagating within $\pm 20^\circ$ of the horizontal axis.

The term 'pathological' is sometimes used in reference to those test cases that prove to be particularly troublesome to models. In the evaluation of propagation models, for example, ocean environments exhibiting double sound speed channels near the surface make especially good pathological test cases.

Ainslie and Harrison (1990) proposed the use of simple analytic algorithms as diagnostic tools for identifying model pathologies. They developed simple analytic expressions for computing the intensity contributions from standard propagation paths such as bottom-reflected paths, bottom-refracted paths, and Lloyd's Mirror interference. Ainslie and Harrison (1990) further demonstrated the utility of these analytical tools in assessing the performance of numerical models of acoustic pro-pagation. They accomplished this by combining the intensity contributions from the individual propagation paths appropriate for any particular ocean environment and sonar system geometry. This approach allowed a path-by-path analysis of numerical model performance and illuminated any modeling pathologies in an instructive fashion. These tools were subsequently organized into a highly interactive model, called INSIGHT, intended for passive sonar applications (Packman *et al.*, 1992). INSIGHT now includes the effects of surface-duct and convergence-zone paths and further permits the specification of arbitrary environmental-acoustic and sonar-system parameters. Related developments include a range-dependent version (called INSTANT). A companion version for active sonar applications (called INSACT) includes calculation of reverberation and echo levels. The ability to alter a variety of environmental or sonar system parameters interactively and then to visualize rapidly the resulting impacts on system performance makes the INSIGHT model especially attractive for instructional purposes.

11.4 QUANTITATIVE ACCURACY ASSESSMENTS

Quantitative accuracy assessments of propagation models can be accomplished according to two procedures: difference techniques and FOM (Figure-of-Merit) techniques (McGirr, 1979).

Difference techniques measure the distance between the model prediction and a standard (which can comprise field measurements or outputs from other models) in terms of dB differences at a given range, or over a set range interval (Fig. 11.3). These techniques are best suited to comparative model evaluations conducted in research environments.

Figure-of-Merit (FOM) techniques are essentially an inverted distance measure wherein a specified dB (or FOM) level is selected and the corresponding ranges (or sonar detection zones) are determined and compared against a standard. The FOM equation was previously discussed in Chapter 10. These techniques are best suited to sensitivity analyses which have as their objective a determination of the operational (versus scientific) impacts of model prediction errors. Such techniques place importance on the location of sonar detection zones relative to a standard. These techniques thus recognize the ultimate application of the model predictions to naval operations. A hypothetical example of this technique, used

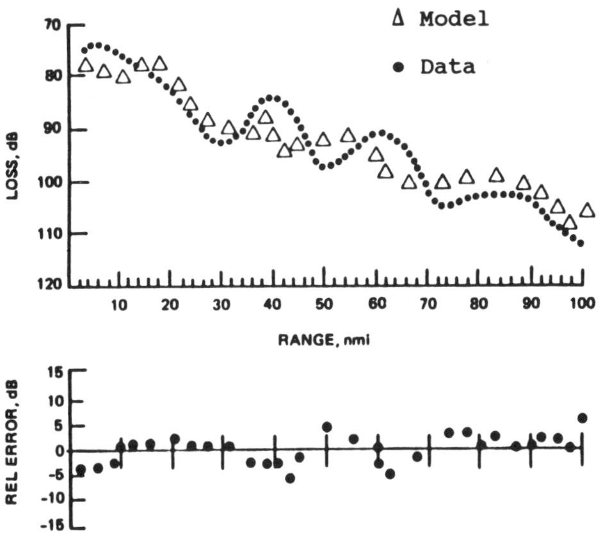

Fig. 11.3 Sample outputs from the AESD model evaluation program (adapted from McGirr, 1979).

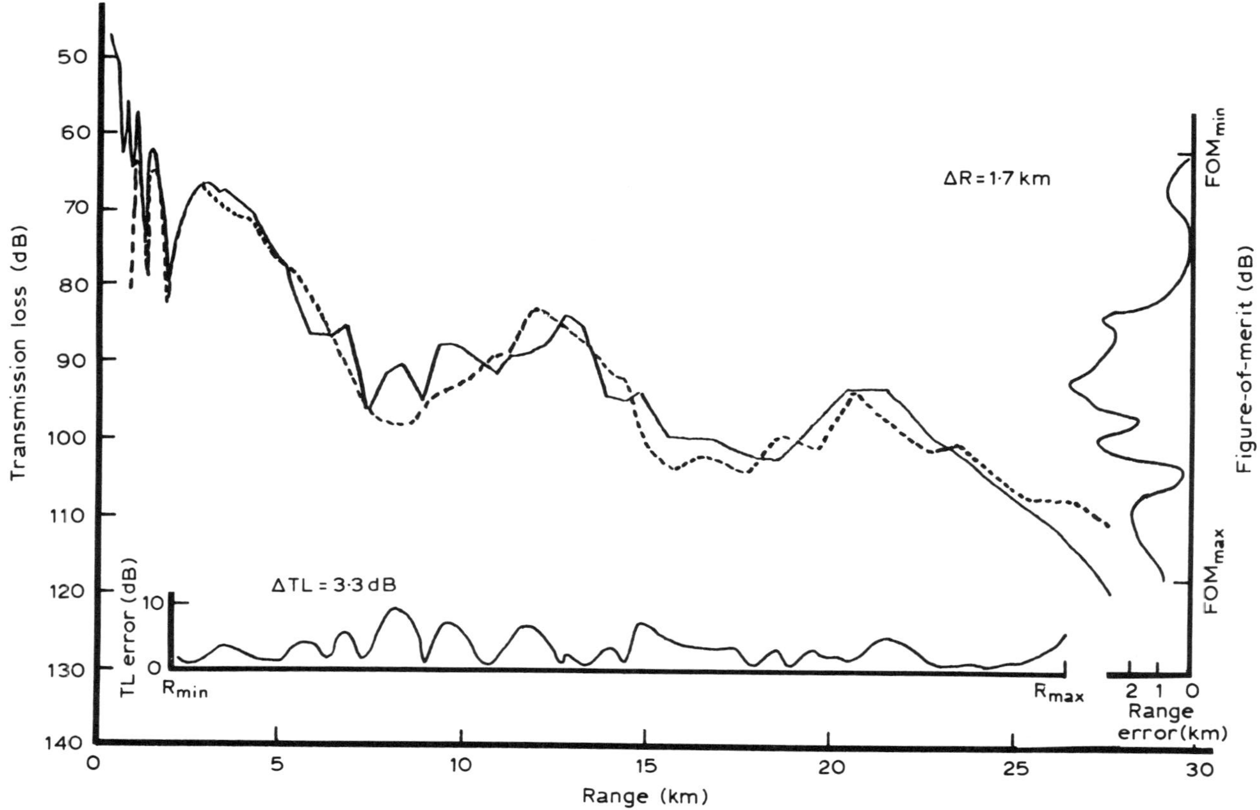

Fig. 11.4 Transmission loss comparison concept proposed by Leibiger (1977) which combines elements of both the difference technique and the FOM technique. This schematic only illustrates the concept and is not meant to be numerically exact.

in combination with the difference technique, is illustrated in Fig. 11.4 (Leibiger, 1977).

11.5 THE POSSM EXPERIENCE – A SPECIFIC EXAMPLE

The deliberations of the Panel on Sonar System Models (POSSM) are considered to be representative of past evaluation efforts, and aspects of these results will be discussed below.

The methodology developed and implemented by POSSM is summarized in Fig. 11.5 (DiNapoli and Deavenport, 1979; Lauer, 1979). Only propagation models were evaluated and one model was selected as the standard against which to compare the other models. More than ten different model versions were evaluated using an ocean environment described by a single sound speed profile and a flat bottom (i.e. range-independent).

All subject models were executed at four frequencies ranging from 35 to 200 Hz. The resulting transmission loss data were modestly smoothed and

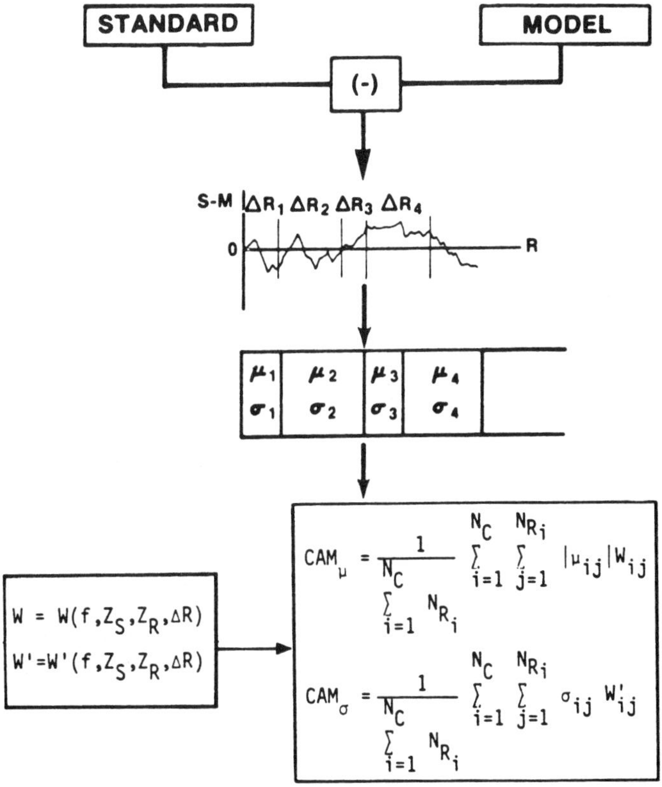

$$CAM_\mu = \frac{1}{\sum_{i=1}^{N_C} N_{R_i}} \sum_{i=1}^{N_C} \sum_{j=1}^{N_{R_i}} |\mu_{ij}| W_{ij}$$

$$CAM_\sigma = \frac{1}{\sum_{i=1}^{N_C} N_{R_i}} \sum_{i=1}^{N_C} \sum_{j=1}^{N_{R_i}} \sigma_{ij} W'_{ij}$$

Fig. 11.5 Summary of the POSSM model evaluation methodology (Lauer, 1979).

then compared with the standard model. These comparisons were conducted within variable range intervals specifically chosen to correspond to common sonar detection zones: direct path (DP), first and second bottom bounce (BB) regions, and first, second and third convergence zones (CZ). The means (μ) and standard deviations (σ) of the differences in transmission loss between the standard and the subject models were then calculated within each interval. A zero mean indicated that the standard model and the subject model produced transmission loss values that were identical when averaged over the interval. The resulting means and standard deviations were appropriately weighted and averaged to obtain the cumulative accuracy measures (CAM). Mathematically, the CAMs for the mean and the standard deviation are expressed as

$$\mathrm{CAM}_{\mu} = \frac{1}{\sum_{i=1}^{N_c} N_{\mathrm{R}_i}} \sum_{i=1}^{N_c} \sum_{j=1}^{N_{\mathrm{R}i}} |\mu_{ij}| W_{ij} \tag{11.1}$$

$$\mathrm{CAM}_{\sigma} = \frac{1}{\sum_{i=1}^{N_c} N_{\mathrm{R}_i}} \sum_{i=1}^{N_c} \sum_{j=1}^{N_{\mathrm{R}i}} \sigma_{ij} W'_{ij} \tag{11.2}$$

where i = case index
j = range interval index
N_c = total number of cases
N_{R_i} = total number of range intervals for the ith case
W_{ij} = weights in each case/range interval bin $(\Delta R)_{ij}$ to be applied to mean values of transmission loss differences
W'_{ij} = weights applied to standard deviations of differences

In the event that all W_{ij}'s and W_{ij}'s are chosen equal to unity, CAM_{μ} becomes the 'grand' mean taken over all cases and range intervals, and CAM_{σ} becomes the 'grand' standard deviation taken over all cases and range intervals.

The weighting functions (W_{ij}, W'_{ij}) are functions of the acoustic frequency (f), source depth (Z_s), receiver depth (Z_R) and range interval (ΔR).

For the purposes of discussions here, only four specific versions of propagation models have been selected. These models represent four of the five different modeling approaches described earlier in Chapter 4. The only modeling approach not investigated by POSSM was PE. At the time of POSSM's activities, PE models were not yet widely used; moreover, FFP models were considered to be the most accurate available. The four models and corresponding approaches represented are:

Model	*Approach*
FACT	Ray theory (with corrections)
RAYMODE	Multipath expansion (hybrid)
NLNM	Normal mode
FFP	Fast field

In the comparisons described by Lauer and Sussman (1979), and by DiNapoli and Deavenport (1979), the FFP model was used as the standard. Table 11.1 summarizes the computer program sizes and execution times for each model in the context of the standard problem selected by POSSM.

A quantitative comparison of model accuracies was reported by Lauer and Sussman (1979) based on the methodology summarized in Fig. 11.5. Only the four particular models identified above will be discussed here. Representative results from one test case are summarized in Table 11.2. It must be stressed that these results pertain only to the configuration of these models as they existed at the time of the tests and are only valid for the particular test case run. Different results might be obtained with more recent configurations of the same models or with other test scenarios and geometries, or even with other measurement standards or host computers.

The cumulative accuracy measures were also computed (Table 11.3). These results support the intuitive notion that model accuracy improves with model complexity and computational intensity. Accuracy alone, however, should not be the sole evaluation criterion. A broader view of the model evaluation process will be presented in the next section.

11.6 EVALUATION GUIDELINES

Model evaluation entails the systematic gathering and promulgation of information about models to determine model limitations and domains of applicability. There is no evaluation procedure appropriate for all models;

Table 11.1 Example program sizes and execution times for selected propagation models (adapted from Lauer and Sussman, 1979)

Model	*Program size (decimal words)*	*Execution time (s)*
FACT	21 855	2·5
RAYMODE	14 115	19·2
NLNM	53 000	30·6
FFP	51 572	373·0

Standard problem: single sound speed profile, flat bottom, low frequency ($\leqslant$200 Hz), UNIVAC 1108 computer.

Table 11.2 Means (μ) and standard deviations (σ) of the differences between the FFP propagation model and candidate model results for one particular case over direct path (DP), bottom bounce (BB) and convergence zone (CZ) regions (adapted from Lauer and Sussman, 1979)

Model	*Range interval*								*Range interval*					
	DP		*BB1*		*CZ*		*BB2*		*to 100 km*		*to 150 km*		*to 200 km*	
	μ	σ	μ	σ	μ	σ	μ	σ	μ	σ	μ	σ	μ	σ
FACT	–	–	1·4	3·6	0·9	2·2	0·1	3·0	−2·2	2·0	−2·0	1·3	−1·9	1·5
RAYMODE	–	–	1·3	3·7	0·5	2·0	−0·4	2·7	−0·3	1·1	−0·4	1·0	−0·8	1·6
NLNM	–	–	0·6	2·0	0·4	2·2	−0·8	1·9	0·7	1·9	−0·3	2·6	−1·0	2·5
FFP	–	–	–	–	–	–	–	–	–	–	–	–	–	–

Table 11.3 Cumulative accuracy measures (CAM): averages of means (μ) and standard deviations (σ) over all cases and range intervals with FFP propagation model results as standard (adapted from Lauer and Sussman, 1979)

Model	$\|\bar{\mu}\|$	$\bar{\sigma}$
FACT	1·3	2·5
RAYMODE	0·9	2·4
NLNM	0·6	2·3
FFP	–	–

the procedures must be tailored on the basis of model structure, documentation and other information available to the evaluators.

Criteria for model evaluation can be segregated into five basic categories (US General Accounting Office, 1979): documentation, verification, validity, maintainability and usability.

11.6.1 Documentation

Documentation of computer models is important for two reasons: to ensure that the model is thoroughly understood and can be operated and maintained in the present as well as in the future, and to facilitate independent evaluation of the model, particularly by someone other than the model developer or initial user.

Elements which should be included in model documentation are (Gass, 1979):

1. a precise statement of what the model is supposed to do;
2. the mathematical and logical definitions, assumptions and formulation of the problem being modeled;
3. a complete set of current input and output, and test cases that have been run;
4. a complete set of flow charts of the computer program;
5. a set of operating instructions for the computer operator;
6. an explanation of the various options available in using the model;
7. the computer program itself (listing), with comments about various operations in the program.

Further guidelines for documentation are provided later in this chapter.

11.6.2 Verification

Varification entails an examination of the model to ensure that the computer program accurately describes the model and that the program runs as expected. In order to do this, the following factors must be examined:

1. consistency of mathematical and logical relationships;
2. accuracy of intermediate numerical results;
3. inclusion of important variables and relationships;
4. proper mechanization and debugging of the program.

Computer hardware selections will sometimes impact model accuracy due to word length and double-precision considerations, among others. Thus, the same model implemented on two different computer systems may produce significantly different results. Related problems concern artifacts which are false features that arise from some quirk of the computer, and which disappear when the software is written differently.

11.6.3 Validity

Validation and critical assessment are required of all theoretical conjectures, hypotheses and models. One of the most difficult but important tasks in model construction is the specification of its limitations: what are its limits, and in what way is it an approximation? The consensus principle in science implies that the evaluation of models must be open, and cannot be accepted on the authority of the model developer alone. The model should not contain adjustable parameters or any hidden variables that have to be invoked to explain discrepancies between theory and experiment. The theoretical properties of the model should be sharply defined, and derived with sufficient mathematical rigor to be compared objectively with the observed phenomena. Elementary errors and misunderstandings will be detected by the independent repetition of experiments and by comparisons of calculations with experimental data, or by theoretical criticism (Ziman, 1978). Even when models are wrong, they can assist in structuring discussions.

The category of validity comprises three factors: theoretical validity, data validity and operational validity.

Theoretical validity entails review of the physics underlying the model and the major stated and implied assumptions. The applicability and restrictiveness of these assumptions must also be examined. In addition, the internal logic of the model should be reviewed.

For empirically based models, data validity is concerned with the accuracy and completeness of the original data and the manner in which the empirical model deals with the transformation of the original data.

Operational validity is concerned with assessing the impact of model errors (i.e. divergences between model predictions and reality) on decision processes. This aspect of model evaluation addresses accuracy. When evaluating model accuracy, it is important to examine the error budget. Errors in model predictions (e_p) are assumed to be the sum of two independent, random variables:

$$e_p = e_d + e_m \tag{11.3}$$

where e_d represents errors related to model input data and e_m represents model errors.

11.6.4 Maintainability

This category includes consideration of the ease of incorporating new data and formulas as well as provisions for reviewing the accuracy of the model as more experimental data become available. A training program must be formalized to ensure that the operators understand how the model should be used and also to make revisions known to the systems personnel.

11.6.5 Usability

Usability addresses the appropriateness of the model for the intended applications. In essence, the model should satisfy the user's specific requirements. This, of course, requires that users articulate their needs very precisely.

It is instructive to distinguish between research and operational models when discussing usability. Specifically, research models are intended to address a wide variety of often ill-posed scientific questions. In order to be responsive to such ambiguous issues, the research models are structured to allow (and often require) a high degree of operator intervention during execution. This permits the researcher the opportunity to adjust parameters as the problem solution evolves. Alternatively, operational models are structured to minimize (if not eliminate) the need for operator intervention; indeed, such intervention is viewed as a nuisance to the operator. The parallel between executive-versus-bundled system architectures and research-versus-operational models is valid (refer to section 10.4.1).

Model usability is heavily influenced by the model's inherent domains of applicability. Factors such as frequency coverage and problem geometry are implicit in these domains. For example, the use of normal mode models in the calculation of high-frequency bistatic reverberation is not practical at present due to excessive computation times. Model output options are also important. For example, transmission loss values in the range–depth plane are more easily generated by wave-theoretical propagation models, while arrival structure information is more easily generated by ray-theoretical propagation models.

11.7 DOCUMENTATION STANDARDS

Documentation standards for computer models have been reviewed by Gass (1979). Both the Department of Defense (DOD) and the National Institute of Standards and Technology (formerly the National Bureau of Standards, or NBS) have issued formal guidelines:

- DOD-STD-7935A – 'Department of Defense standard: Automated data systems (ADS) documentation'
- DOD-STD-2167A – 'Military standard: Defense system software development'
- FIPS PUB 38 – 'Guidelines for documentation of computer programs and automated data systems'

According to the FIPS (Federal Information Processing Standards), model documentation is keyed to the various phases and stages of the software life cycle (Table 11.4). The major phases are:

Initiation phase – During this phase, the objectives and general definition of the requirements for the software are established. Feasibility studies, cost–benefit analyses and the documentation prepared within this phase are determined by agency procedures and practices.

Development phase – During this phase, the requirements for the software are determined, and software is then defined, specified, programmed and tested. Ten major documents are prepared in this phase to provide an adequate record of the technical information developed (Table 11.4).

Operation phase – During this phase, the software is maintained, evaluated and changed as additional requirements are identified. The documentation is maintained and updated accordingly.

The development phase of the software life cycle is further subdivided into four main stages, as follows:

Definition stage – when the requirements for the software and documentation are determined.

Design stage – when the design alternatives, specific requirements and functions to be performed are analyzed and a design is specified.

Programming stage – when the software is coded and debugged.

Test stage – when the software is tested and related documentation is reviewed. The software and documentation are then evaluated in terms of readiness for implementation.

Table 11.4 Software life cycle and documentation types according to the Federal Information Processing Standards (Gass, 1979)

Initiation phase	*Development phase*				*Operation phase*
	Definition stage	*Design stage*	*Programming stage*	*Test stage*	
	Functional requirements document	System/ subsystem specification	User's manual		
		Program specification	Operations manual		
	Data requirements document	Database specification	Program maintenance manual	Test plan	
				Test analysis report	

Functional requirements document. The purpose of the functional requirements document is to provide a basis for the mutual understanding between users and designers of the initial definition of the software, including the requirements, operating environment and development plan.

Data requirements document. The purpose of the data requirements document is to provide, during the definition stage of software development, a data description and technical information about data collection requirements.

System/subsystem specification. The purpose of the system/subsystem specification is to specify for analysts and programmers the requirements, operating environment, design characteristics and program specifications (if desired) for a system or subsystem.

Program specification. The purpose of the program specification is to specify for programmers the requirements, operating environment and design characteristics of a computer program.

Database specification. The purpose of the database specification is to specify the identification, logical characteristics and physical characteristics of a particular database.

User's manual. The purpose of the user's manual is to sufficiently describe the functions performed by the software in non-ADP terminology, such that the user organization can determine its applicability and when and how to use it. It should serve as a reference document for preparation of input data and parameters, and for interpretation of results.

Operations manual. The purpose of the operations manual is to provide computer operation personnel with a description of the software and of the operational environment so that the software can be run.

Program maintenance manual. The purpose of the program maintenance manual is to provide the maintenance programmer with the information necessary to understand the programs, their operating environment, and their maintenance procedures.

Test plan. The purpose of the test plan is to provide a plan for the testing of software; detailed specifications, descriptions and procedures for all tests; and test data reduction and evaluation criteria.

Test analysis report. The purpose of the test analysis report is to document the test analysis results and findings, present the demonstrated capabilities and deficiencies for review, and provide a basis for preparing a statement of software readiness for implementation.

Formal documentation guidelines are frequently amended or even superseded by newer guedelines. Care should be taken to consult the latest governing instructions. Efforts are presently underway to consolidate DOD-STD-7935A and DOD-STD-2167A into one MIL-STD-SDD (tentatively designated 498). This consolidation is intended, in part, to implement governing ISO/IEC standards (e.g. DIS 12207 – 'software life-cycle processes').

Appendix A: Abbreviations and acronyms

AAIW	Antarctic Intermediate Water
ACCURAY	Range-Dependent Shallow-Water Ray Propagation Model
ADCP	Acoustic Doppler Current Profiler
ADIAB	Adiabatic Normal Mode Model
ADONIS	Acoustic Daylight Ocean Noise Imaging System
ADP	Automated Data Processing
ADS	Automated Data Systems
AEAS	Advanced Environmental Acoustic Support
AESD	Acoustic Environmental Support Detachment
AIAA	American Institute of Aeronautics and Astronautics
AMBENT	Ambient Noise Model
AMEC	Acoustic Model Evaluation Committee
AMODE	Acoustic Mid-Ocean Dynamics Experiment
AMOS	Acoustic, Meteorological and Oceanographic Survey
AN	Ambient Noise
ANDES	Ambient Noise Directionality Estimation System
AP	Arbitrary Profile Model
APL	Applied Physics Laboratory
APP	Acoustic Performance Prediction
ARA	Acoustic Rainfall Analysis
ARL	Applied Research Laboratory
ARSRP	Acoustic Reverberation Special Research Project
ASEPS	Automated Signal Excess Prediction System

ASERT	ASTRAL System for Estimation of Radial Transmission
ASPM	Acoustic System Performance Model
ASTRAL	ASEPS Transmission Loss Model
ASW	Anti-Submarine Warfare
ATOC	Acoustic Thermometry of Ocean Climate
AUAMP	Advanced Underwater Acoustic Modeling Program
AUTO OCEAN	Automated Oceanographic Database
AUTO SHIPS	Automated Shipping Density Database
AUV	Autonomous Underwater Vehicle
AVHRR	Advanced Very High Resolution Radiometer
BAM	Bistatic Acoustic Model
BB	Bottom Bounce
BBN	Bolt, Beranek and Newman, Inc.
BEAMPL	Beam Program Library
BEM	Boundary Element Method
BISAPP	Bistatic Acoustic Performance Prediction
BISSM	Bistatic Scattering Strength Model
BLUG	Bottom Loss Upgrade
BN	Beam Noise
BP	Beam Pattern
BS	Beam Signal
BT	Bathythermograph
BTL	Bell Telephone Laboratories
CAM	Cumulative Accuracy Measures
CAPARAY	Broadband Eigenray Model
CASP	Canadian Atlantic Storms Program
CASS	Comprehensive ASW Sonar Simulation
CASTAR	Computer Aided Sonar Tactical Recommendations
CENTRO	Normal Mode Model with Shear Wave Effects
CFM	Calculation Frequency Method
CMM3D	Three-Demensional Coupled Normal Mode Model
CNOISE	Noise Model
COMODE	Normal Mode Model
CONGRATS	Continuous Gradient Ray-Tracing System
COUPLE	Coupled Mode Model
C-SALT	Caribbean Sheets and Layers Transects Program
CST	Critical Sea Test
CTD	Conductivity, Temperature, Depth Sensor
CW	Continuous Wave

CZ	Convergence Zone
DANES	Directional Ambient Noise Estimation System
dB	Decibel
DBDB	Digital Bathymetric Database
DI	Directivity Index
DIS	Draft International Standard
DMA	Defense Mapping Agency
DOD	Department of Defense
DODGE	Normal Mode Model
DOP	Doppler Content of Reverberation Model
DOSL	Diurnal Ocean Surface Layer Model
DP	Direct Path
DREA	Defence Research Establishment Atlantic (Canada)
DREP	Defence Research Establishment Pacific (Canada)
DSBN	Discrete Shipping Beam Noise Model
DSC	Deep Sound Channel
DSL	Deep Scattering Layer
DT	Detection Threshold
DUNES	Directional Underwater Noise Estimates Model
EIGEN/REVERB	Eigenray/Reverberation Model
EOF	Empirical Orthogonal Function
ERL	Environmental Research Laboratory
ESL	Echo Signal Level
EVA	Environmental Acoustics
FACT	Fast Asymptotic Coherent Transmission Model
FACTEX	FACT Extended to Range-Dependent Environments
FAME	Fast Multipath Expansion Model
FANM	Fast Ambient Noise Model
FE	Finite Element
FEPE	Finite Element PE Model
FEPES	Finite Element PE Model with Shear Wave Effects
FFP	Fast-Field Program
FFRAME	Finite-Element, Full-Wave Range-Dependent, Acoustic Marching Element Model
FFT	Fast Fourier Transform
FIPS	Federal Information Processing Standards
FLIP	Floating Instrument Platform

FLIRT	Fast Linear Intermediate Range Transmission Model
FNMOC	Fleet Numerical Meteorology and Oceanography Center (formerly FNOC)
FNMSS	Fast Normal Mode with Surface Scattering Integrals Model
FNOC	Fleet Numerical Oceanography Center (now FNMOC)
FOAM	Finite Element Ocean Acoustic Model
FOM	Figure of Merit
FOR3D	Finite Difference Methods, Ordinary Differential Equations, and Rational Function Approximations to Solve the LSS 3D Wave Equation
GAMARAY	Broadband Ray Propagation Model
GDEM	Generalized Digital Environmental Model
GEOSECS	Geochemical Ocean Section Study
GRAB	Gaussian Ray Bundles (Propagation Model in CASS)
GRASS	Germinating Ray-Acoustics Simulation System
GSM	Generic Sonar Model
HAPE	High-Angle PE Model
HARPO	Hamiltonian Acoustic Raytracing Program – Ocean
HFBL	High-Frequency Bottom Loss
HIE	Historical Ice Edge
HIFT	Heard Island Feasibility Test
HITS	Historical Temporal Shipping Database
HOOM	Harvard Open Ocean Model
HWS	Historical Wind Speed
HYPER	Hybrid PE–Ray Model
IAPSO	International Association for the Physical Sciences of the Ocean
ICAPS	Integrated Command ASW Prediction System
ICECAP	Interactive Computer Environment for Comprehensive Arctic Predictions
ICERAY	Under-Ice Ray Propagation Model
IEC	International Electrotechnical Commission
IEEE	Institute of Electrical and Electronics Engineers
IFD	Implicit Finite Difference Model
Im	Imaginary

IMACS	International Association for Mathematics and Computers in Simulation
INSACT	Analytical Active/Passive Sonar Model
INSIGHT	Analytical Range-Independent Propagation Model
INSTANT	Analytical Range-Dependent Propagation Model
ISO	International Organization for Standardization
ISVRFEM	Institute of Sound and Vibration Research Finite Element Model
IUSS	Integrated Undersea Surveillance System
KRAKEN	Adiabatic/Coupled Normal Mode Model
LFBL	Low-Frequency Bottom Loss (formerly BLUG)
LIRA	Low-Frequency Intermediate-Surveillance-Range Active Model
LOFAR	Low-Frequency Analysis and Recording
LOGPE	PE Model Using Logarithmic Expression for Index of Refraction
LORA	Long Range Active Model
LRAPP	Long Range Acoustic Propagation Project
LSS	Lee–Saad–Schultz Method
MaCh1	Broadband, Range-Dependent Propagation Model
MAM	Monostatic Acoustic Model
MATE	Mid-Ocean Acoustic Transmission Experiment
MBT	Mechanical Bathythermograph
MEDUSA	Propagation Model
MEP	Model Evaluation Program
MGS	Marine Geophysical Survey
MIL	Military
MIPE	(University of) Miami PE Model
MIT	Massachusetts Institute of Technology
MIW	Mediterranean Intermediate Water
MIZ	Marginal Ice Zone
MLD	Mixed Layer Depth
MOATL	Modal Acoustic Transmission Loss Model
MOCASSIN	Monte Carlo Schall–Strahlen Intensitäten; Monte Carlo Sound Ray Intensities
MOODS	Master Oceanographic Observation Data Set
MOREPE	Modified Refraction PE Model
MOS	Model-Operating System
MOST	Mobile Sonar Technology

MPC	Multiple Profile Configuration Model
MPP	Multiple Profile Program
MSPFFP	Multiple Scattering Pulse FFP Model
MST	Moving Ship Tomography
MTS	Marine Technology Society
MULE	Multilayer Expansion Model
N	Noise Sources
NABTAM	Narrow Beam Towed Array Model
NADC	Naval Air Development Center (now NAWC)
NAVDAB	Navy Ocean Environmental Acoustic Data Bank
NAVOCEANO	Naval Oceanographic Office
NAWC	Naval Air Warfare Center
NBS	National Bureau of Standards (now National Institute of Standards and Technology)
NCCOSC	Naval Command Control and Ocean Surveillance Center
NCDC	National Climatic Data Center
NDRC	National Defense Research Committee
NEMESIS/PLMODE	Propagation Model
NEONS	Naval Environmental Operational Nowcasting System
NEPBR	Numerable Energy Paths by RAYMODE
NGDC	National Geophysical Data Center
NISSM	Navy Interim Surface Ship Model
NL	Noise Level
NLNM	N-Layer Normal Mode Model
NML	Noise-Masking Level
NOAA	National Oceanic and Atmospheric Administration
NOARL	Naval Oceanographic and Atmospheric Research Laboratory (now part of NRL)
NODC	National Oceanographic Data Center
NODDS	Navy/NOAA Oceanographic Data Distribution System
NORDA	Naval Ocean Research and Development Activity
NORM 2L	Normal Mode 2-Leyer Model
NORMOD 3	Normal Mode Model
NOSC	Naval Ocean Systems Center (now NCCOSC)
NPE	Nonlinear Progressive-Wave Equation
NRL	Naval Research Laboratory
NUSC	Naval Underwater Systems Center (now NUWC)

NUWC	Naval Undersea Warfare Center
OAML	Oceanographic and Atmospheric Master Labrary
OASES	SAFARI Extended
ODE	Ordinary Differential Equation
OE	Ocean Environment
OMS	Optimum Mode Selection System
ONR	Office of Naval Research
ORBS	Ocean Refraction and Bathymetric Scattering Model
OS2IFD	Implicit Finite Difference PE Model
OTIS	Optimum Thermal Interpolation System
PAREQ	Parabolic Equation Model
PE	Parabolic Equation
PEREV	PE Reverberation Model
PERUSE	PE Rough Surface
PESOGEN	PE Solution Generator
PLRAY	Ray Propagation Loss Model
POSSM	Panel on Sonar System Models
PRISM	Program for Integrated Sonar Modeling
PROLOS	Propagation Loss Model
PROTEUS	Propagation Model
psu	Practical Salinity Unit
PWE	Progressive Wave Equation
RANDI	Research Ambient Noise Directionality Model
RANGER	Range-Invariant Ray Model
RAP	Reliable Acoustic Path
RASP	Range-Dependent Active System Performance Model
RAYMODE	Ray/Normal Mode Model
RAYWAVE	Ray/Wave Propagation Model
RBR	Refracted-Bottom-Reflected
RD	Recognition Differential; Range-Dependent
RDFFP	Range-Dependent FFP Model
Re	Real
re	Relative to
REVGEN	Reverberation Generator
REVMOD	Reverberation Spectrum Model
REVSIM	Reverberation Simulation
RI	Range-Independent
RL	Reverberation Level
RML	Reverberation-Masking Level
rms	Root Mean Square
RP-70	Ray-Tracing Program – 1970

RR	Refracted-Refracted
RSR	Refracted-Surface-Reflected
RSRBR	Refracted-Surface-Reflected-Bottom-Reflected
RUMBLE	Reverberation Underwater Model, Bistatic Level Estimation
SACLANTCEN	Supreme Allied Commander, Atlantic (SACLANT) Undersea Research Centre
SAFARI	Seismo-Acoustic Fast-Field Algorithm for Range-Independent Environments
SAFE	Seismo-Acoustic Finite Element Model
SAFRAN	Hybrid BEM + WI Model
SALT	Sound Angle, Level, and Travel Time
SC	System Chracteristics
SD	Surface Duct; Source Depth
SDD	Software Design Document; Software Development and Documentation
SEARAY	Mine-Hunting Sonar Performance Model
SHALFACT	Shallow-Water FACT
SHEAR2	Normal Mode Model With Shear Wave Effects
SIAM	Simulated Ambient Noise Model; Society for Industrial and Applied Mathematics
SIMAS	Sonar *In-Situ* Mode Assessment System
SIO	Scripps Institution of Oceanography
SL	Source Level; Signal Level
SLD	Sonic Layer Depth
SN	Self-Noise; Shipping Noise
SNAP	SACLANTCEN Normal-Mode Acoustic Propagation Model
SNR	Signal-to-Noise Ratio
SOFAR	Sound Fixing and Ranging
SONAR	Sound Navigation and Ranging
SP	Sonar Performance
SPAWAR	US Navy Space and Naval Warfare Systems Command
Spectral PE	PE Model with 3D Backscattered Energy
SRS	Software Requirements Specification
SST	Sonar Simulation Toolset; Sea-Surface Temperature
STD	Salinity, Temperature, Depth Sensor; Software Test Description; Standard
SYNBAPS	Synthetic Bathymetric Profiling System
TA	Transmission Anomaly
TAMU	Texas A & M University

TDA	Tactical Decision Aid
TDPA	Time-Domain Parabolic Approximation Model
TDPE	Time-Domain Parabolic Equation
TENAR	Target Echo, Noise and Reverberation
TESS	Tactical Environmental Support System
TKE	Turbulent Kinetic Energy
TL	Transmission Loss
TLM	Transmission-Line Matrix Modeling
TOPS	Thermodynamical Ocean Prediction System
TOTS	Tactical Ocean Thermal Structure
TRIMAIN	Propagation Model
TS	Target Strength
Two-Way PE	PE Model with Backscattered Energy
ULETA	Propagation Model
ULF	Ultralow Frequency
UMPE	University of Miami PE Model
UNIMOD	Propagation Model
USI	Underwater Systems, Inc.
VLA	Vertical Line Array
VLF	Very Low Frequency
VSS	Volume Scattering Strength
WEDGE	Range-Dependent Shallow-Water Normal Mode Model
WI	Wavenumber Integration
WKB	Wentzel, Kramers and Brillouin
WKBJ	Wentzel, Kramers, Brillouin and Jeffreys
WOTAN	Weather Observation Through Ambient Noise
WPL	Wave Propagation Laboratory
WRAP	Wide-Area Rapid Acoustic Prediction
WRN	Wind and Residual Noise
XBT	Expendable Bathythermograph
XSV	Expendable Sound Velocimeter

Appendix B: Glossary of terms

Absorption – loss of acoustic energy due to conversion to heat.

Acoustic daylight – imaging of underwater objects using the ambient noise field.

Acoustic impedance – quantitatively equal to the product of density and sound speed of the medium.

Acoustic tomography – inverse technique which uses acousic signals to sample the interior of a water body.

Adiabatic process – thermodynamic change in which there is no transfer of heat or mass across the system boundaries.

Advection – movement of oceanic properties by currents.

Ambient noise – background level of unwanted sound in the sea, exclusive of occasional noise sources.

Ambient noise models – mathematical models that predict the levels and directionality of noise in the ocean due to surface weather and shipping sources.

Analog models – controlled acoustic experiments in water tanks that utilize appropriate oceanic scaling factors.

Analytic models – same as physical models.

Anticyclonic – gyral pattern of motion induced by the rotation of the Earth which is clockwise in the northern hemisphere but counter-clockwise in the southern hemisphere.

Archipelagic apron – a gentle slope with a smooth surface on the sea floor.

Arrival structure – relationship between vertical angle of arrival of acoustic rays and horizontal distance traveled by rays.

Attenuation – loss of acoustic energy due to combined effects of absorption and scattering.

Backscattering – scattering of sound in the direction of the source.

Bank – an elevation of the sea floor located on a shelf.

Baroclinic – state of stratification in the ocean in which surfaces of constant pressure intersect surfaces of constant density.

Barotropic – state of stratification in the ocean in which surfaces of constant pressure and surfaces of constant density are parallel.

Basic acoustic models – category of models containing underwater acoustic propagation, noise and reverberation models.

Basin – a depression of variable extent, generally in a circular or oval form.

Bathythermograph – instrument used to measure water temperature versus depth.

Beam displacement – lateral displacement of an acoustic beam, of finite width, undergoing reflection at a water–sediment interface.

Beam noise statistics models – mathematical models that predict the levels and directionality of low-frequency shipping noise in the ocean for application to large-aperture, narrow-beam passive sonar systems.

Bistatic – geometry in which the acoustic source and receiver are not at the same position.

Borderland – a region adjacent to a continent that is highly irregular with depths in excess of those typical of a shelf.

Bottom bounce – ray paths which have reflected off the sea floor.

Bottom limited – ocean environment characterized by a water depth and sound speed profile that will not support long-range refracted paths.

Bottom loss – reflection loss at the sea floor.

Boundary conditions – constraints imposed on possible solutions of the wave equation by adjacent surfaces.

Buoyancy frequency – frequency at which a water parcel displaced from equilibrium will oscillate.

Canonical profile – standard sound speed profile applicable to ocean areas having a deep sound channel; the profile normally assumes an exponential form in the region of the sound channel axis.

Canyon – a narrow, deep depression with steep slopes.

Caustic – envelope formed by set of tangential rays.

Cell-scattering models – mathematical models of reverberation based on the assumption that the scatterers are uniformly distributed throughout the ocean.

Chaotic – a deterministic process for which certain ranges of parameters are unpredictable.

Composite roughness model – model that partitions treatment of boundary scattering into two regimes according to large-scale and small-scale surface roughness.

Conjugate depth – depth below the sound channel axis at which the value of sound speed equals that at a shallower depth.

Continental margin – a zone separating the continent from the deeper sea bottom, generally consisting of the rise, slope and shelf.

Continental rise – a gentle slope rising toward the foot of the continental slope.

Continental shelf – zone adjacent to a continent or island from the waterline to the depth at which there is usually a marked increase of slope to greater depth (shelf break).

Continental slope – zone between the continental shelf and the continental rise.

Convergence zone – ray paths formed by long-range refraction in deep water environments having both a critical depth and sufficient depth excess.

Cordillera – an entire underwater mountain system including all the subordinate ranges, interior plateaus and basins.

Coriolis force – an apparent force acting on moving particles resulting from the Earth's rotation; the force is proportional to the speed and latitude of the particle, and results in a deflection to the right of motion in the northern hemisphere, or to the left of motion in the southern hemisphere.

Critical angle – angle in the vertical plane separating the angular region of total reflection from that of partial reflection and partial transmission at a boundary.

Critical depth – depth at which the value of sound speed equals that of the near-surface maximum sound speed.

Cusped caustic – intersection of two (smooth) caustics.

Cyclonic – gyral pattern of motion induced by the rotation of the Earth which is counter-clockwise in the northern hemisphere but clockwise in the southern hemisphere.

Cylindrical spreading – form of geometrical spreading which is confined between two parallel planes.

Decibel – unit of measure of acoustic intensity based on a logarithmic scale.

Deep scattering layer – layer of biological organisms associated with an increased scattering of sound.

Deep sound channel axis – location of the deep sound speed minimum in the ocean.

Deterministic – state of being completely predictable (as distinguished from stochastic or chaotic).

Diagnostic information – information used to analyze the past or present state of a system, particularly with reference to identifying sonar system pathologies.

Diel – characterized by a 24-hour period.

Diffraction – frequency-dependent interference effects.

Direct path – ray paths directly connecting source and receiver over short ranges without boundary interaction.

Dispersion – separation of sound into component frequencies; condition in which the phase velocity is frequency dependent.

Diurnal – having a daily cycle.

Domains of applicability – the spatial, temporal or spectral ranges over which a model's output can be considered valid; in general, such limitations are imposed by the model's physical or mathematical basis.

Doppler – frequency shift resulting from relative motion of source and receiver.

Downwelling – downward-directed vertical water motion in the ocean.

Ducted precursors – acoustic energy that leaks out of a surface duct, travels via a convergence zone or bottom bounce path, or both, before coupling back into the surface duct down range.

Earth curvature correction – correction applied to sound speed profile to adjust for propagation of sound over the curved Earth surface rather than a flat surface.

Eddy – isolated pattern of gyral motion in the ocean.

Eigenray – ray which connects the source and receiver.

Eigenvalue – solution to the normal mode equation.

Eikonal – acoustic path length as a function of the endpoints.

Ekman drift current – surface current produced by the wind but directed perpendicular to the wind direction.

Empirical models – mathematical models based on observations.

Environmental models – empirical algorithms used to quantify the boundary conditions and volumetric effects of the ocean environment.

Escarpment – an elongated and comparatively steep slope of the sea floor separating flat or gently sloping areas.

Fan – a gently sloping, fan-shaped feature normally located near the lower termination of a canyon.

Fast-field models – mathematical models of acoustic propagation based on a variation of the normal-mode technique.

Fathometer returns – multiply reflected echoes arriving from the sea surface directly above or from the sea floor directly below a monostatic sonar.

Feature model – statistical representation of a common synoptic structure in the ocean such as a front or an eddy.

Figure of Merit (FOM) – quantitative measure of sonar performance; value equals allowable one-way transmission loss in passive sonars, or two-way transmission loss in noise-limited active sonars.

Finite difference – mathematical technique for solving differential equations.

Finite element – mathematical technique for solving complex problems; in underwater acoustics, this method is implemented by dividing the range–depth plane into a gridded pattern of small elements.

Fracture zone – an extensive linear zone of unusually irregular topography of the sea floor characterized by large seamounts, steep-sided ridges, troughs or escarpments.

Front – boundary between two different water masses.

Gap – a depression cutting transversely across a ridge or rise.

Gaussian beam tracing – method which associates each acoustic ray with a beam having a Gaussian intensity profile normal to the ray.

Geodesic path – shortest distance between two points on the surface of an elliptical Earth; the shortest distance between two points on the surface of a spherical Earth is referred to as a great circle path.

Grazing angle – angle in the vertical plane measured from the horizontal.

Half-channel – ducted environment defined by the equivalent sound speed profile either above or below the deep-sound-channel axis (the deep sound channel is considered to be a full channel).

Harmonic solution – solution to the wave equation based on a single frequency.

Hidden depths – concept that regards as unimportant the deep ocean sediment structure below the ray turning point.

Hill – a small elevation rising less than about 200 m above the sea floor.

Hindcast – estimate of a previous state of a system.

Hole – a small depression on the sea floor.

Hybrid models – models based on multiple physical or mathematical approaches in order to broaden the domains of applicability.

Incidence angle – angle in the vertical plane measured from the vertical.

Inertial motion – periodic, quasi-circular oceanic motions influenced by the rotation of the Earth.

Internal wave – subsurface gravity waves which propagate along density interfaces in the ocean.

Isotherm – line connecting values of equal temperature.

Knoll – an elevation rising less than about 1 km above the sea floor and of limited extent across the summit.

Levee – an embankment bordering either one or both sides of a sea channel, or the low-gradient seaward part of a canyon or valley.

Limiting ray angle – angle of ray measured relative to a horizontal surface which leaves the source tangent to the horizontal plane.

Littoral zone–region between the shore and water depths of approximately 200 meters. (The usage and interpretation of this term can vary widely.)

Lloyd mirror effect – near-field interference patterns associated with a shallow acoustic source and receiver.

Marginal ice zone (MIZ) – area of ice cover separating pack ice from the open ocean.

Mathematical models – category of models containing empirical models and numerical models.

Meso-scale features – dynamic features of the ocean with characteristic length scales on the order of 100 m to 100 km; examples include eddies and internal waves.

Mixed layer – surface layer of uniform temperature which is well mixed by wind or wave action or thermohaline convection.

Moat – a depression located at the base of many seamounts or islands.

Modified data banks – data banks containing modified or extrapolated data derived from a primary data bank.

Monostatic – geometry in which the acoustic source and receiver are at the same position.

Mountains – a well-delineated subdivision of a large and complex feature, generally part of a cordillera.

Multipath – propagation conditions characterized by a combination of different types of parths.

Multipath expansion models – mathematical models of acoustic propagation based on a variation of the normal-mode technique.

Noise models – mathematical models of noise levels and directionality in the ocean consisting of two categories: ambient noise and beam-noise statistics models.

Normal mode models – mathematical models of acoustic propagation based on the normal mode technique.

Normal modes – natural frequencies at which a medium vibrates in response to an excitation.

Nowcast – estimate of the present state of a system.

Numerical models – mathematical models based on the governing physics.

Ocean impulse response function – time-variant and space-variant Green's function that describes the response of the ocean medium to a unit impulse.

Parabolic equation models – mathematical models of acoustic propagation based on a solution of the parabolic versus the elliptic-reduced wave equation.

Pathological test case – specification of inputs to a model that prove to be particularly troublesome; a test case that produces disorders in an otherwise well-functioning model.

Physical models – theoretical or conceptual representations of the physical and acoustical processes occurring within the ocean.

Plain – a flat, gently sloping or nearly level region of the sea floor.

Plateau – a comparatively flat-topped elevation of the sea floor of considerable extent across the summit and usually rising more than 200 m above the sea floor on at least one side.

Point-scattering models – mathematical models of reverberation based on the assumption that the scatterers are randomly distributed throughout the ocean.

Primary data banks – data banks containing original or modestly processed environmental acoustic data.

Prognostic information – information used to forecast the future state of a system, particularly with reference to predicting sonar system performance.

Propagation – transmission of acoustic energy through the ocean medium.

Propagation models – mathematical models of underwater acoustic transmission loss consisting of five categories: ray theory, normal mode, multipath expansion, fast field and parabolic equation models.

Province – a region composed of a group of similar bathymetric features whose characteristics are markedly in contrast with the surrounding areas.

Range – a series of generally parallel ridges or seamounts.

Rayleigh parameter – measure of the acoustic roughness of a surface.

Ray theory models – mathematical models of acoustic propagation based on ray-tracing techniques.

Reef – an offshore consolidation with a depth below the sea surface of less than about 20 m.

Reflection – return of a portion of the incident energy in the forward direction after an encounter with a boundary.

Refraction – directional changes in acoustic propagation caused by density discontinuities.

Reliable acoustic path – ray paths formed when a source or receiver is located at the critical depth.

Reverberation – backscattered sound.

Reverberation models – mathematical models of reverberation based on boundary and volumetric scattering processes consisting of two categories: cell-scattering and point-scattering models.

Ridge – a long, narrow elevation of the sea floor with steep sides.

Ring – same as eddy.

Saddle – a low part on a ridge, or between seamounts.

Salinity – a measure of the quantity of dissolved salts in sea water.

Scattering – random dispersal of sound after encounters with boundaries or with volumetric inhomogeneities.

Seachannel – a long, narrow, shallow depression of the sea floor, usually occurring on a gently sloping plain or fan, with either a U-shaped or V-shaped cross-section.

Seamount – an elevation rising 1 km or more above the sea floor, and of limited extent across the summit.

Self-noise – noise background originating from own ship or sonar structure.

Shadow zone – region of the range–depth plane into which little, if any, acoustic energy penetrates owing to the refractive properties of the water column.

Shoal – an offshore hazard to navigation with a depth below the sea surface of 20 m or less, usually composed of unconsolidated material.

Signal-processing models – mathematical models of signal detection processes.

Significant wave height – the average height of the one-third highest waves of a given wave group.

Sill – the low part of a ridge or rise separating ocean basins from one another or from the adjacent sea floor.

Simulation – dynamic execution of models to generate prognostic or diagnostic information.

Skip distance – distance traveled by a ray that just grazes the sea surface, the sea floor or some intermediate layer in the ocean such as the sonic layer depth.

Sonar performance models – mathematical models organized to solve specific sonar applications problems; contain environmental models, basic acoustic models and signal processing models.

Sonic layer – acoustic equivalent of mixed layer; surface layer characterized by near iso-velocity conditions.

Spherical spreading – form of geometrical spreading which is unbounded.

Spur – a subordinate elevation, ridge or rise projecting from a larger feature.

Stochastic – state of being random; describable in probabilistic terms.

Surface duct – propagation channel bounded by the sea surface and the base of the sonic layer.

Surface loss – reflection loss at the sea surface.

Tablemount – a seamount having a comparatively smooth, flat top (also called a guyot).

Terrace – a bench-like feature bordering an undersea feature (also called a bench).

Thermocline – region of water column characterized by a negative temperature gradient.

Thermohaline – of or relating to the effects of water temperature (thermo) and salinity (haline).

Transmission anomaly – difference between the observed transmission loss and the loss expected from spherical spreading alone.

Transmission loss – measure of acoustic energy loss (in dB) versus range.

Trench – a long, narrow and deep depression of the sea floor, with relatively steep sides.

Trough – a long depression of the floor, normally wider and shallower than a trench.

Turning point – position in the range–depth plane at which an upgoing or a downgoing ray reverses direction.

Upwelling – upward-directed vertical motion in the ocean.

Valley – a relatively shallow, wide depression with gentle slopes, the bottom of which grades continually downward (as opposed to a canyon).

Virtual mode – mode associated with acoustic propagation in the surface duct.

Water mass – body of water characterized by a unique temperature–salinity relationship.

Wave equation – equation describing motion of acoustic waves in the ocean.

Wispering gallery – phenomenon relating to the propagation of sound along the curved walls of a gallery, especially those of a hemispherical dome.

References

Abramowitz, M. and Stegun. I.A. (1964) *Handbook of Mathematical Functions with Formulas, Graphs, and Mathematical Tables.* Applied Mathematics Series 55, National Bureau of Standards, Washington, DC.

Ackerman, C.L. and Kesser, R.L. (1973) Reverberation spectrum model for matched filter homing systems. Appl. Res. Lab., Pennsylvania State Univ., Tech. Memo. 73–285.

Ainslie, M.A. and Harrison, C.H. (1990) Diagnostic tools for the ocean acoustic modeller. *In Computational Acoustics. Vol. 3: Seismo-Ocean Acoustics and Modeling,* ed. D. Lee, A. Cakmak and R. Vichnevetsky, North-Holland, Amsterdam, pp. 107–30.

Akal, T. and Berkson, J.M. (eds) (1986) *Ocean Seismo-Acoustics* (Low-Frequency Underwater Acoustics). NATO Conf. Series IV, Marine Sciences, Vol. 16. Plenum, New York.

Alapati, N.K., Kirklin, R.H. and Etter, P.C. (1993) Analysis of Chaotic Waveforms for Application to Active Sonar Systems. Radix Systems, Inc., TR-93-081.

Albers, V.M. (1965) *Underwater Acoustics Handbook – II.* The Pennsylvania State Univ. Press, University Park, USA.

Almeida, R.J. Jr and Medeiros, R.C. (1985) RAYMODE Accuracy, Limitations and Transportability. Nav. Underwater Syst. Ctr, Tech. Memo. 841032.

Amato, I. (1993) A sub surveillance network becomes a window on whales. *Science,* **261**, 549–50.

Ammicht, E. and Stickler, D.C. (1984) Uniform asymptotic evaluation of the continuous spectrum contribution for a stratified ocean. *J. Acoust. Soc. Amer.,* **76**, 186–91.

Andersen, N.R. and Zahuranec, B.J. (eds) (1977) *Oceanic Sound Scattering Prediction.* Plenum, New York.

Anderson, A.L. and Hampton, L.D. (1980a) Acoustics of gas-bearing sediments I. Background. *J. Acoust. Soc. Amer.,* **67**, 1865–89.

Anderson, A.L. and Hampton, L.D. (1980b) Acoustics of gas-bearing sediments II. Measurements and models. *J. Acoust. Soc. Amer.,* **67**, 1890–903.

Anderson, V.C. (1958) Arrays for the investigation of ambient noise in the ocean. *J. Acoust. Soc. Amer.,* **30**, 470–7.

Anderson, V.C. (1979) Variation of the vertical directionality of noise with depth in the North Pacific. *J. Acoust. Soc. Amer.,* **66**, 1446–52.

Apel, J.R. (1987) *Principles of Ocean Physics*, International Geophysics Series, Vol. 38. Academic Press, Orlando, USA.

Arnold, J.M. and Felsen, L.B. (1983) Rays and local modes in a wedge-shaped ocean. *J. Acoust. Soc. Amer.*, **73**, 1105–19.

Arrhenius, G. (1963) Pelagic sediments. In *The Sea*, Vol. 3, The Earth Beneath the Sea, History, ed. M.N. Hill. Wiley Interscience, New York, pp. 655–727.

Arvelo, J.I. and Überall, H. (1990) Adiabatic normal-mode theory of sound propagation including shear waves in a range-dependent ocean floor. *J. Acoust. Soc. Amer.*, **88**, 2316–25.

Au, W.W.L., Penner, R.H. and Turl, C.W. (1987) Propagation of Beluga echolocation signals. *J. Acoust. Soc. Amer.*, **82**, 807–13.

Auer, S.J. (1987) Five-year climatological survey of the Gulf Stream system and its associated rings. *J. Geophys. Res.*, **92**, 11709–26.

Axelrod, E.H., Schoomer, B.A. and Von Winkle, W.A. (1965) Vertical directionality of ambient noise in the deep ocean at a site near Bermuda. *J. Acoust. Soc. Amer.*, **37**, 77–83.

Baer, R.N. (1981) Propagation through a three-dimensional eddy including effects on an array. *J. Acoust. Soc. Amer.*, **69**, 70–5.

Baer, R.N., Berman, D.H., Perkins, J.S. and Wright, E.B. (1985). A three-dimensional model for acoustic scattering from rough ocean bathymetry. *Comp. Maths Applic.*, **11**, 863–71.

Baggeroer, A.B., Kuperman, W.A. and Mikhalevsky, P.N. (1993) An overview of matched field methods in ocean acoustics. *IEEE J. Oceanic Engr.*, **18**, 401–24.

Baker, C.L. (1976) FANIN preprocessor for the FANM-I ambient noise model. Ocean Data Syst., Inc.

Baker, C.L. and Spofford, C.W. (1974) The FACT model, Vol. II. Acoustic Environmental Support Detachment, Off. Nav. Res., Tech. Note TN-74-04.

Bannister, R.W. (1986) Deep sound channel noise from high latitude winds. *J. Acoust. Soc. Amer.*, **79**, 41–8.

Bannister, R.W., Kewley, D.J. and Burgess, A.S. (1989) Directional underwater noise estimates – The DUNES model. Australia Defence Science and Technology Organisation, Weapons Systems Research Laboratory, WSRL-TN-34/89.

Barkhatov, A.N. (1978) Modeling of sound propagation in the sea. Gidrometeorologicheskoe Izdatel'stvo. (Translated from Russian by J.S. Wood, Consultants Bureau, a Division of Plenum Publishing Corp., New York, 1971.)

Bartberger, C.L. (1965) Lecture notes on underwater acoustics. Nav. Air Devel. Ctr, Rept NADC-WR-6509.

Bartberger, C.L. (1978a). PLRAY–A ray propagation loss program. Nav. Air Devel. Ctr, Rept NADC-77296-30.

Bartberger, C.L. (1978b). AP-2 normal mode program. Nav. Air Devel. Ctr, Rept.

Bartberger, C.L. (1985) The NADC bistatic active sonar model. Nav. Air Devel. Ctr, Rept.

Bartberger, C.L. (1991a) The physics of the bistatic acoustic model. Nav. Air Devel. Ctr, Rept NADC-91020-50.

Bartberger, C.L. (1991b) The NADC monostatic sonar model (version 1.0). Nav. Air Devel. Ctr, TN-5044-7-91.

Bathen, K.H. (1972) On the seasonal changes in the depth of the mixed layer in the North Pacific Ocean. *J. Geophys. Res.*, **77**, 7138–50.

Becken, B.A. (1961) The directional distribution of ambient noise in the ocean. Scripps Inst. Oceanogr., Rept 61–4.

Beckmann, P. and Spizzichino, A. (1963) *The Scattering of Electromagnetic Waves from*

Rough Surfaces, International Series of Monographs on Electromagnetic Waves, Vol. 4. Macmillan, New York.

Beebe, J.H. McDaniel, S.T. and Rubano, L.A. (1982) Shallow-water transmission loss prediction using the Biot sediment model. *J. Acoust. Soc. Amer.*, **71**, 1417–26.

Bennett, A.F. (1992) *Inverse Methods in Physical Oceanography*, Cambridge University Press, New York.

Berger, M.D., Boucher, C.E., Daley, E.M., Renner, W.W., Pastor, V.L., Haines, L.C. and Eller, A.I. (1994) Acoustic system performance model (ASPM) 4.0A: User's guide. Sci. Appl. Inter. Corp., SAIC-94/1000.

Berman, D.H., Wright, E.B. and Baer, R.N. (1989) An optimal PE-type wave equation. *J. Acoust. Soc. Amer.*, **86**, 228–33.

Bertuccelli, H.C. (1975) Digital computer programs for analyzing acoustic search performance in refractive waters, Vol. 3. Nav. Undersea Ctr, Tech. Pub. 164.

Bialek, E.L. (1966) *Handbook of Oceanographic Tables.* Nav. Oceanogr. Off., Special Pub. SP-68, Washington, DC.

Biot, M.A. (1956a) Theory of propagation of elastic waves in a fluid-saturated porous solid. I. Low-frequency range. *J. Acoust. Soc. Amer.*, **28**, 168–78.

Biot, M.A. (1956b) Theory of propagation of elastic waves in a fluid-saturated porous solid. II. Higher frequency range. *J. Acoust. Soc. Amer.*, **28**, 179–91.

Bishop, G.C. (1987) A bistatic, high-frequency, under-ice, acoustic scattering model. *J. Acoust. Soc. Amer.*, **82**, (Suppl. 1), S30 (abstract).

Bishop, G.C. (1989a) A bistatic high-frequency, under-ice, acoustic scattering model. I: Theory. *J. Acoust. Soc. Amer.*, **85**, 1903–11.

Bishop, G.C. (1989b) A bistatic, high-frequency, under-ice, acoustic scattering model. II. Applications: *J. Acoust. Soc. Amer.*, **85**, 1912–24.

Bishop, G.C., Mellberg, L.E. and Ellison, W.T. (1986) A simulation model for high-frequency, under-ice reverberation. Nav. Underwater Syst. Ctr, Tech. Rept 6268.

Bishop, G.C., Ellison, W.T. and Mellberg, L.E. (1987) A simulation model for high-frequency under-ice reverberation. *J. Acoust. Soc. Amer.*, **82**, 275–86.

Blatstein, I.M. (1974) Comparisons of normal mode theory, ray theory, and modified ray theory for arbitrary sound velocity profiles resulting in convergence zones. Nav. Ordnance Lab., NOLTR 74–95.

Blumen, L.S. and Spofford, C.W. (1979) The ASTRAL model, Vol. II: Software implementation. Sci. Appl., Inc., SAI-79-743-WA.

Bogart, C.W. and Yang, T.C. (1992) Comparative performance of matched-mode and matched-field localization in a range-dependent environment. *J. Acoust. Soc. Amer.*, **92**, 2051–68.

Bom, N. (1969) Effect of rain on underwater noise level. *J. Acoust. Soc. Amer.*, **45**, 150–6.

Botseas, G. Lee, D. and Gilbert, K.E. (1983) IFD: Wide angle capability. Nav. Underwater Syst. Ctr, Tech. Rept 6905.

Botseas, G. Lee, D. and King, D. (1987) FOR3D: A computer model for solving the LSS three-dimensional, wide angle wave equation. Nav. Underwater Syst. Ctr, Tech. Rept 7943.

Botseas, G. Lee, D. and Siegmann, W.L. (1989) IFD: Interfaced with Harvard open ocean model forecasts. Nav. Underwater Syst. Ctr, Tech. Rept 8367.

Bowditch, N. (1977) *American Practical Navigator*, Vol. I. Defense Mapping Agency Hydrographic Center, Pub. 9, Washington, DC.

Boyles, C.A. (1984) *Acoustic Waveguides. Applications to Oceanic Science.* John Wiley and Sons, New York.

Bradley, M.R. and Bradley, B.W. (1984) Computer program performance specification for the geophysics ambient noise model. Planning Syst., Inc., TR-S310018.

Breeding, J.E., Jr (1993) Description of a noise model for shallow water: RANDI-III. *J. Acoust. Soc. Amer.*, **94**, 1820 (abstract).

Brekhovskikh, L. and Lysanov, Yu. (1991) *Fundamentals of Ocean Acoustics*, 2nd edn. Springer-Verlag, New York.

Brock, H.K. (1978) The AESD parabolic equation model. Nav. Ocean Res. Devel. Activity, Tech. Note 12.

Brock, H.K., Buchal, R.N. and Spofford, C.W. (1977) Modifying the sound-speed profile to improve the accuracy of the parabolic equation technique. *J. Acoust. Soc. Amer.*, **62**, 543–52.

Brown, J.R. (1964) Reverberation under Arctic ice. *J. Acoust. Soc. Amer.*, **36**, 601–3.

Brown, M.G. (1994) Global acoustic propagation modeling using MaCh1. *J. Acoust. Soc. Amer.*, **95**, 2880 (abstract).

Brown, M.G., Tappert, F.D. and Goñi, G. (1991a) An investigation of sound ray dynamics in the ocean volume using an area preserving mapping. *Wave Motion*, **14**, 93–9.

Brown, M.G., Tappert, F.D., Goñi, G.J. and Smith, K.B. (1991b). Chaos in underwater acoustics. In *Ocean Variability and Acoustic Propagation*, ed. J. Potter and A. Warn–Varnas. Kluwer Academic Publishers, Dordrecht, The Netherlands, pp. 139–60.

Browning, D.G., Christian, R.J. and Petitpas, L.S. (1994) A global survey of the impact on acoustic propagation of deep water warm core ocean eddies. *J. Acoust. Amer.*, **95**, 2880 (abstract).

Brutzman, D.P., Kanayama, Y. and Zyda, M.J. (1992) Integrated simulation for rapid development of autonomous underwater vehicles. Proc. 1992 IEEE Symposium on Autonomous Underwater Vehicle Technology, pp. 3–10.

Bryan, G.M. (1967) Underwater sound in marine geology. In *Underwater Acoustics*, Vol. 2, ed. V.M. Albers. Plenum, New York, pp. 351–61.

Buck, B.M. (1968) Arctic acoustic transmission loss and ambient noise. In *Arctic Drifting Stations*, ed. J.E. Sater. Arctic Inst. of North Amer., Calgary, Canada, pp. 427–38.

Buck, B.M. (1981) Preliminary underice propagation models based on synoptic ice roughness. Polar Res. Lab., TR-30.

Buck, B.M. (1985) Long term statistical measurements of environmental acoustics parameters in the Arctic (AEAS Report No. 2 – Low frequency transmission loss measurements in the central Arctic Ocean). Polar Res. Lab., TR-55.

Bucker, H.P. (1971) Some comments on ray theory with examples from current NUC ray trace models. SACLANTCEN Conf. Proc. No. 5, Part I, pp. 32–6.

Bucker, H.P. (1976) Use of calculated sound fields and matched-field detection to locate sound sources in shallow water. *J. Acoust. Soc. Amer.*, **59**, 368–73.

Bucker, H.P. (1983) An equivalent bottom for use with the split-step algorithm. *J. Acoust. Soc. Amer.*, **73**, 486–91.

Bucker, H.P. (1986) Use of SALT tables for rapid calculation of sound angle, level and travel time. *J. Acoust. Soc. Amer.*, **80**, (Suppl. 1), S63 (Abstract).

Bucker, H.P. (1994). A simple 3-D Gaussian beam sound propagation model for shallow water. *J. Acoust. Soc. Amer.*, **95**, 2437–40.

Bucker, H.P. and Morris, H.E. (1968) Normal-mode reverberation in channels or ducts. *J. Acoust. Soc. Amer.*, **44**, 827–8.

Buckingham, M.J. (1991) On acoustic transmission in ocean-surface waveguides. *Phil. Trans. R. Soc. Lond.*, (A) **335**, 513–55.

Buckingham, M.J. (1992) Ocean-acoustic propagation models. *J. Acoustique*, No. 3, 223–87.

Buckingham, M.J. (1993) Theory of acoustic imaging in the ocean with ambient noise. *J. Comput. Acoust.*, **1**, 117–40.

Buckingham, M.J., Berkhout, B.V. and Glegg, S.A.L. (1992) Imaging the ocean with ambient noise. *Nature*, **356**, 327–9.

Buckingham, M.J. and Tolstoy, A. (1990) An analytical solution for benchmark problem 1: The 'ideal' wedge. *J. Acoust. Soc. Amer.*, **87**, 1511–3.

Burdic, W.S. (1991) *Underwater Acoustic System Analysis*, 2nd edn. Prentice-Hall, Englewood Cliffs, USA.

Burridge, R. and Weinberg, H. (1977) Horizontal rays and vertical modes. In *Wave Propagation and Underwater Acoustics*, ed. J.B. Keller and J.S. Papadakis. Lecture Notes in Physics, Vol. 70. Springer-Verlag, New York, pp. 86–152.

Candy, J.V. and Sullivan, E.J. (1992) Ocean acoustic signal processing: A model-based approach. *J. Acoust. Soc. Amer.*, **92**, 3185–201.

Candy, J.V. and Sullivan, E.J. (1993) Sound velocity profile estimation: A system theoretic approach. *IEEE J. Oceanic Engr.*, **18**, 240–52.

Carey, W.M. and Wagstaff, R.A. (1986) Low-frequency noise fields. *J. Acoust. Soc. Amer.*, **80**, 1523–6.

Carey, W.M. Evans, R.B. and Davis, J.A. (1987) Downslope propagation and vertical directionality of wind noise. *J. Acoust. Soc. Amer.*, **82**, (Suppl. 1), S63 (abstract).

Carey, W.M., Evans, R.B., Davis, J.A. and Botseas, G. (1990). Deep-ocean vertical noise directionality. *IEEE J. Oceanic Engr.*, **15**, 324–34.

Carlson, J.T. (1994) IUSS community adopts collateral uses. *Sea Technology*, **35** (5), 25–8.

Caruthers, J.W. and Novarini, J.C. (1993) Modeling bistatic bottom scattering strength including a forward scatter lobe. *IEEE J. Oceanic Engr.*, **18**, 100–7.

Caruthers, J.W. and Yoerger, E.J. (1993) ARSRP reconn results and BISSM modeling of direct path backscatter. In *Ocean Reverberation*, ed. D.D. Ellis, J.R. Preston and H.G. Urban. Kluwer Academic Publishers, Dordrecht, The Netherlands, pp. 203–8.

Caruthers, J.W., Sandy, R.J. and Novarini, J.C. (1990) Modified bistatic scattering strength model (BISSM2). Nav. Oceanogr. Atmos. Res. Lab., SP 023:200:90.

Cavanagh, R.C. (1974a) Fast Ambient Noise Model I (FANM I). Acoustic Environmental Support Detachment, Off. Nav. Res.

Cavanagh, R.C. (1974b) Fast Ambient Noise Model II (FANM II). Acoustic Environmental Support Detachment, Off. Nav. Res.

Cavanagh, R.C. (1978a) Acoustic fluctuation modeling and system performance estimation, Vol. I. Sci. Appl., Inc., SAI-79-737-WA.

Cavanagh, R.C. (1978b) Acoustic fluctuation modeling and system performance estimation, Vol. II. Sci. Appl., Inc., SAI-79-738-WA.

Chaika, E.D., Granum, R.M. and Marusich, R.B. (1979) Program for Integrated Sonar Modeling (PRISM) user's guide. Nav. Ocean Syst. Ctr, Tech. Note 851.

Chamberlain, S.G. and Galli, J.C. (1983) A model for numerical simulation of nonstationary sonar reverberation using linear spectral prediction. *IEEE J. Oceanic Engr.*, **OE-8**, 21–36.

Chapman, N.R. and Ebbeson, G.R. (1983) Acoustic shadowing by an isolated seamount. *J. Acoust. Soc. Amer.*, **73**, 1979–84.

Chapman, R.P. and Harris, J.H. (1962) Surface backscattering strengths measured with explosive sound sources. *J. Acoust. Soc. Amer.*, **34**, 1592–7.

Chapman, R.P. and Scott, H.D. (1964) Surface backscattering strengths measured over an extended range of frequencies and grazing angles. *J. Acoust. Soc. Amer.*, **36**, 1735–7.

Chen, C.-T. and Millero, F.J. (1977) Speed of sound in seawater at high pressures. *J. Acoust. Soc. Amer.*, **62**, 1129–35.

Cheney, R.E. and Winfrey, D.E. (1976) Distribution and classification of ocean fronts. Nav. Oceanogr. Off., Tech. Note 3700-56-76.

Chin-Bing, S.A. and Murphy, J.E. (1987) Two numerical models – one simplistic, the other sophisticated–applied to under-ice acoustic propagation and scattering. *J. Acoust. Soc. Amer.*, **82** (Suppl. 1), S31 (abstract).

Chin-Bing, S.A. and Murphy, J.E. (1988) Long-range, range-dependent, acoustic propagation simulation using a full-wave, finite-element model coupled with a one-way parabolic equation model. *J. Acoust. Soc. Amer.*, **84** (Suppl. 1), S90 (abstract).

Chin-Bing, S.A., Davis, J.A. and Evans, R.B. (1982) Nature of the lateral wave effect on bottom loss measurements. *J. Acoust. Soc. Amer.*, **71**, 1433–7.

Chin-Bing, S.A., Gilbert, K.E., Evans, R.B., Werby, M.F. and Tango, G.J. (1986) Research and development of acoustic models at NORDA. In *The NORDA Review* (March 31, 1976-March 31, 1986). Stennis Space Center, USA, pp. 39–44.

Chin-Bing, S.A., King, D.B. and Boyd, J.D. (1994) The effects of ocean environmental variability on underwater acoustic propagation forecasting. In *Oceanography and Acoustics: Prediction and Propagation Models*, ed. A.R. Robinson and D. Lee. American Institute of Physics, New York, Chapter 2, pp. 7–49.

Chin-Bing, S.A., King, D.B. and Murphy, J.E. (1993a) Numerical simulations of lower-frequency acoustic propagation and backscatter from solitary internal waves in a shallow water environment. In *Ocean Reverberation*, ed. D.D. Ellis, J.R. Preston and H.G. Urban. Kluwer Academic Publishers, Dordrecht, The Netherlands, pp. 113–18.

Chin-Bing, S.A., King, D.B., Davis, J.A. and Evans, R.B. (eds) (1993b) *PE workshop II. Proceedings of the Second Parabolic Equation Workshop*. Naval Research Laboratory NRL/BE/7181-93-0001. US Government Printing Office, Washington, DC.

Chiu, C.-S. and Ehret, L.L. (1990) Computation of sound propagation in a three-dimensionally varying ocean: A coupled normal mode approach. In *Computational Acoustics. Vol. 1: Ocean-Acoustic Models and Supercomputing*, ed. D. Lee, A. Cakmak and R. Vichnevetsky. North-Holland, Dordrecht, The Netherlands, pp. 187–202.

Chiu, C.-S. and Ehret, L.L. (1994) Three-dimensional acoustic normal mode propagation in the Gulf Stream. In *Oceanography and Acoustics: Prediction and Propagation Models*, ed. A.R. Robinson and D. Lee. American Institute of Physics, New York, Chapter 8, pp. 179–97.

Claerbout, J.F. (1976) *Fundamentals of Geophysical Data Processing with Applications to Petroleum Prospecting*, International Series in the Earth and Planetary Sciences. McGraw-Hill, New York.

Clancy, R.M. and Martin, P.J. (1979) The NORDA/FLENUMOCEANCEN Thermodynamical Ocean Prediction System (TOPS): A technical description. Nav. Ocean Res. Devel. Activity, Tech. Note 54.

Clancy, R.M. and Pollack, K.D. (1983) A real-time synoptic ocean thermal analysis/forecast system. *Prog. Oceanogr.*, **12**, 383–424.

Clancy, R.M., Cummings, J.A. and Ignaszewski, M.J. (1991a) New ocean thermal structure model operational at FNOC. *Naval Oceanography Command News*, **11** (5 and 6), 1–5.

Clancy, R.M., Martin, P.J., Piacsek, S.A. and Pollak, K.D. (1981) Test and evaluation of an operationally capable synoptic upper-ocean forecast system. Nav. Ocean Res. Devel. Activity, Tech. Note 92.

Clancy, R.M., Price, J.F. and Hawkins, J.D. (1991b). The FNOC diurnal ocean surface layer (DOSL) model. *Naval Oceanography Command News*, **11** (7), 8–10.

Clay, C.S. and Medwin, H. (1964) High-frequency acoustical reverberation from a rough-sea surface. *J. Acoust. Soc. Amer.*, **36**, 2131–4.

Clay, C.S. and Medwin, H. (1977) *Acoustical Oceanography: Principles and Applications*. Wiley Interscience, New York.

Coates, R., de Cogan, D. and Willison, P.A. (1990) Transmission line matrix modeling applied to problems in underwater acoustics. Proc. IEEE Oceans 90 Conf., pp. 216–20.

Cohen, J.S. and Einstein, L. T. (1970) Continuous gradient ray tracing system (CONGRATS) II: Eigenray processing programs. Navy Underwater Sound Lab., Rept 1069.

Cohen, J.S. and Weinberg, H. (1971) Continuous gradient ray-tracing system (CONGRATS) III: Boundary and volume reverberation. Nav. Underwater Syst. Ctr, Rept 4071.

Colilla, R.A. (1970) Improvement in the efficiency of the present model and documentation of the program RP-70. Ocean Data Syst., Inc.

Collins, M.D. (1988a) FEPE user's guide. Nav. Ocean Res. and Devel. Activity, Tech. Note 365.

Collins, M.D. (1988b) The time-domain solution of the wide-angle parabolic equation including the effects of sediment dispersion. *J. Acoust. Soc. Amer.*, **84**, 2114–25.

Collins, M.D. (1990a) The rotated parabolic equation and sloping ocean bottoms. *J. Acoust. Soc. Amer.*, **87**, 1035–7.

Collins, M.D. (1990b) Benchmark calculations for higher-order parabolic equations. *J. Acoust. Amer.*, **87**, 1535–8.

Collins, M.D. (1991) Higher-order Padé approximations for accurate and stable elastic parabolic equations with application to interface wave propagation. *J. Acoust. Soc. Amer.*, **89**, 1050–7.

Collins, M.D. (1992) A self-starter for the parabolic equation method. *J. Acoust. Soc. Amer.*, **92**, 2069–74.

Collins, M.D. (1993a) A two-way parabolic equation for elastic media. *J. Acoust. Soc. Amer.*, **93**, 1815–25.

Collins, M.D. (1993b) An energy-conserving parabolic equation for elastic media. *J. Acoust. Soc. Amer.*, **94**, 975–82.

Collins, M.D. and Chin-Bing, S.A. (1990) A three-dimensional parabolic equation model that includes the effects of rough boundaries. *J. Acoust. Soc. Amer.*, **87**, 1104–9.

Collins, M.D. and Evans, R.B. (1992) A two-way parabolic equation for acoustic backscattering in the ocean. *J. Acoust. Soc. Amer.*, **91**, 1357–68.

Collins, M.D. and Kuperman, W.A. (1994) Overcoming ray chaos. *J. Acoust. Soc. Amer.*, **95**, 3167–70.

Condron, T.P., Onyx, P.M. and Dickson, K.R. (1955) Contours of propagation loss and plots of propagation loss vs range for standard conditions at 2, 5, and 8 kc. Navy Underwater Sound Lab., Tech. Memo. 1110-14-55.

Coppens, A.B. (1981) Simple equations for the speed of sound in Neptunian waters. *J. Acoust. Soc. Amer.*, **69**, 862–3.

Coppens, A.B. and Sanders, J.V. (1980) Propagation of sound from a fluid wedge into a fast fluid bottom. In *Bottom-Interacting Ocean Acoustics*, ed. W.A. Kuperman and F.B. Jensen. Plenum, New York, pp. 439–50.

Cornuelle, B., Wunsch, C., Behringer, D., Birdsall, T., Brown, M., Heinmiller, R., Knox, R., Metzger, K., Munk, W., Spiesberger, J., Spindel, R., Webb, D. and Worcester, P. (1985) Tomographic maps of the ocean mesoscale, Part 1: Pure acoustics. *J. Phys. Oceangor.*, **15**, 133–52.

Cornyn, J.J. (1973a) GRASS: A digital-computer ray-tracing and transmission-loss- prediction system. Vol. I – overall description. Nav. Res. Lab., Rept 7621.

Cornyn, J.J. (1973b) GRASS: A digital-computer ray-tracing and transmission-loss-prediction system. Vol. II – user's manual. Nav. Res. Lab., Rept 7642.

Cox, H. (1989) Fundamentals of bistatic active sonar. In *Underwater Acoustic Data Processing*, NATO Advanced Science Institutes Series E: Applied Sciences, Vol. 161, ed. Y.T. Chan. Kluwer Academic Publishers, Dordrecht, The Netherlands, pp. 3–24.

Crum, L.A. Roy, R.A. and Prosperetti, A. (1992) The underwater sounds of precipitation. *Nav. Res. Rev.*, **44**(2), 2–12.

Culkin, F. and Ridout, P. (1989) Salinity: Definitions, determinations, and standards. *Sea T echnology*, **30**(10), 47–9.

Cummings, W.C. and Holliday, D.V. (1987) Sounds and source levels from Bowhead whales off Pt Barrow, Alaska. *J. Acoust. Soc. Amer.*, **82**, 814–21.

Dantzler, H.L. Jr, Sides, D.J. and Neal, J.C. (1993) An automated tactical oceanographic monitoring system. *Johns Hopkins APL T ech. Dig.*, **14**, 281–92.

Dashen, R. and Munk, W. (1984) Three models of global ocean noise. *J. Acoust. Soc. Amer.*, **76**, 540–54.

Davis, J.A., White, D. and Cavanagh, R.C. (1982) NORDA Parabolic Equation Workshop, 31 March–3 April 1981. Nav Ocean Res. Devel. Activity, Tech. Note 143.

Deavenport, R.L. and DiNapoli, F.R. (1982) Evaluation of Arctic transmission loss models. Nav. Underwater Syst. Ctr, Tech. Memo. 82–1160.

Del Balzo, D.R., Feuillade, C. and Rowe, M.M. (1988) Effects of water-depth mismatch on matched-field localization in shallow water. *J. Acoust. Soc. Amer.*, **83**, 2180–5.

Del Grosso, V.A. (1974) New equation for the speed of sound in natural waters (with comparisons to other equations). *J. Acoust. Soc. Amer.*, **56**, 1084–91.

DeSanto, J.A. (1979) Derivation of the acoustic wave equation in the presence of gravitational and rotational effects. *J. Acoust. Soc. Amer.*, **66**, 827–30.

Diachok, O.I. (1976) Effects of sea-ice ridges on sound propagation in the Arctic Ocean. *J. Acoust. Soc. Amer.*, **59**, 1110–20.

Diachok, O. (1980) Arctic hydroacoustics. *Cold Regions Sci. Tech.*, **2**, 185–201.

Diachok, O.I. and Winokur, R.S. (1974) Spatial variability of underwater ambient noise at the Arctic ice-water boundary. *J. Acoust. Soc. Amer.*, **55**, 750–3.

DiNapoli, F.R. (1971) Fast field program for multilayered media. Nav. Underwater Syst. Ctr, Rept 4103.

DiNapoli, F.R. and Deavenport, R.L. (1979) Numerical models of underwater acoustic propagation. In *Ocean Acoustics*, Topics in Current Physics, Vol. 8, ed. J.A. DeSanto. Springer-Verlag, New York, pp. 79–157.

DiNapoli, F.R. and Deavenport, R.L. (1980) Theoretical and numerical Green's function field solution in a plane multilayered medium. *J. Acoust. Soc. Amer.*, **67**, 92–105.

Donn, W.L. and Rind, D. (1971) Natural infrasound as an atmospheric probe. *Geophys. J. Roy. Astron. Soc.*, **26**, 111–33.

Doolittle, R., Tolstoy, A. and Buckingham, M. (1988) Experimental confirmation of horizontal refraction of CW acoustic radiation from a point source in a wedge-shaped ocean environment. *J. Acoust. Soc. Amer.*, **83**, 2117–25.

Dosso, S.E., Yeremy, M.L., Ozard, J.M. and Chapman, N.R. (1993) Estimation of ocean-bottom properties by matched-field inversion of acoustic field data. *IEEE J. Oceanic Engr.*, **18**, 232–9.

Dozier, L.B. (1984) PERUSE: A numerical treatment of rough surface scattering for the parabolic wave equation. *J. Acoust. Soc. Amer.*, **75**, 1415–32.

Dozier, L.B. and Cavanagh, R.C. (1993) Overview of selected underwater acoustic propagation models. Office of Naval Research, Advanced Environmental Acoustic Support Program, OAML Rept 93–101.

Dozier, L.B. and White, D. (1988) Software requirements specification for the ASTRAL model. Sci. Appl. Inter. Corp., OAML-SRS-23.

Dushaw, B.D., Worcester, P.E., Cornuelle, B.D. and Howe, B.M. (1993) On equations for the speed of sound in seawater. *J. Acoust. Soc. Amer.*, **93**, 255–75.

Duston, M.D., Gilbert, R.P. and Wood, D.H. (1986) A wave propagation computation technique using function theoretic representation. In *Numerical Mathematics and Applications*, ed. R. Vichnevetsky and J. Vignes. Elsevier Science Publishers, London, pp. 281–8.

Dyer, I (1973) Statistics of distant shipping noise. *J. Acoust. Soc. Amer.*, **53**, 564–70.

Dyer, I. (1984) The song of sea ice and other Arctic melodies. In *Arctic Technology and Policy*, ed. I. Dyer and C. Chryssostomidis. Hemisphere Publishing Corp., Philadelphia, pp. 11–37.

Dyer, I. (1988) Arctic ambient noise: ice source mechanisms. Physics News in 1987 – Acoustics. *Physics Today*, **41** (1), S5–S6.

Earle, M.D. and Bishop, J.M. (1984) *A Practical Guide to Ocean Wave Measurement and Analysis*. ENDECO, Inc., Marion, USA.

Ebbeson, G.R. and Turner, R.G. (1983) Sound propagation over Dickins Seamount in the northeast Pacific Ocean. *J. Acoust. Soc. Amer.*, **73**, 143–52.

Eckart, C. (1953) The scattering of sound from the sea surface. *J. Acoust. Soc. Amer.*, **25**, 566–70.

Einstein, P.A. (1975) Underwater sonic ray tracing in three dimensions. *J. Sound Vib.*, **42**, 503–8.

Eller, A.I. (1984a) Findings and recommendations of the Surface Loss Model Working Group: Final report, 1984. Nav. Ocean Res. Devel. Activity, Tech. Note 279.

Eller, A.I. (1984b) Acoustics of shallow water: A status report. Nav. Res. Lab., Memo. Rept 5405.

Eller, A.I. (1985) Findings and recommendations of the Under-Ice Scattering Loss Model Working Group. Nav. Ocean Res. Devel. Activity, Tech. Note 255.

Eller, A.I. (1986) Shallow water sonar performance prediction assessment. Nav. Ocean Res. Devel. Activity, Tech. Note 330.

Eller, A.I. (1990) Automated long range acoustic modeling. Proc. IEEE Oceans 90 Conf., pp. 205–8.

Eller, A.I. and Haines, L. (1987) Identification and acoustic characterization of seamounts. Sci. Appl. Inter. Corp., SAIC-87/1722.

Eller, A.I. and Venne, H.J. Jr (1981) Evaluation procedure for environmental acoustic models. IEEE Inter. Conf. Acoust. Speech Signal Process., pp. 1026–9.

Eller, A.I., Venne, H.J. and Hoffman, D.W. (1982) Evaluation of ocean acoustic reverberation models. Proc. MTS/IEEE Oceans 82 Conf., pp. 206–10.

Ellis, D.D. (1980) NORM2L: An interactive computer program for acoustic normal mode calculations for the Pekeris model. Defence Research Establishment Atlantic (Canada), Tech. Memo. 80/K.

Ellis, D.D. (1985) Two-ended shooting technique for calculating normal modes in underwater acoustic propagation. Defence Research Establishment Atlantic (Canada), DREA Rept 85/105.

Ellis, D.D. (1993) Shallow water reverberation: Normal-mode model predictions compared with bistatic towed-arrary measurements. *IEEE J. Oceanic Engr.*, **18**, 474–82.

Ellis, D.D. and Chapman, D.M.F. (1985) A simple shallow water propagation model including shear wave effects. *J. Acoust. Soc. Amer.*, **78**, 2087–95.

Ellis, D.D., Preston, J.R. and Urban, H.G. (eds) (1993) *Ocean Reverberation*. Kluwer Academic Publishers Dordrecht, The Netherlands.

Emery, W.J. and Meincke, J. (1986) Global water masses: Summary and review. *Oceanol. Acta*, **9**, 383–91.

Estalote, E. (1984) NORDA acoustic models and data bases. Nav. Ocean Res. Devel. Activity, Tech. Note 293.

Estalote, E.A., Kerr, G.A. and King, D.B. (1986) Recent advances in application of acoustic models. In *The NORDA Review* (March 31, 1976-March 31, 1986). Stennis Space Center, USA, pp. 45–50.

Etter, P.C. (1981) Underwater acoustic modeling techniques. *Shock Vib. Dig.*, **13**(2), 11–20.

Etter, P.C. (1984) Underwater acoustic modeling techniques. *Shock Vib. Dig.*, **16**(1), 17–23.

Etter, P.C. (1987a) Underwater acoustic modeling techniques. *Shock Vib. Dig.*, **19**(2),3–10.

Etter, P.C. (1987b). Numerical modeling techniques in underwater acoustics. *J. Acoust. Soc. Amer.*, **82** (Suppl. 1), S102 (abstract).

Etter, P.C. (1987c) Arctic oceanography and meteorology review. Nav. Training Syst. Ctr, NTSC-TR-87-032.

Etter, P.C. (1989) Underwater acoustic modeling for antisubmarine warfare. *Sea Technology*, **30**(5), 35–7.

Etter, P.C. (1990) Underwater acoustic modeling techniques. *Shock Vib. Dig.*, **22**(5), 3–12.

Etter, P.C. and Cochrane, J.D. (1975) Water temperature on the Texas–Louisiana Shelf. Texas A & M Univ., TAMU-SG-75-604.

Etter, P.C. and Flum, R.S. Sr (1978) A survey of marine environmental/acoustic data banks and basic acoustic models with potential application to the Acoustic Performance Prediction (APP) program. ASW Syst. Proj. Off., ASWR-78-117.

Etter, P.C. and Flum, R.S. Sr (1979) An overview of the state-of-the-art in naval underwater acoustic modeling. *J. Acoust. Soc. Amer.*, **65** (Suppl. 1), S42 (abstract).

Etter, P.C. and Flum, R.S. Sr (1980) A survey of underwater acoustic models and environmental-acoustic data banks. ASW Syst. Proj. Off., ASWR-80-115.

Etter, P.C., Deffenbaugh, R.M. and Flum, R.S. Sr (1984) A survey of underwater acoustic models and environmental-acoustic data banks. ASW Syst. Prog. Off., ASWR-84-001.

Etter, P.C., Lamb, P.J. and Portis, D.H. (1987) Heat and freshwater budgets of the Caribbean Sea with revised estimates for the Central American seas. *J. Phys. Oceanogr.*, **17**, 1232–48.

Evans, R.B. (1983) A coupled mode solution for acoustic propagation in a waveguide with stepwise depth variations of a penetrable bottom. *J. Acoust. Soc. Amer.*, **74**, 188–95.

Evans, R.B. (1986) COUPLE: A user's manual. Nav. Ocean Res. Devel. Activity. Tech. Note 332.

Evans, R.B. and Gilbert, K.E. (1985) Acoustic propagation in a refracting ocean waveguide with an irregular interface. *Comp. Maths. Applic.*, **11**, 795–805.

Ewart, T.E. and Reynolds, S.A. (1984) The mid-ocean acoustic transmission experiment, MATE. *J. Acoust. Soc. Amer.*, **75**, 785–802.

Ewing. M. and Worzel, J.L. (1948) Long-range sound transmission. *Geol. Soc. Amer. Mem.*, **27**(Part III), 1–35.

Farquhar, G.B. (ed.) (1970) *Proceedings of an International Symposium on Biological Sound Scattering in the Ocean*. US Government Printing Office, Washington, DC.

Felsen, L.B. (1986) Benchmarks: Are they helpful, diversionary, or irrelevant? *J. Acoust. Soc. Amer.*, **80** (Suppl. 1), S36 (abstract).

Felsen, L.B. (1987) Chairman's introduction to session on numerical solutions of two benchmark problems. *J. Acoust. Soc. Amer.*, **81** (Suppl. 1), S39 (abstract).

Felsen, L.B. (1990) Benchmarks: An option for quality assessment. *J. Acoust. Soc. Amer.*, **87**, 1497–8.

Ferla, M.C., Jensen, F.B. and Kuperman, W.A. (1982) High-frequency normal-mode calculations in deep water. *J. Acoust. Soc. Amer.*, **72**, 505–9.

Ferris, R.H. (1972) Comparison of measured and calculated normal-mode amplitude functions for acoustic waves in shallow water. *J. Acoust. Soc. Amer.*, **52**, 981–8.

Fisher, A., Jr and Pickett, R. (1973) Evaluation of methods for merging BT-derived and deep climatological sound speeds. Nav. Oceanogr. Off., Tech. Note 7700-12-73.

Fisher, F.H. and Simmons, V.P. (1977) Sound absorption in sea water. *J. Acoust. Soc. Amer.*, **62**, 558–64.

Fizell, R.G. (1987) Application of high-resolution processing to range and depth estimation using ambiguity function methods. *J. Acoust. Soc. Amer.*, **82**, 606–13.

Fizell, R.G. and Wales, S.C. (1985) Source localization in range and depth in an Arctic environment. *J. Acoust. Soc. Amer.*, **78**, (Suppl. 1), S57 (abstract).

Flanagan, R.P., Weinberg, N.L. and Clark, J.G. (1974) Coherent analysis of ray propagation with moving source and fixed receiver. *J. Acoust. Soc. Amer.*, **56**, 1673–80.

Flatté, S.M. (ed.) (1979) *Sound Transmission through a Fluctuating Ocean*. Cambridge Univ. Press, Cambridge, UK.

Focke, K.C. (1984) Acoustic attenuation in ocean sediments found in shallow water regions. Appl. Res. Lab., Univ. Texas at Austin, USA, ARL-TR-84-6.

Fofonoff, N.P. (1985) Physical properties of seawater: A new salinity scale and equation of state for seawater. *J. Geophys. Res.*, **90**, 3332–42.

Forbes, A. (1994) Acoustic monitoring of global ocean climate. *Sea Technology*, **35**(5), 65–7.

Foreman, T.L. (1977) Acoustic ray models based on eigenrays. Appl. Res. Lab., Univ. Texas at Austin, USA, ARL-TR-77-1.

Foreman, T.L. (1982) The ARL:UT range-dependent ray model MEDUSA. Appl. Res. Lab., Univ. Texas at Austin, USA, ARL-TR-82-64.

Foreman, T.L. (1983) Ray modeling methods for range dependent ocean environments. Appl. Res. Lab., Univ. Texas at Austin, USA, ARL-TR-83-41.

Fortuin, L. (1970) Survey of literature on reflection and scattering of sound waves at the sea surface. *J. Acoust. Soc. Amer.*, **47**, 1209–28.

Franchi, E.R., Griffin, J.M. and King, J.B. (1984) NRL reverberation model: A computer program for the prediction and analysis of medium-to long-range boundary reverberation. Nav. Res. Lab., Rept 8721.

Franz, G.J. (1959) Splashes as sources of sound in liquids. *J. Acoust. Soc. Amer.*, **31**, 1080–96.

Frisk, G.V. (1994) *Ocean and Seabed Acoustics: A Theory of Wave Propagation*. Prentice-Hall, Englewood Cliffs, USA.

Frye, H.W. and Pugh, J.D. (1971) A new equation for the speed of sound in seawater. *J. Acoust. Soc. Amer.*, **50**, 384–6.

Gabrielson, T.B. (1982) Mathematical foundations for normal mode modeling in waveguides. Nav. Air Devel. Ctr, Rept NADC-81284-30.

Gaillard, F. (1985) Ocean acoustic tomography with moving sources or receivers. *J. Geophys. Res.*, **90**, 11891–8.

Garon, H.M. (1975) FACTEX: FACT extended to range-dependent environments. Acoustic Environmental Support Detachment, Off. Nav. Res.

Garon, H.M. (1976) SHALFACT: A shallow water transmission loss model. Acoustic Environmental Support Detachment, Off. Nav. Res.

Garrett, C. and Munk, W. (1979) Internal waves in the ocean. *Ann. Rev. Fluid Mech.*, **11**, 339–69.

Garwood, R.W. Jr (1979) Air – sea interaction and dynamics of the surface mixed layer. *Rev. Geophys. Space Physics*, **17**, 1507–24.

Gass, S.I. (1979) Computer model documentation: A review and an approach. National Bureau of Standards Special Pub. 500–39.

Gemmill, W. and Khedouri, E. (1974) A note on sound ray tracing through a Gulf Stream eddy in the Sargasso Sea. Nav. Oceanogr. Off., Tech. Note 6150-21-74.

Georges, T.M., Jones, R.M. and Lawrence, R.S. (1990) A PC version of the HARPO ocean acoustic ray-tracing program. Wave Propagation Lab., NOAA Tech. Memo. ERL WPL-180.

Gerstoft, P. and Schmidt, H. (1991) A boundary element approach to ocean seismoacoustic facet reverberation. *J. Acoust. Soc. Amer.*, **89**, 1629–42.

Gilbert, K.E. and Evans, R.B. (1986) A Green's function method for one-way wave propagation in a range-dependent ocean environment. In *Ocean Seismo-Acoustics*, ed. T. Akal and J.M. Berkson. Plenum, New York, pp. 21–8.

Gilbert, K.E., Evans, R.B., Chin-Bing, S.A., White, D. and Kuperman, W.A. (1983). Some new models for sound propagation in bottom-limited ocean environments. Conf. Proc., Acoustics and the Sea Bed, Bath Univ. Press, England, pp. 243–50.

Gilbert, R.P. and Wood, D.H. (1986) A transmutation approach to underwater sound propagation. *Wave Motion*, **8**, 383–97.

Gill, A.E. (1982) *Atmosphere – Ocean Dynamics*, International Geophysics Series, Vol. 30. Academic Press, New York.

Goddard, R.P. (1989) The sonar simulation toolset. Proc. IEEE Oceans 89 Conf., pp. 1217–22.

Goddard, R.P. (1993) Simulating ocean reverberation: A review of methods and issues. Appl. Physics Lab., Univ. Washington, APL-UW TR 9313.

Goddard, R.P. (1994) Sonar simulation toolset software description. Release 2.5. Appl. Physics Lab., Univ. Washington, APL-UW TR 9211.

Gold, B.A. and Renshaw, W.E. (1978) Joint volume reverberation and biological measurements in the tropical Western Atlantic. *J. Acoust. Soc. Amer.*, **63**, 1809–19.

Goldberg, E.D., McCave, I.N., O'Brien, J.J. and Steele, J.H. (eds) (1977) *The Sea*, Vol. 6, Marine Modeling. Wiley Interscience, New York.

Goldman, J. (1974) A model of broadband ambient noise fluctuations due to shipping. Bell Telephone Labs Rept OSTP-31 JG.

Goldstein, C.I. (1984) The numerical solution of underwater acoustic propagation problems using finite difference and finite element methods. Nav. Res. Lab. Rept 8820.

Gonzalez, R. and Payne, S.G. (1980) User's manual for NEMESIS and PLMODE. Appl. Res. Lab., Univ. Texas at Austin, USA, ARL-TM-80-6.

Goodman, R.R. and Farwell, R.W. (1979) A note on the derivation of the wave equation in an inhomogeneous ocean. *J. Acoust. Soc. Amer.*, **66**, 1895–6.

Gordon, D.F. (1964) Extension of the ray intensity procedure for underwater sound based on a profile consisting of curvilinear segments. Nav. Electron. Lab., Res. Rept 1217.

Gordon, D.F. (1979) Underwater sound propagation-loss program. Computation by normal modes for layered oceans and sediments. Nav. Ocean Syst. Ctr, Tech. Rept 393.

Gordon, D.F. and Bucker, H.P. (1984) Arctic acoustic propagation model with ice scattering. Nav. Ocean Syst. Ctr, Tech. Rept 985.

Gragg, R.F. (1985) The broadband normal mode model PROTEUS. Appl. Res. Lab., Univ. Texas at Austin, USA, ARL-TM-85-6.

Green, D.M. (1994) Sound's effects on marine mammals need investigation. *EOS, Trans. Amer. Geophys. Union*, **75**, 305–6.

Greene, C.H. and Wiebe, P.H. (1988) New developments in bioacoustical oceanography. *Sea Technology*, **29**(8), 27–9.

Greene, R.R. (1984) The rational approximation to the acoustic wave equation with bottom interaction. *J. Acoust. Soc. Amer.*, **76**, 1764–73.

Haines, L.C., Renner, W.W. and Eller, A.I. (1988) Prediction system for acoustic returns from ocean bathymetry. Proc. MTS/IEEE Oceans 88 Conf., pp. 295–7.

Hall, H.R. and Watson, W.H. (1967) An empirical bottom reflection loss expression for use in sonar range prediction. Nav. Undersea Warfare Ctr, Tech. Note 10.

Hall, M.V. (1989) A comprehensive model of wind-generated bubbles in the ocean and predictions of the effects on sound propagation at frequencies up to 40 kHz. *J. Acoust. Soc. Amer.*, **86**, 1103–17.

Hall, M.V. and Irving, M.A. (1989) Application of adiabatic mode theory to the calculation of horizontal refraction through a mesoscale eddy. *J. Acoust. Soc. Amer.*, **86**, 1465–77.

Hamilton, E.L. (1980) Geoacoustic modeling of the sea floor. *J. Acoust. Soc. Amer.*, **68**, 1313–40.

Hamson, R.M. and Wagstaff, R.A. (1983) An ambient-noise model that includes coherent hydrophone summation for sonar system performance in shallow water. SACLANT ASW Res. Ctr, Rept SR-70.

Hanna, J.S. (1976) Example of acoustic model evaluation and data interpretation. *J. Acoust. Soc. Amer.*, **60**, 1024–31.

Hanna, J.S. and Rost, P.V. (1981) Parabolic equation calculations versus North Pacific measurement data. *J. Acoust. Soc. Amer.*, **70**, 504–15.

Hardin, R.H. and Tappert, F.D. (1973) Applications of the split-step Fourier method to the numerical solution of nonlinear and variable coefficient wave equations. *SIAM Rev.*, **15**, 423.

Harlan, J.A., Georges, T.M. and Jones, R.M. (1991a) PROFILE – A program to generate profiles from HARPO/HARPA environmental models. Wave Propagation Lab., NOAA Tech. Memo. ERL WPL-198.

Harlan, J.A., Jones, R.M. and Georges, T.M. (1991b) PSGRAPH – A plotting program for PC-HARPO, PROFILE, CONPLT, and EIGEN. Wave Propagation Lab., NOAA Tech. Memo. ERL WPL-203.

Harrison, C.H. (1989) Ocean propagation models. *Appl. Acoust.*, **27**, 163–201.

Harrison, C.H., Ainslie, M.A. and Packman, M.N. (1990) INSIGHT: A fast, robust propagation loss model providing clear understanding. Proc. Undersea Defence Technology Conf. (London, England), pp. 317–22.

Hassab, J.C. (1989) *Underwater Signal and Data Processing*. CRC Press, Boca Raton, Florida, USA.

Hawker, K.E. (1980) The use and evaluation of acoustic transmission loss models. Appl. Res. Lab., Univ. Texas at Austin, USA, ARL-TR-80-1.

Hawkins, J.D. (1992) TOTS: Three dimensional ocean thermal structure analysis. *Sea Technology*, **33**(1), 62–3.

Hawkins, J.D., Clancy, R.M. and Price, J.F. (1993) Use of AVHRR data to verify a system for forecasting diurnal sea surface temperature variability. *Int. J. Remote Sensing*, **14**, 1347–57.

Heaney, K.D., Kuperman, W.A. and McDonald, B.E. (1991) Perth-Bermuda sound propagation (1960): Adiabatic mode interpretation. *J. Acoust. Soc. Amer.*, **90**, 2586–94.

Heathershaw, A.D., Stretch, C.E. and Maskell, S.J. (1991) Coupled ocean-acoustic model studies of sound propagation through a front. *J. Acoust. Soc. Amer.*, **89**, 145–55.

Heindsmann, T.E., Smith, R.H. and Arneson, A.D. (1955) Effect of rain upon underwater noise levels. *J. Acoust. Soc. Amer.*, **27**, 378–9.

Henrick, R.F. (1983) A cautionary note on the use of range-dependent propagation models in underwater acoustics. *J. Acoust. Soc. Amer.*, **73**, 810–12.

Hodges, P. (1987) Three decades by the numbers. *Datamation*, **33**(18) 77–87.

Hodgkiss, W.S. (1980) Reverberation model: I. Technical description and user's guide. Marine Physical Lab., Scripps Inst. Oceanogr., Tech. Memo. 319.

Hodgkiss, W.S. Jr (1984) An oceanic reverberation model. *IEEE J. Oceanic Engr.*, **9**, 63–72.

Hodgkiss, W.S. Jr and Alexandrou, D. (1985) Under-ice reverberation rejection. *IEEE J. Oceanic Engr.*, **10**, 285–9.

Hodgkiss, W.S., Jr and Fisher, F.H. (1990) Vertical directionality of ambient noise at 32°N as a function of longitude and wind speed. *IEEE J. Oceanic Engr.*, **15**, 335–9.

Hoffman, D.W. (1976) LORA: A model for predicting the performance of long-range active sonar systems. Nav. Undersea Ctr, Tech. Pub. 541.

Hoffman, D.W. (1979) LIRA: A model for predicting the performance of low-frequency active-sonar systems for intermediate surveillance ranges. Nav. Ocean Syst. Ctr, Tech. Doc. 259.

Holland, C.W. and Brunson, B.A. (1988) The Biot-Stoll sediment model: An experimental assessment. *J. Acoust. Soc. Amer.*, **84**, 1437–43.

Holmes, E.S. (1988) Software requirements specification for the parabolic equation model. Sci. Appl. Inter. Corp., OAML-SRS-22.

Holmes, E.S. and Gainey, L.A. (1992a) Software requirements specification for the parabolic equation model version 3.4. Anal. and Technol., Inc. and Sci. Appl. Inter. Corp., OAML-SRS-22.

Holmes, E.S. and Gainey, L.A. (1992b) Software test description for the parabolic equation model version 3.4. Anal. and Technol., Inc. and Sci. Appl. Inter. Corp., OAML-STD-22.

Holmes, E.S. and Gainey, L.A. (1992c) Software design document for the parabolic equation model version 3.4. Anal. and Technol., Inc. and Sci. Appl. Inter. Corp., OAML-SDD-22.

Holmes, E.S., Miller, E.C. and Stephens, R.H. (1990) A PC-based acoustic model operating system. Proc. IEEE Oceans 90 Conf., pp. 227–31.

Holt, R.M. (1985) FACT-10B version description document. ODSI Defense Syst., Inc., DSIR-PU-85-0124.

Hovem, J.M. (1993) Mechanisms of Bottom Loss in Underwater Acoustics. In *Acoustic Signal Processing for Ocean Exploration*, ed. J.M.F. Moura and I.M.G. Lourtie. Kluwer Academic Publishers, Dordrecht, The Netherlands, pp. 21–40.

Hovem, J.M., Richardson, M.D. and Stoll, R.D. (eds) (1991) *Shear Waves in Marine Sediments*. Kluwer Academic Publishers, Dordrecht, The Netherlands.

Hovland, M. (1988) Organisms: The only cause of scattering layers? *EOS, Trans. Amer. Geophys. Union*, **69**, 760.

Ingenito, F., Ferris, R.H., Kuperman, W.A. and Wolf, S.N. (1978) Shallow water acoustics summary report (first phase). Nav. Res. Lab., Rept 8179.

Jackson, D.R., Winebrenner, D.P. and Ishimaru, A. (1986) Application of the composite roughness model to high-frequency bottom back scattering. *J. Acoust. Soc. Amer.*, **79**, 1410–22.

Jacobs, G. (1974) Multiple profile restructuring and supplemental plot programs. Ocean Data Syst., Inc.

Jacobs, G. (1982) FACT 9-H: Version description document. Nav. Ocean Res. Devel. Activity, Tech. Note 133T.

Jennette, R.L., Sander, E.L. and Pitts, L.E. (1978) The USI array noise model, Version I documentation. Underwater Syst., Inc., USI-APL-R-8.

Jensen, F.B. (1982) Numerical models of sound propagation in real oceans. Proc. MTS/IEEE Oceans 82 Conf., pp. 147–54.

Jensen, F.B. (1984) Numerical models in underwater acoustics. In *Hybrid Formulation of Wave Propagation and Scattering*, ed. L.B. Felsen. Martinus Nijhoff, Dordrecht, The Netherlands, pp. 295–335.

Jensen, F.B. (1986) The art of generating meaningful results with numerical codes. *J. Acoust. Soc. Amer.*, **80**, (Suppl. 1), S20 (abstract).

Jensen, F.B. (1988) Wave theory modeling: A convenient approach to CW and pulse propagation modeling in low-frequency acoustics. *IEEE J. Oceanic Engr.*, **13**, 186–97.

Jensen, F. (1993) CW and Pulse Propagation Modeling in Ocean Acoustics. In *Acoustic Signal Processing for Ocean Exploration*, ed. J.M.F. Moura and I.M.G. Lourtie. Kluwer Academic Publishers, Dordrecht, The Netherlands, pp. 3–20.

Jensen, F.B. and Ferla, M.C. (1979) SNAP: The SACLANTCEN normal-mode acoustic propagation model. SACLANT ASW Res. Ctr, Memo. SM-121.

Jensen, F.B. and Ferla, C.M. (1988) Numerical solutions of range-dependent benchmark problems in ocean acoustics. SACLANT Undersea Res. Ctr, Rept SR-141.

Jensen, F.B. and Ferla, C.M. (1990) Numerical solutions of range-dependent benchmark problems in ocean acoustics. *J. Acoust. Soc. Amer.*, **87**, 1499–1510.

Jensen, F.B. and Krol, H. (1975) The use of the parabolic equation method in sound propagation modelling. SACLANT ASW Res. Ctr, Memo. SM-72.

Jensen, F.B. and Kuperman, W.A. (1979) Environmental acoustical modelling at SACLANTCEN. SACLANT ASW Res. Ctr, Rept SR-34.

Jensen, F.B. and Kuperman, W.A. (1983) Optimum frequency of propagation in shallow water environments. *J. Acoust. Soc. Amer.*, **73**, 813–19.

Jensen, F.B. and Martinelli, G. (1985) Accurate PE calculations for range-dependent environments based on c_0 updates. *J. Acoust. Soc. Amer.*, **78**, (Suppl. 1), S23 (abstract).

Jensen, F.B. and Schmidt, H. (1984) Review of numerical models in underwater acoustics, including recently developed fast-field program. SACLANT ASW Res. Ctr, Rept SR-83.

Jensen, F.B. and Schmidt, H. (1987) Subcritical penetration of narrow Gaussian beams into sediments. *J. Acoust. Soc. Amer.*, **82**, 574–9.

Jensen, F.B., Kuperman, W.A., Porter, M.B. and Schmidt, H. (1994) *Computational Ocean Acoustics*. American Institute of Physics, New York.

Johnson, H.R., Backus, R.H., Hersey, J.B. and Owen, D.M. (1956) Suspended echo-sounder and camera studies of midwater sound scatterers. *Deep-Sea Res.*, **3**, 266–72.

Johnson, O.G. (1984) Three-dimensional wave equation computations on vector computers. *Proc. IEEE*, **72**, 90–5.

Jones, R.M. (1982) Algorithms for reflecting rays from general topographic surfaces in a ray tracing program. Wave Propagation Lab., NOAA Tech. Memo. ERL WPL-98.

Jones, R.M. (1983) A survey of underwater-acoustic ray tracing techniques. Wave Propagation Lab., NOAA Tech. Memo. ERL WPL-111.

Jones, R.M., Riley, J.P. and Georges, T.M. (1982) A versatile three-dimensional Hamiltonian ray-tracing computer program for acoustic waves in the atmosphere. Wave Propagation Lab., NOAA Tech. Memo. ERL WPL-103.

Jones, R.M. and Georges, T.M. (1991) HARPX – A program to extend HARPO or HARPA ray paths in horizontally uniform media. Wave Propagation Lab., NOAA Tech. Memo. ERL WPL-201.

Jones, R.M., Riley, J.P. and Georges, T.M. (1986) HARPO–A versatile three-dimensional Hamiltonian ray-tracing program for acoustic waves in an ocean with irregular bottom. Wave Propagation Lab., NOAA Tech. Rept.

Jones, R.M., Georges, T.M., Nesbitt, L. and Weickmann, A. (1991) Ocean acoustic tomography inversion in the adiabatic-invariant approximation. Wave Propagation Lab., NOAA Tech. Memo. ERL WPL-217.

Kanabis, W.G. (1975) A shallow water acoustic model for an ocean stratified in range and depth, Vol. I. Nav. Underwater Syst. Ctr, Tech. Rept 4887-I.

Kanabis, W.G. (1976) A shallow water acoustic model for an ocean stratified in range and depth, Vol. II. Nav. Underwater Syst. Ctr, Tech. Rept 4887-II.

Keiffer, R.S. and Novarini, J.C. (1990) A wedge assemblage method for 3-D acoustic scattering from sea surfaces: Comparison with a Helmholtz–Kirchhoff method. In *Computational Acoustics. Vol. 1: Ocean-Acoustic Models and Supercomputing*, ed. D.Lee, A Cakmak and R. Vichnevetsky. North-Holland, Dordrecht, The Netherlands. pp. 67–81.

Keller, J.B. and Papadakis, J.S. (eds) (1977) *Wave Propagation and Underwater Acoustics*, Lecture Notes in Physics, Vol. 70. Springer-Verlag, New York.

Kerman, B.R. (ed.) (1988) *Sea Surface Sound. Natural Mechanisms of Surface Generated Noise in the Ocean*, NATO Advanced Science Institutes Series C: Mathematical and Physical Sciences 238. Kluwer Academic Publishers, Dordrecht, The Netherlands.

Kerman, B.R. (ed.) (1993) *Natural Physical Sources of Underwater Sound*, Sea Surface Sound (2). Kluwer Academic Publishers, Dordrecht, The Netherlands.

Kerr, D.E. (ed.) (1951) *Propagation of Short Radio Waves*, MIT Radiation Laboratory Series, Vol. 13. McGraw-Hill, New York.

Kerr, R.A. (1977) Oceanography: A closer look at Gulf Stream rings. *Science*, **198**, 387–9, 430.

Kewley, D.J. and Bucker, H.P. (1987) A fast bistatic reverberation and systems model. *J. Acoust. Soc. Amer.*, **82** (Suppl. 1), S75 (abstract).

Kibblewhite, A.C. and Ewans, K.C. (1985) Wave-wave interactions, microseisms and infrasonic ambient noise in the ocean. *J. Acoust. Soc. Amer.*, **73**, 981–94.

Kibblewhite, A.C., Shooter, J.A. and Watkins, S.L. (1976) Examination of attenuation at very low frequencies using the deep-water ambient noise field. *J. Acoust. Soc. Amer.*, **60**, 1040–7.

King, D. and White, D. (1986a) FACT version 10B Volume 1: Physics description Nav. Ocean Res. Devel. Activity, Tech. Note 319.

King, D. and White, D. (1986b) FACT version 10B. Volume 2: Software description. Nav. Ocean Res. Devel. Activity, Tech. Note 320.

Kinsler, L.E., Frey, A.R., Coppens, A.B. and Sanders, J.V. (1982) *Fundamentals of Acoustics*, 3rd edn. John Wiley and Sons, New York.

Knight, W.C., Pridham, R.G. and Kay, S.M. (1981) Digital signal processing for sonar. *Proc. IEEE*, **69**, 1451–1506.

Knobles, D.P. and Vidmar, P.J. (1986) Simulation of bottom interacting waveforms. *J. Acoust. Soc. Amer.*, **79**, 1760–6.

Knudsen, V.O., Alford, R.S. and Emling, J.W. (1948). Underwater ambient noise. *J. Mar. Res.*, **7**, 410–29.

Koch, R.A., Rutherford, S.R. and Payne, S.G. (1983) Slope propagation: Mechanisms and parameter sensitivities. *J. Acoust. Soc. Amer.*, **74**, 210–8.

Kraus, E.B. (ed.) (1977) *Modelling and Prediction of the Upper Layers of the Ocean*. Pergamon Press, New York.

Kuperman, W.A. (1985) Models of sound propagation in the ocean. *Naval Research Reviews*, **37**(3), 32–41.

Kuperman, W.A. and Ingenito, F. (1980). Spatial correlation of surface generated noise in a stratified ocean. *J. Acoust. Soc. Amer.*, **67**, 1988–96.

Kuperman, W.A., Porter, M.B. and Perkins, J.S. (1987) Three-dimensional oceanographic acoustic modeling of complex environments. *J. Acoust. Soc. Amer.*, **82**, (Suppl 1), S42 (abstract).

Kuperman, W.A., Porter, M.B., Perkins, J.S. and Piacsek, A.A. (1988) Rapid three-dimensional ocean acoustic modeling of complex environments. In *Proceedings of the IMACS 12th World congress on Scientific Computation*, ed. R. Vichnevetsky, P. Borne and J. Vignes. GERFIDN – Cite Scientifique, Villeneuve d'Ascq, France, pp. 231–3.

Kuperman, W.A., Porter, M.B., Perkins, J.S. and Evans, R.B. (1991) Rapid computation of acoustic fields in three-dimensional ocean environments. *J. Acoust. Soc. Amer.*, **89**, 125–33.

Kutschale, H.W. (1973) Rapid computation by wave theory of propagation loss in the Arctic Ocean. Lamont – Doherty Geol. Obs., CU-8-73.

Kutschale, H.W. (1984) Arctic marine acoustics. Lamont – Doherty Geol. Obs., Final Rept, ONR Contr, N00014-80-C-0021.

Labianca, F.M. (1973) Normal modes, virtual modes, and alternative representations in the theory of surface-duct sound propagation. *J. Acoust. Soc. Amer.*, **53**, 1137–47.

LaFond, E.C. (1962) Internal waves. Part I. In *T he Sea*, Vol. 1, Physical Oceanography, ed. M.N. Hill. Wiley Interscience, New York, pp. 731–51.

Lai, D.Y. and Richardson, P.L. (1977) Distribution and movement of Gulf Stream rings. *J. Phys. Oceanogr.*, 7, 670–83.

Lai, C.-C.A., Qian, W. and Glenn, S.M. (1994) Data assimilation and model evaluation experiment datasets. *Bull. Amer. Meteor. Soc.*, **75**, 793–809.

Lamb, P.J. (1984) On the mixed-layer climatology of the North and tropical Atlantic. *Tellus*, **36A**, 292–305.

Lasky, M. and Colilla, R. (1974) FANM-I fast ambient noise model. Program documentation and user's guide. Ocean Data Syst., Inc.

Lau, R.L., Lee, D. and Robinson, A.R. (eds) (1993) *Computational Acoustics. Vol. 1: Scattering, Supercomputing and Propagation*. North-Holland, Amsterdam.

Lauer, R.B. (1979) Acoustic model evaluation: Issues and recommendations incorporating the experience of the Panel on Sonar System Models (POSSM). Nav. Underwater Syst. Ctr, Tech. Rept 6025.

Lauer, R.B. and Sussman, B. (1976) A methodology for the comparison of models for sonar system applications, Vol. I. Nav. Sea Syst. Command, SEA 06H1/036-EVA/MOST-10.

Lauer, R.B. and Sussman, B. (1979) A methodology for the comparison of models for sonar system applications – Results for low frequency propagation loss in the

Mediterranean Sea, Vol. II. Nav. Sea Syst. Command, SEA 06H1/036-EVA/MOST-11.

Laville, F., Abbott, G.D. and Miller, M.J. (1991) Underwater sound generated by rainfall. *J. Acoust. Soc. Amer.*, **89**, 715–21.

Lee, D. (1983) Effective methods for predicting underwater acoustic wave propagation. Proc. AIAA 8th Aeroacoustics Conf. Paper No. 83-0683.

Lee, D. and Botseas, G. (1982) IFD: An implicit finite-difference computer model for solving the parabolic equation. Nav. Underwater Syst. Ctr, Tech. Rept 6659.

Lee, D. and McDaniel, S.T. (1983) Wave field computations on the interface: An ocean acoustic model. *Math. Modelling*, **4**, 473–88.

Lee, D. and McDaniel, S.T. (1987) Ocean acoustic propagation by finite difference methods. *Comp. Maths Applic.*, **45**(5), special hardcover issue.

Lee, D. and Papadakis, J.S. (1980) Numerical solutions of the parabolic wave equation: An ordinary-differential-equation approach. *J. Acoust. Soc. Amer.*, **68**, 1482–8.

Lee, D. and Siegmann, W.L. (1986) A mathematical model for the 3-dimensional ocean sound propagation. *Math. Modelling*, **7**, 143–62.

Lee, D., Botseas, G. and Papadakis, J.S. (1981) Finite-difference solution to the parabolic wave equation. *J. Acoust. Soc. Amer.*, **70**, 795–800.

Lee, D., Botseas, G. and Siegmann, W.L. (1992) Examination of three-dimensional effects using a propagation model with azimuth-coupling capability (FOR3D). *J. Acoust. Soc. Amer.*, **91**, 3192–202.

Lee, D., Cakmak, A. and Vichnevetsky, R. (eds) (1990a) *Computational Acoustics. Vol. 1: Ocean-Acoustic Models and Supercomputing*. North-Holland, Amsterdam.

Lee, D., Cakmak, A. and Vichnevetsky, R. (eds) (1990b) *Computational Acoustics. Vol. 2: Scattering, Gaussian Beams, and Aeroacoustics*. North-Holland, Amsterdam.

Lee, D., Cakmak, A. and Vichnevetsky, R. (eds) (1990c) *Computational Acoustics. Vol. 3: Seismo-Ocean Acoustics and Modeling*. North-Holland, Amsterdam.

Lee, D., Saad, Y. and Schultz, M.H. (1988) An efficient method for solving the three-dimensional wide angle wave equation. In *Computational Acoustics, Vol. 1, Wave Propagation*, ed. D. Lee, R.L. Sternberg and M.H. Schultz. North-Holland, Amsterdam, pp. 75–89.

Lee, D., Vichnevetsky, R. and Robinson, A.R. (eds) (1993) *Computational Acoustics. Vol. 2: Acoustic Propagation*. North-Holland, Amsterdam.

Leibiger, G.A. (1968) Wave propagation in an inhomogeneous medium with slow spatial variation. Ph.D. Dissertation, Stevens Institute of Technology.

Leibiger, G.A. (1977) Criteria for propagation loss model assessment for APP application. Nav. Underwater Syst. Ctr, Tech. Memo. 771245.

Lemon, D.D. and Duddridge, G. (1987) A numerical model for the calibration of CASP WOTAN wind data. Proc. MTS/IEEE Oceans 87 Conf., pp. 167–71.

LePage, K. and Schmidt, H. (1994) Modeling of low-frequency transmission loss in the central Arctic. *J. Acoust. Soc. Amer.*, **96**, 1783–95.

Leroy, C.C. (1969) Development of simple equations for accurate and more realistic calculation of the speed of sound in seawater. *J. Acoust. Soc. Amer.*, **46**, 216–26.

Levitus, S. (1982) Climatological atlas of the world ocean. NOAA professional Paper 13.

Lewis, E.L. (1980) The practical salinity scale 1978 and its antecedents. *IEEE J. Oceanic Engr.*, **OE-5**, 3–8.

Lewis, J. K. and Denner, W.W. (1988) Higher frequency ambient noise in the Arctic Ocean. *J. Acoust. Soc. Amer.*, **84**, 1444–55.

Locklin, J.H. and Etter, P.C. (1988) Sonar model operating system functional requirement. SYNTEK Engr. Comp. Syst., Inc.

Locklin, J. and Webster, J. (1980) NORDA model operating system functional description and FY 80 implementation plan. Ocean Data Syst., Inc.

Long, D. (1979) FANM/SIAM noise plot program. Ocean Data Syst., Inc.

Love, R.H. (1975) Predictions of volume scattering strengths from biological trawl data. *J. Acoust. Soc. Amer.*, **57**, 300–6.

Love, R.H. (1978) Resonant acoustic scattering by swimbladder-bearing fish. *J. Acoust. Soc. Amer.*, **64**, 571–80.

Love, R.H. (1993) A comparison of volume scattering strength data with model calculations based on quasisynoptically collected fishery data. *J. Acoust. Soc. Amer.*, **94**, 2255–68.

Lovett, J.R. (1978) Merged seawater sound-speed equations. *J. Acoust. Soc. Amer.*, **63**, 1713–18.

Luby, J.C. and Lytle, D.W. (1987) Autoregressive modeling of nonstationary multibeam sonar reverberation. *IEEE J. Oceanic Engr.*, **OE-12**, 116–29.

Ludwing, D. (1966) Uniform asymptotic expansions at a caustic. *Commun Pure Appl. Math.*, **19**, 215–50.

Lukas, I.J., Hess, C.A. and Osborne, K.R. (1980a) ASERT/ASEPS version 4.1 FNOC user's manual. Ocean Data Syst., Inc.

Lukas, I.J., Hess, C.A. and Osborne, K.R. (1980b) DANES/ASEPS version 4.1 FNOC user's manual. Ocean Data Syst., Inc.

Lynch, J.F., Newhall, A.E., Chiu, C.-S. and Miller, J.H. (1994) Three-dimensional ray acoustics in a realistic ocean. In *Oceanography and Acoustics: Prediction and Propagation Models*, ed. A.R. Robinson and D. Lee. American Institute of Physics, New York, Chapter 9, pp 198–232.

Lyons, A.P., Anderson, A.L. and Dwan, F.S. (1994) Acoustic scattering from the seafloor: Modeling and data comparison. *J. Acoust. Soc. Amer.*, **95**, 2441–51.

McCabe, B.J. (1976) Ambient noise effects in the modeling of detection by a field of sensors. Daniel H. Wagner Assoc.

McCammon, D.F. (1988) Fundamental relationships between geoacoustic parameters and predicted bottom loss using a thin layer model. *J. Geophys. Res.*, **93**, 2363–9.

McCammon, D.F. (1991) Underwater acoustic modeling. *Sea Technology*, **32**(8), 53–5.

McCammon, D.F. and Crowder, D.C. (1981) NEPBR – numerable energy paths by RAYMODE computer program. Appl. Res. Lab., Pennsylvania State Univ., Tech. Memo. TM 81–50.

McCammon, D.F. and McDaniel, S.T. (1985) The influence of the physical properties of ice on reflectivity. *J. Acoust. Soc. Amer.*, **77**, 499–507.

McConnell, S.O. (1983) Remote sensing of the air-sea interface using microwave acoustics. Proc. MTS/IEEE Oceans 83 Conf., pp. 85–92.

McDaniel, S.T. (1993) Sea surface reverberation: A review. *J. Acoust. Soc. Amer.*, **94**, 1905–22.

McDonald, B.E. and Kuperman, W.A. (1987) Time domain formulation for pulse propagation including nonlinear behaviour at a caustic. *J. Acoust. Soc. Amer.*, **81**, 1406–17.

McGirr, R.W. (1979) Acoustic model evaluation procedures: A review. Nav. Ocean Syst. Ctr, Tech. Doc. 287.

McGirr, R.W. (1980) Applications of statistical distributions in acoustic model evaluation. Nav. Ocean Syst. Ctr, Tech. Doc. 387.

McGirr, R.W. and Hall, J.C. (1974) FLIRT, a fast linear intermediate-range transmission loss model. Nav. Undersea Ctr, Tech. Note 1282.

McGirr, R., White, D. and Bartberger, C. (1984) Technical evaluation of FACT 10A. Nav. Ocean Res. Devel. Activity, Rept 70.

Mackenzie, K.V. (1961) Bottom reverberation for 530- and 1030-cps sound in deep water. *J. Acoust. Soc. Amer.*, **33**, 1498–504.

Mackenzie, K.V. (1981) Nine-term equation for sound speed in the oceans. *J. Acoust. Soc. Amer.*, **70**, 807–12.

McKinney, C.M. and Anderson, C.D. (1964) Measurements of backscattering of sound from the ocean bottom. *J. Acoust. Soc. Amer.*, **36**, 158–63.

Macpherson, J.D. and Daintith, M.J. (1967) Practical model of shallow-water acoustic propagation. *J. Acoust. Soc. Amer.*, **41**, 850–4.

Mahler, J.I., Sullivan, F.J.M. and Moll, M. (1975) Statistical methodology for the estimation of noise due to shipping in small sectors and narrow bands. Bolt, Beranek and Newman, Inc., Tech. Memo. W273.

Makris, N.C. (1993) Imaging ocean-basin reverberation via inversion. *J. Acoust. Soc. Amer.*, **94**, 983–93.

Makris, N.C. and Dyer, I. (1986) Environmental correlates of pack ice noise. *J. Acoust. Soc. Amer.*, **79**, 1434–40.

Makris, N.C. and Dyer, I. (1991) Environmental correlates of Arctic ice-edge noise. *J. Acoust. Soc. Amer.*, **90**, 3288–98.

Marsh, H.W. (1950) Theory of the anomalous propagation of acoustic waves in the ocean. Navy Underwater Sound Lab., Rept 111.

Marsh, H.W. (1963) Sound reflection and scattering from the sea surface. *J. Acoust. Soc. Amer.*, **35**, 240–4.

Marsh, H.W. and Schulkin, R.H. (1963) Underwater sound propagation in the Arctic ocean. *J. Acoust. Soc. Amer.*, **35**, 552–63.

Marsh, H.W. Jr and Mellen, M. (1955) Report on the status of Project AMOS (Acoustic Meteorological, and Oceanographic Survey) (1 January 1953–31 December 1954). Navy Underwater Sound Lab., Res. Rept 225.

Marsh, H.W. and Schulkin, M. (1962a) Underwater sound transmission. AVCO Marine Electronics Office Rept.

Marsh, H.W. and Schulkin, M. (1962b) Shallow-water transmission. *J. Acoust. Soc. Amer.*, **34**, 863–4.

Marsh, H.W., Schulkin, M. and Kneale, S.G. (1961) Scattering of underwater sound by the sea surface. *J. Acoust. Soc. Amer.*, **33**, 334–40.

Marsh, P. (1976) A computer program for studying the Doppler content of reverberation. Nav. Sea Syst. Command, OD 52258.

Marsh, P. and Poynter, A.B. (1969) Digital computer programs for analyzing acoustic search performance in refractive waters, Vols 1 and 2. Nav. Undersea Ctr, Tech. Pub. 164.

Martin, P.J. (1993) Sensitivity of acoustic transmission loss prediction to mixed-layer hindcasts calculated with data from Ocean Weather Station Papa. Nav. Res. Lab., NRL/FR/7322-93-9426.

Medeiros, R.C. (1982a) RAYMODE passive propagation loss program performance specification. New England Tech. Services, Doc. 8205.

Medeiros, R.C. (1982b) Active acoustic performance prediction using RAYMODE integration. New England Tech. Services, Doc. 8203.

Medeiros, R.C. (1985a) 1985 baseline RAYMODE passive propagation loss program performance specification. New England Tech. Services, Doc. 8501 Rev. 1.

Mederios, R.C. (1985b) 1985 baseline active RAYMODE propagation loss prediction computer program performance specification. New England Tech. Services, Doc. 8502 Rev. 1.

Medwin, H. (1975) Speed of sound in water: A simple equation for realistic parameters. *J. Acoust. Soc. Amer.*, **58**, 1318–19.

Medwin, H., Childs, E., Jordan, E.A. and Spaulding, R.A. Jr (1984) Sound scatter and shadowing at a seamount: Hybrid physical solutions in two and three dimensions. *J. Acoust. Soc. Amer.*, **75**, 1478–90.

Medwin, H., Browne, M.J., Johnson, K.R. and Denny, P.L. (1988) Low-frequency backscatter from Arctic leads. *J. Acoust. Soc. Amer.*, **83**, 1794–803.

Medwin, H., Nystuen, J.A., Jacobus, P.W., Ostwald, L.H. and Snyder, D.E. (1992) The anatomy of underwater rain noise. *J. Acoust. Soc. Amer.*, **92**, 1613–23.

Mellen, R.H. and Marsh, H.W. (1965) Underwater sound in the Arctic Ocean. AVCO Marine Electronics Office, Rept MED-65-1002.

Mellen, R.H., Scheifele, P.M. and Browning, D.G. (1987a) Global model for sound absorption in sea water. Nav. Underwater Syst. Ctr, Tech. Rept 7923.

Mellen, R.H., Scheifele, P.M. and Browning, D.G. (1987b) Global model for sound absorption in sea water. Part II: GEOSECS pH data analysis. Nav. Underwater Syst. Ctr, Tech. Rept 7925.

Mellen, R.H., Scheifele, P.M. and Browning, D.G. (1987c) Global model for sound absorption in sea water. Part III: Arctic regions. Nav. Underwater Syst. Ctr, Tech. Rept 7969.

Mellor, G.L. and Yamada. T. (1974) A hierarchy of turbulence closure models for planetary boundary layers. *J. Atmos. Sci.*, **31**, 1791–806.

Menke, W. (1989) *Geophysical Data Analysis: Discrete Inverse Theory* (Rev. Edn), International Geophysics Series, Vol. 45. Academic Press, Orlando, USA.

Merklinger, H.M. (ed.) (1987) *Progress in Underwater Acoustics*. Proc. Twelfth International Congress on Acoustics Associated Symposium on Underwater Acoustics, July 16–18, 1986, Halifax, Nova Scotia, Canada. Plenum Press, New York.

Miller, J.F. (1982) Range-dependent ocean acoustic transmission loss calculations in a real-time framework. 4th Interservice/Industry Training Equip. Conf., pp. 287–99.

Miller, J.F. (1983) Acoustic propagation loss calculations in a complex ocean environment for simulator-based training. IEE Inter. Conf. Simulators. (Brighton, England), Conf. Pub. 226, pp. 215–21.

Miller, J.F. and Ingenito, F. (1975) Normal mode FORTRAN programs for calculating sound propagation in the ocean. Nav. Res. Lab., Memo. Rept 3071.

Miller, J.F. and Wolf. S.N. (1980) Modal acoustic transmission loss (MOATAL): A transmission loss computer program using a normal-mode model of the acoustic field in the ocean. Nav. Res. Lab., Rept 8429.

Miller, J.F., Nagl, A. and Überall, H. (1986) Upslope sound propagation through the bottom of a wedge-shaped ocean beyond cutoff. *J. Acoust. Soc. Amer.*, **79**, 562–5.

Miller, J.H., Lynch, J.F. and Chiu, C.-S. (1989) Estimation of sea surface spectra using acoustic tomography. *J. Acoust. Soc. Amer.*, **86**, 326–45.

Milne, A.R. (1964) Underwater backscattering strengths of Arctic pack ice. *J. Acoust. Soc. Amer.*, **36**, 1551–6.

Milne, A.R. (1967) Sound propagation and ambient noise under sea ice. In *Underwater Acoustics*, Vol. 2, ed. V.M. Albers. Plenum, New York, pp. 103–38.

Moler, C.B. and Solomon, L.P. (1970) Use of splines and numerical integration in geometrical acoustics. *J. Acoust. Soc. Amer.*, **48**, 739–44.

Moll, M., Zeskind, R.M. and Sullivan, F.J.M. (1977) Statistical measures of ambient noise: Algorithms, program, and predictions. Bolt, Beranek and Newman, Inc., Rept 3390.

Moll, M., Zeskind, R.M. and Scott, W.L. (1979) An algorithm for beam noise prediction. Bolt, Beranek and Newman, Inc., Rept 3653.

Mooers, C.N.K., Piacsek, S.A. and Robinson, A.R. (eds). (1982) Ocean prediction:

The scientific basis and the Navy's needs. A status and prospectus report. Proc. Ocean Prediction Workshop, Monterey, California, USA, May 1981.

Moore-Head, M.E., Jobst, W. and Showalter, J.A. (1989) The calculation-frequency method for fast parabolic-equation modeling at high frequencies. *J. Acoust. Soc. Amer.*, **85**, 1527–30.

Moritz, E. (1982) Reverberation research overview. Nav. Coastal Syst. Ctr, NCSC-TM-362-82.

Morris, G.B. (1978) Depth dependence of ambient noise in the northeastern Pacific Ocean. *J. Acoust. Soc. Amer.*, **64**, 581–90.

Moskowitz, L. (1964) Estimates of the power spectrums for fully developed seas for wind speeds of 20 to 40 knots. *J. Geophys. Res.*, **69**, 5161–79.

Munk, W.H. (1974) Sound channel in an exponentially stratified ocean, with application to SOFAR. *J. Acoust. Soc. Amer.*, **55**, 220–6.

Munk, W.H. (1980) Horizontal deflection of acoustic paths by mesoscale eddies. *J. Phys. Oceanogr.*, **10**, 596–604.

Munk, W.H. (1991) Bermuda shadow. *Atmos.-Ocean*, **29**, 183–96.

Munk, W. (1994) Heard Island and beyond. *J. Acoust. Soc. Amer.*, **95**, 567.

Munk, W. and Baggeroer, A. (1994) The Heard Island papers: A contribution to global acoustics. *J. Acoust. Soc. Amer.*, **96**, 2327–9.

Munk, W.H. and Forbes, A.M.G. (1989) Global ocean warming: An acoustic measure? *J. Phys. Oceanogr.*, **19**, 1765–78.

Munk, W. and Wunsch, C. (1979) Ocean acoustic tomography: A scheme for large scale monitoring. *Deep-Sea Res.*, **26**, 123–61.

Munk, W.H. and Zachariasen, F. (1991) Refraction of sound by islands and seamounts.*J. Atmos. Ocean. Technol.*, **8**, 554–74.

Murphy, J.E. and Chin-Bing, S.A. (1988) A finite element model for ocean acoustic propagation. *Mathl Comput. Modelling*, **11**, 70–4.

Murphy, J.E. and Chin-Bing, S.A. (1989) A finite-element model for ocean acoustic propagation and scattering. *J. Acoust. Soc. Amer.*, **86**, 1478–83.

Murphy, J.E. and Chin-Bing, S.A. (1991) A seismo-acoustic finite element model for underwater acoustic propagation. In *Shear Waves in Marine Sediments*, ed. J.M. Hovem, M.D. Richardson and R.D. Stoll. Kluwer Academic Publishers, Dordrecht, The Netherlands, pp. 463–70.

National Defense Research Committee (1946) *Physics of Sound in the Sea*, Summary Technical Report of Division 6, NDRC, Vol. 8. Reprinted by Dept of the Navy, Naval Material Command, 1969, NAVMAT P-9675, Washington DC.

National Oceanographic Data Center (1992) *Users Guide, Key to Oceanographic Records Documentation No. 14*, 2nd edn. Washington, DC.

National Research Council (1994) Low-frequency sound and marine mammals: Current knowledge and research needs. National Academy Press, Washington, DC.

Naval Oceanographic Office (1967) *Oceanographic Atlas of the North Atlantic Ocean. Section II – Physical Properties*. Pub. No. 700, US Dept of the Navy, Washington, DC.

Naval Oceanographic Office (1972) *Environmental-Acoustics Atlas of the Caribbean Sea and Gulf of Mexico. Volume II – Marine Environment*. SP-189II, US Dept of the Navy, Washington, DC.

Naval Oceanographic Office (1991a) Software design document for the 1985 baseline passive RAYMODE computer program. OAML-SDD-01A, US Dept of the Navy, Stennis Space Center, Mississippi.

Naval Oceanographic Office (1991b) Software design document for the 1985 baseline active RAYMODE computer program. OAML-SDD-02A, US Dept of the Navy, Stennis Space Center, Mississippi.

Naval Oceanographic Office (1991c) Test specification for the 1985 baseline passive RAYMODE computer program. OAML-TS-01G, US Dept of the Navy, Stennis Space Center, Mississippi.

Naval Oceanographic Office (1991d) Test specification for the 1985 baseline active RAYMODE computer program. OAML-TS-02F, US Dept of the Navy, Stennis Space Center, Mississippi.

Naval Oceanographic Office (1993) Oceanographic and atmospheric master library summary. OAML-SUM-21C, US Dept of the Navy, Stennis Space Center, Mississippi.

Neumann, G. and Pierson, W.J. Jr (1966) *Principles of Physical Oceanography*. Prentice-Hall, Englewood Cliffs, USA.

Newhall, A.E., Lynch, J.F., Chiu, C.-S. and Daugherty, J.R. (1990) Improvements in three-dimensional raytracing codes for underwater acoustics. In *Computational Acoustics. Vol. 1: Ocean-Acoustic Models and Supercomputing*, ed. D. Lee, A. Cakmak and R. Vichnevetsky. North-Holland, Dordrecht, The Netherlands, pp. 169–85.

Nghiem-Phu, L., Daubin, S.C. and Tappert, F. (1984) A high-speed, compact, and interactive parabolic equation solution generator (PESOGEN) system. *J. Acoust. Soc. Amer.*, **75**, (Suppl. 1), S26 (abstract).

Nielsen, R.O. (1991) *Sonar Signal Processing*. Artech House, Boston.

Nystuen, J.A. (1986) Rainfall measurements using underwater ambient noise. *J. Acoust. Soc. Amer.*, **79**, 972–82.

Nystuen, J.A. (1994) Acoustical rainfall analysis. *J. Acoust. Soc. Amer.*, **95**, 2882 (abstract).

Officer, C.B. (1958) *Introduction to the Theory of Sound Transmission with Application to the Ocean*. McGraw-Hill, New York.

Ogden, P.M. and Erskine, F.T. (1994) Surface scattering measurements using broadband explosive charges in the Critical Sea Test experiments. *J. Acoust. Soc. Amer.*, **95**, 746–61.

Oğuz, H.N. (1994) A theoretical study of low-frequency oceanic ambient noise. *J. Acoust. Soc. Amer.*, **95**, 1895–912.

Orchard, B.J., Siegmann, W.L. and Jacobson, M.J. (1992) Three-dimensional time-domain paraxial approximations for ocean acoustic wave propagation. *J. Acoust. Soc. Amer.*, **91**, 788–801.

Orcutt, J.A. (1988) Ultralow- and very-low-frequency seismic and acoustic noise in the Pacific. *J. Acoust. Soc. Amer.*, **84** (Suppl. 1), S194 (abstract).

Orme, E.A., Johns, P.B. and Arnold, J.M. (1988) A hybrid modelling technique for underwater acoustic scattering. *Int. J. Numer. Model.: Electron. Netw. Devices Fields*, **1**, 189–206.

Orris, G.J. and Collins, M.D. (1994) The spectral parabolic equation and three-dimensional backscattering. *J. Acoust. Soc. Amer.*, **96**, 1725–31.

Osborne, K.R. (1979) DANES – a directional ambient noise prediction model for FLENUMOCEANCEN. Ocean Data Syst., Inc.

Pack, P.M.W. (1986) The finite element method in underwater acoustics. Ph.D. Thesis, Institute of Sound and Vibration Research, Univ. Southampton.

Packman, M.N., Harrison, C.H. and Ainslie, M.A. (1992) INSIGHT – A fast, robust, propagation loss model with physical intuition. *Acoust. Bull.*, **17**(4), 21–4.

Palmer, L.B. and Fromm, D.M. (1992) The range-dependent active system performance prediction model (RASP). Nav. Res. Lab. Rept. NRL/FR/5160-92-9383.

Palmer, D.R., Brown, M.G., Tappert, F.D. and Bezdek, H.F. (1988a) Classical chaos in nonseparable wave propagation problems. *Geophys. Res. Lett.*, **15**, 569–72.

Palmer, D.R., Lawson. L.M., Daneshzadeh, Y.-H. and Behringer, D.W. (1988b).

Computational studies of the effect of an El Niño/southern oscillation event on underwater sound propagation. In *Computational Acoustics, Vol. 2, Algorithms and Applications*, ed. D. Lee, R.L. Sternberg and M.H. Schultz. North-Holland, Amsterdam, pp. 335–56.

Pedersen, M.A. and Gordon, D.F. (1965) Normal-mode theory applied to short-range propagation in an underwater acoustic surface duct. *J. Acoust. Soc. Amer.*, **37**, 105–18.

Pedersen, M.A. and Gordon, D.F. (1967) Comparison of curvilinear and linear profile approximation in the calculation of underwater sound intensities by ray theory. *J. Acoust. Soc. Amer.*, **41**, 419–38.

Pedersen, M.A. and McGirr, R.W. (1982) Use of theoretical controls in underwater acoustic model evaluation. Nav. Ocean Syst. Ctr, Tech. Rept 758.

Pedersen, M.A., Gordon, D.F. and Hosmer, R.F. (1984) Comparison of experimental detailed convergence-zone structure with results of uniform asymptotic ray theory. Nav. Ocean Syst. Ctr, Tech. Rept 997.

Pedersen, M.A., Gordon, D.F. and Keith, A.J. (1962) A new ray intensity procedure for underwater sound based on a profile consisting of curvilinear segments. Nav. Electron. Lab., Res. Rept. 1105.

Peixoto, J. P. and Oort, A.H. (1992) *Physics of Climate*. American Institute of Physics, New York.

Pekeris, C.L. (1946) Theory of propagation of sound in a half-space of variable sound velocity under conditions of formation of a shadow zone. *J. Acoust. Soc. Amer.*, **18**, 295–315.

Pekeris, C.L. (1948) Theory of propagation of explosive sound in shallow water. *Geol. Soc. Amer. Mem.*, **27**(Part II), 1–117.

Peloquin, R.A. (1992) The Navy ocean modeling and prediction program. *Oceanography*, **5**(1), 4–8.

Penrose, J.D., Pauly, T.J., Arcus, W.R., Duncan, A.J. and Bush, G. (1993) Remote sensing with underwater acoustics. *Acoust. Aust.*, **21**(1), 19–21.

Perkin, R.G. and Lewis, E.L. (1980) The practical salinity scale 1978: Fitting the data. *IEEE J. Oceanic Engr.*, **OE-5**, 9–16.

Perkins, J.S. and Baer, R.N. (1978) A corrected parabolic-equation program package for acoustic propagation. Nav. Res. Lab., Memo. Rept 3688.

Perkins, J.S. and Baer, R.N. (1982) An approximation to the three-dimensional parabolic-equation method for acoustic propagation. *J. Acoust. Soc. Amer.*, **72**, 515–22.

Perkins, J.S., Baer, R.N., Wright, E.B. and Roche, L.F. (1982) Solving the parabolic equation for underwater acoustic propagation by the split-step algorithm. Nav. Res. Lab., Rept 8607.

Perkins, J.S., Baer, R.N., Roche, L.F. and Palmer, L.B. (1983) Three-dimensional parabolic-equation-based estimation of the ocean acoustic field. Nav. Res. Lab., Rept 8685.

Perkins, J.S., Adams, B.B. and McCoy, J.J. (1984) Vertical coherence along a macroray path in an inhomogeneous anisotropic ocean. Nav. Res. Lab., Rept 8792.

Perkins, J.S., Kuperman, W.A., Ingenito, F., Fialkowski, L.T. and Glattetre, J. (1993) Modeling ambient noise in three-dimensional ocean environments. *J. Acoust. Soc. Amer.*, **93**, 739–52.

Perkins, J.S., Williamson, M., Kuperman, W.A. and Evans, R.B. (1990) Sound propagation through the Gulf Stream: Current status of two three-dimensional models. In *Computational Acoustics. Vol. 1: Ocean-Acoustic Models and Supercomputing*, ed. D. Lee, A. Cakmak and R. Vichnevetsky. North-Holland, Dordrecht, The Netherlands, pp. 203–15.

Peters, H. and Gregg, M.C. (1987) Equatorial turbulence: Mixed layer and thermocline. In *Dynamics of the Oceanic Surface Mixed Layer*, ed. P. Muller and D. Henderson. Hawaii Inst. Geophys. Special Pub., Honolulu, USA, pp. 25–45.

Pickard, G.L. (1963) *Descriptive Physical Oceanography*. Pergamon Press, New York.

Pickard, G.L. and Emery, W.J. (1990) *Descriptive Physical Oceanography: An Introduction*, 5th edn. Pergamon Press, New York.

Pierce, A.D. (1982) Guided mode disappearance during upslope propagation in variable depth shallow water overlying a fluid bottom. *J. Acoust. Soc. Amer.*, **72**, 523–31.

Pierce, A.D. and Thurston, R.N. (eds) (1993) *Underwater Scattering and Radiation, Physical Acoustics XXII*. Academic Press, San Diego, USA.

Pierson, W.J., Jr (1964) The interpretation of wave spectrums in terms of the wind profile instead of the wind measured at a constant height. *J. Geophys. Res.*, **69**, 5191–203.

Pierson, W.J., Jr (1991) Comment on 'Effect of Sea Maturity on Satellite Altimeter Measurements' by Roman E. Glazman and Stuart H. Pilorz. *J. Geophys. Res.*, **96**, 4973–7.

Pierson, W. J., Jr and Moskowitz, L. (1964) A proposed spectral form for fully developed wind seas based on the similarity theory of S.A. Kitaigorodskii. *J. Geophys. Res.*, **69**, 5181–90.

Podeszwa, E.M. (1969) Computer program for COLOSSUS II shallow-water transmission loss equations. Navy Underwater Sound Lab., Tech. Memo. 2040–85–69.

Pomerenk, K. and Novick, A. (1987) BISAPP user's manual. Mission Sci. Corp.

Porter, M.B. (1990) The time-marched fast-field program (FFP) for modeling acoustic pulse propagation. *J. Acoust. Soc. Amer.*, **87**, 2013–23.

Porter, M. (1991) The KRAKEN normal mode program. SACLANT Undersea Res. Ctr, Memo. SM-245.

Porter, M.B. (1993) Acoustic models and sonar systems. *IEEE J. Oceanic Engr.*, **18**, 425–37.

Porter, M.B. and Bucker, H.P. (1987) Gaussian beam tracing for computing ocean acoustic fields. *J. Acoust. Soc. Amer.*, **82**, 1349–59.

Porter, M.B. and Jensen, F.B. (1993) Anomalous parabolic equation results for propagation in leaky surface ducts. *J. Acoust. Soc. Amer.*, **94**, 1510–6.

Porter, M. and Reiss, E.L. (1984) A numerical method for ocean-acoustic normal modes. *J. Acoust. Soc. Amer.*, **76**, 244–52.

Porter, M.B. and Reiss, E.L. (1985) A numerical method for bottom interacting ocean acoustic normal modes. *J. Acoust. Soc. Amer.*, **77**, 1760–7.

Potter, J.R. (1994) Acoustic imaging using ambient noise: Some theory and simulation results. *J. Acoust. Soc. Amer.*, **95**, 21–33.

Primack, H. and Gilbert, K.E. (1991) A two-dimensional downslope propagation model based on coupled wedge modes. *J. Acoust. Soc. Amer.*, **90**, 3254–62.

Princehouse, D.W. (1977) REVGEN, a real-time reverberation generator. IEEE Inter. Conf. Acoust. Speech Signal Process., pp. 827–35.

Pritchard, R.S. (1990) Sea ice noise-generating processes. *J. Acoust. Soc. Amer.*, **88**, 2830–42.

Prosperetti, A. and Oğuz, H.N. (1993) The impact of drops on liquid surfaces and the underwater noise of rain. *Ann. Rev. Fluid Mech.*, **25**, 577–602.

Pumphrey, H.C. (1994) Underwater rain noise – The initial impact component. *Acoust. Bull.*, **19**(2), 19–28.

Pumphrey, H.C. and Crum, L.A. (1990) Free oscillations of near-surface bubbles as a source of the underwater noise of rain. *J. Acoust. Soc. Amer.*, **87**, 142–8.

Rajan, S.D. (1992) Determination of geoacoustic parameters of the ocean bottom – data requirements. *J. Acoust. Soc. Amer.*, **92**, 2126–40.

Ramsdale, D.J. and Posey, J.W. (1987) Understanding underwater acoustics under the Arctic ice canopy. *Sea Technology*, **28**(7), 22–8.

Rasband, S.N. (1990) *Chaotic Dynamics of Nonlinear Systems*. John Wiley and Sons, New York.

Rayleigh, J.W.S. (1945) *The Theory of Sound*, Vols I and II. Dover Publications, New York.

Renner, W.W. (1986a) Ambient noise directionaligy estimation system (ANDES) user's guide. Sci. Appl. Inter. Corp., SAIC-86-1705.

Renner, W.W. (1986b) Ambient noise directionality estimation system (ANDES) technical description. Sci. Appl. Inter. Corp., SAIC-86/1645.

Renner, W.W. (1988) Ambient noise directionality estimation system (ANDES) II user's guide (VAX 11/78X installations). Sci. Appl. Inter. Corp., SAIC-88/1567.

Roberts, B.G. (1974) Horizontal-gradient acoustical ray-trace program TRIMAIN. Nav. Res. Lab., Rept 7827.

Robertson, J.S. (1989) A classification scheme for computational ocean acoustic benchmark problems. *Appl. Acoust.*, **27**, 65–8.

Robertson, J.S., Arney, D.C., Jacobson, M.J. and Siegmann, W.L. (1989) An efficient enhancement of finite-difference implementations for solving parabolic equations. *J. Acoust. Soc. Amer.*, 86, 252–60.

Robertson, J.S., Siegmann, W.L. and Jacobson, M.J. (1991) OS2IFD: A microcomputer implementation of the parabolic equation for predicting underwater sound propagation. *Comput. and Geosciences*, **17**, 731–57.

Robinson, A.R. (1987) Predicting open ocean currents, fronts and eddies. In *Three-Dimensional Models of Marine and Estuarine Dynamics*, ed. J. C. J. Nihoul and B.M. Jamart. Elsevier Oceanography Series, 45, pp. 89–111.

Robinson, A.R. (1992) Shipboard prediction with a regional forecast model. *Oceanography*, **5**(1), 42–8.

Robinson, A.R., Carman, J.C. and Glenn, S.M. (1994) A dynamical system for acoustic applications. In *oceanography and Acoustics: Prediction and Propagation Models*, ed. A.R. Robinson and D. Lee. American Institute of Physics, New York, Chapter 4, pp. 80–117.

Robinson, A.R., Carton, J.A., Mooers, C.N.K., Walstad, L.J., Carter, E.F., Rienecker, M.M., Smith, J.A. and Leslie, W.G. (1984) A real-time dynamical forecast of ocean synoptic/mesoscale eddies. *Nature*, **309**, 781–3.

Robinson, E.R. and McConnell, S.O. (1983) Sensitivity of high frequency surface-generated noise to sonar and environmental parameters. Proc. MTS/IEEE Oceans 83 Conf., pp. 11–15.

Rogers, P.H. (1981) Onboard prediction of propagation loss in shallow water. Nav. Res. Lab., Rept 8500.

Rohr, J. and Detsch, R. (1992) A low sea-state study of the quieting effect of monomolecular films on the underlying ambient-noise field. *J. Acoust. Soc. Amer.*, **92**, 365–83.

Ross, D. (1976) *Mechanics of Underwater Noise*. Pergamon Press, New York.

Ross, D. (1993) On ocean underwater ambient noise. *Acoust. Bull.*, **18**(1), 5–8.

Rubenstein, D. and Greene, R. (1991) Modeling the acoustic scattering by under-ice-ridge keels. *J. Acoust. Soc. Amer.*, 89, 666–72.

Runyan, L. (1991) 40 Years on the frontier. *Datamation*, **37**(6), 34–57.

Sachs, D.A. and Silbiger, A. (1971) Focusing and refraction of harmonic sound and transient pulses in stratified media. *J. Acoust. Soc. Amer.*, **49**, 824–40.

Sadowski, W., Katz, R. and McFadden, K. (1984) Ambient noise standards for acoustic modeling and analysis. Nav. Underwater Syst. Ctr, Tech. Doc. 7265.

Saenger, R.A. (1984) Volume scattering strength algorithm: A first generation model. Nav. Underwater Syst. Ctr, Tech. Memo. 841193.

Sagen, H., Johannessen, O.M. and Sandven, S. (1990) The influence of sea ice on ocean ambient sound. In *Ice Technology for Polar Operations*, ed. T.K.S. Murthy, J.G. Paren, W.M. Sackinger and P. Wadhams. Second Inter. Conf. Ice Technology, Cambridge, England, Computational Mechanics Publications, pp. 415–26.

Schmidt, H. (1988) SAFARI. Seismo-acoustic fast field algorithm for range-independent environments. SACLANT Undersea Res. Ctr, Rept SR-113.

Schmidt, H. (1991) Numerical modeling in ocean seismo-acoustics. Proc. IEEE Oceans 91 Conf., pp. 84–92.

Schmidt, H. and Glattetre, J. (1985) A fast field model for three-dimensional wave propagation in stratified environments based on the global matrix method. *J. Acoust. Soc. Amer.*, **78**, 2105–14.

Schmitt, R.W. (1987) The Caribbean sheets and layers transects (C-SALT) program. *EOS, Trans. Amer. Geophys. Union.*, **68**, 57–60.

Schmitt, R.W., Perkins, H., Boyd, J.D. and Stalcup, M.C. (1987) C-SALT: an investigation of the thermohaline staircase in the western tropical North Atlantic. *Deep-Sea Res.*, **34**, 1655–65.

Schneider, H.G. (1990) MOCASSIN – Sound Propagation and Sonar Range Prediction Model for Shallow Water Environments, User's Guide. Tech. Rept 1990–9, Forschungsanstalt der Bundeswehr für Wasserschall- und Geophysik, Kiel, Germany.

Schneider, H.G. (1993) Surface loss, scattering, and reverberation with the split-step parabolic wave equation model. *J. Acoust. Soc. Amer.*, **93**, 770–81.

Schramm, W.G. (1993) NEONS: A database management system for environmental data. *Earth System Monitor*, **3**(4), 7–8. (Published by Dept of Commerce, NOAA, Silver Spring, Maryland, USA.)

Schreiner, H.F., Jr (1990) The RANDI-PE noise model. Proc. IEEE Oceans 90 Conf., pp. 576–77.

Schulkin, M. and Mercer, J.A. (1985) Colossus revisited: A review and extension of the Marsh-Schulkin shallow water transmission loss model (1962) Appl. Physics Lab., Univ. Washington, APL-UW 8508.

Schurman, I.W., Siegmann, W.L. and Jacobson, M.J. (1991) An energy-conserving parabolic equation incorporating range refraction. *J. Acoust. Soc. Amer.*, **89**, 134–44.

Scrimger, J.A., Evans, D.J., McBean, G.A., Farmer, D.M. and Kerman, B.R. (1987) Underwater noise due to rain, hail and snow. *J. Acoust. Soc. Amer.*, **81**, 79–86.

Scully-Power, P.D. and Lee, D. (eds) (1984) Recent progress in the development and application of the parabolic equation. Nav. Underwater Syst. Ctr, Tech. Doc. 7145.

Selsor, H.D. (1993) Data from the sea: Navy drifting buoy program. *Sea Technology*, **34**(12), 53–8.

Seong, W. (1990) Hybrid Galerkin boundary element – wavenumber integration method for acoustic propagation in laterally inhomogeneous media. Ph.D. Thesis, Massachusetts Institute of Technology.

Shaw, P.T., Watts, D.R. and Rossby, H.T. (1978) On the estimation of oceanic wind speed and stress from ambient noise measurements. *Deep-Sea Res.*, **25**, 1225–33.

Siegmann, W.L., Kriegsmann, G.A. and Lee, D. (1985) A wide-angle three-dimensional parabolic wave equation. *J. Acoust. Soc. Amer.*, **78**, 659–64.

Siegmann, W.L., Jacobson, M.J. and Law, L.D. (1987) Effects of bottom attenuation on acoustic propagation with a modified ray theory. *J. Acoust. Soc. Amer.*, **81**, 1741–51.

Sienkiewicz, C.G., Boyd, M.L., Rosenzweig, J.R. and Pelton, R.M. (1975) Computer-Programmed models used at the University of Washington's Applied Physics Laboratory for evaluating torpedo sonar performance. Appl. Physics Lab., Univ. Washington, APL-UW 7611.

Smith, E.M. (1982) Active RAYMODE program user's guide. Nav. Underwater Syst. Ctr, Tech., Memo. 821081.

Smith, K.B. and Tappert, F.D. (1993) UMPE: The University of Miami parabolic equation model, version 1.0. Marine Physical Lab., Scripps Inst. Oceanogr., Tech. Memo. 432.

Smith, K.B., Brown, M.G. and Tappert, F.D. (1992a) Ray chaos in underwater acoustics. *J. Acoust. Soc. Amer.*, **91**, 1939–49.

Smith, K.B., Brown, M.G. and Tappert, F.D. (1992b) Acoustic ray chaos induced by mesoscale ocean structure. *J. Acoust. Soc. Amer.*, **91**, 1950–9.

Smith, K.B., Tappert, F.D. Hodgkiss, W.S. (1993) Comparison of UMPE/PEREV bistatic reverberation predictions with observations in the ARSRP natural lab. *J. Acoust. Soc. Amer.*, **94**, 1766 (abstract).

Solomon, L.P., Ai, D.K.Y. and Haven, G. (1968) Acoustic propagation in a continuously refracted medium. *J. Acoust. Soc. Amer.*, **44**, 1121–9.

Solomon, L.P., Barnes, A.E. and Lunsford, C.R. (1977) Ocean route envelopes. Planning Syst., Inc., TR–036049.

Spiesberger, J.L. (1989) Remote sensing of western boundary currents using acoustic tomography. *J. Acoust. Soc. Amer.*, **86**, 346–51.

Spiesberger, J.L. and Metzger, K. (1991a) Basin-scale tomography: A new tool for studying weather and climate. *J. Geophys. Res.*, **96**, 4869–89.

Spiesberger, J.L. and Metzger, K. (1991b) A new algorithm for sound speed in seawater. *J. Acoust. Soc. Amer.*, **89**, 2677–88.

Spiesberger, J.L., Metzger, K. and Furgerson, J.A. (1992) Listening for climatic temperature change in the northeast Pacific: 1983–1989. *J. Acoust. Soc. Amer.*, **92**, 384–96.

Spindel, R.C. (1985) Sound transmission in the ocean. *Ann. Rev. Fluid. Mech.*, **17**, 217–37.

Spindel, R.C. and Worcester, P.F. (1990a) Ocean acoustic tomography programs: Accomplishments and plans. Proc. IEEE Oceans 90 Conf., pp. 1–10.

Spindel, R.C. and Worcester, P.F. (1990b) Ocean acoustic tomography. *Scientific American*, **263**(4), 94–9.

Spindel, R.C. and Worcester, P.F. (1991) Ocean acoustic tomography: A decade of development. *Sea Technology*, **32**(7), 47–52.

Spofford, C.W. (1973a) The Bell Laboratories multiple-profile ray-tracing program. Bell Telephone Labs.

Spofford, C.W. (1973b) A synopsis of the AESD Workshop on Acoustic-Propagation Modeling by Non-Ray-Tracing Techniques, 22–25 May 1973, Washington, DC. Acoustic Environmental Support Detachment. Off. Nav. Res. Tech. Note TN-73-05.

Spofford, C.W. (1974) The FACT model, Vol. I. Acoustic Environmental Support Detachment, Off. Nav. Res., Maury Ctr Rept 109.

Spofford, C.W. (1979) The ASTRAL model Volume I: Technical description. Sci. Appl., Inc., SAI-79-742-WA.

Spofford, C.W. and Haynes, J.M. (eds) (1983) Stochastic Modeling Workshop. Proc. Workshop, Nav. Res. Lab., Oct. 26–28, 1982, Washington, DC.

Stephen, R.A. (1990) Solutions to range-dependent benchmark problems by the finite-difference method. *J. Acoust. Soc. Amer.*, **87**, 1527–34.

Stickler, D.C. (1975) Normal-mode program with both the discrete and branch line contributions. *J. Acoust. Soc. Amer.*, **57**, 856–61.

Stickler, D.C. and Ammicht, E. (1980) Uniform asymptotic evaluation of the continuous spectrum contribution for the Pekeris model. *J. Acoust. Soc. Amer.*, **67**, 2018–24.

Stoll, R.D. (1974) Acoustic waves in saturated sediments. In *Physics of Sound in Marine Sediments*, ed. L. Hampton. Plenum, New York, pp. 19–39.

Stoll, R.D. (1980) Theoretical aspects of sound transmission in sediments. *J. Acoust. Soc. Amer.*, **68**, 1341–50.

Stoll, R.D. (1989) *Sediment Acoustics*, Lecture Notes in Earth Sciences, Vol. 26. Springer-Verlag, New York.

Stotts, S.A. and Bedford, N.R. (1991) Development of an Arctic ray model. *J. Acoust. Soc. Amer.*, **90**, 2299 (abstract).

Stotts, S.A., Koch, R.A. and Bedford, N.R. (1994) Development of an Arctic ray model. *J. Acoust. Soc. Amer.*, **95**, 1281–98.

Sverdrup, H.U. Johnson, M.W. and Fleming, R.H. (1942) *The Oceans. Their Physics, Chemistry, and General Biology*. Prentice-Hall, Englewood Cliffs, USA.

Swift, S.A. and Stephen, R.A. (1994) The scattering of a low-angle pulse beam from seafloor volume heterogeneities. *J. Acoust. Soc. Amer.*, **96**, 991–1001.

Talham, R.J. (1964) Ambient-sea-noise model. *J. Acoust. Soc. Amer.*, **36**, 1541–4.

Tango, G.J. (1988) Numerical models for VLF seismic-acoustic propagation prediction: A review. *IEEE J. Oceanic Engr.*, **13**, 198–214.

Tappert, F.D. (1977) The parabolic approximation method. In *Wave Propagation and Underwater Acoustics*, ed. J.B. Keller and J.S. Papadakis. Lecture Notes in Physics, Vol. 70. Springer-Verlag, New York, pp. 224–87.

Tappert, F.D. and Lee, D. (1984) A range refraction parabolic equation. *J. Acoust. Soc. Amer.*, **76**, 1797–803.

Tappert, F.D., Brown, M.G. and Goñi, G. (1991) Weak chaos in an area-preserving mapping for sound ray propagation. *Phys. Lett.* A, **153**, 181–5.

Tappert, F., Lee, D. and Weinberg, H. (1984) High-frequency propagation modeling using HYPER. *J. Acoust. Soc. Amer.*, **75** (Suppl. 1), S30 (abstract).

Tappert, F.D., Brown, M.G., Palmer, D.R. and Bezdek, H. (1988) Chaos in underwater acoustics. *J. Acoust. Soc. Amer.*, **83** (Suppl. 1), S36 (abstract).

Teague, W.J., Carron, M.J. and Hogan, P.J. (1990) A comparison between the generalized digital environmental model and Levitus climatologies. *J. Geophys. Res.*, **95**, 7167–83.

The Acoustic Mid-Ocean Dynamics Experiment Group (1994) Moving ship tomography in the North Atlantic. *EOS, Trans. Amer. Geophys. Union*, **75**, 17–23.

The Ring Group (1981) Gulf Stream cold-core rings: Their physics, chemistry, and biology. *Science*, **212**, 1091–100.

Thomson, D.J. (1990) Wide-angle parabolic equation solutions to two range-dependent benchmark problems. *J. Acoust. Soc. Amer.*, **87**, 1514–20.

Thomson, D.J. and Chapman, N.R. (1983) A wide-angle split-step algorithm for the parabolic equation. *J. Acoust. Soc. Amer.*, **74**, 1848–54.

Thomson, D.J. and Mayfield, M.E. (1994) An exact radiation condition for use with the a posteriori PE method. *J. Comput. Acoust.*, **2**, 113–32.

Thomson, D.J. and Wood, D.H. (1987) A posteriori phase corrections to the parabolic equation. In *Progress in Underwater Acoustics*, ed. H.M. Merklinger. Plenum Press, New York, pp. 425–31.

Thomson, D.J., Brooke, G.H. and DeSanto, J.A. (1990) Numerical implementation of a modal solution to a range-dependent benchmark problem. *J. Acoust. Soc. Amer.*, **87**, 1521–6.

Thorp, W.H. (1967) Analytic description of the low-frequency attenuation coefficient. *J. Acoust. Soc. Amer.*, **42**, 270.

Thorsos, E.I. (1990) Acoustic scattering from a 'Pierson–Moskowitz' sea surface. *J. Acoust. Soc. Amer.*, **88**, 335–49.

Tindle, C.T. and Bold, G.E.J. (1981) Improved ray calculations in shallow water. *J. Acoust. Soc. Amer.*, **70**, 813–9.

Tolstoy, A. (1992) Review of matched field processing for environmental inverse problems. *Inter. J. Modern Phys. C*, **3**, 691–708.

Tolstoy, A. (1993) *Matched Field Processing for Underwater Acoustics*. World Scientific Publishing, Singapore.

Tolstoy, A., Berman, D.H. and Franchi, E.R. (1985) Ray theory versus the parabolic equation in a long-range ducted environment. *J. Acoust. Soc. Amer.*, **78**, 176–89.

Tolstoy, I. and Clay, C.S. (1966) *Ocean Acoustics*. McGraw-Hill, New York (reprinted 1987 by the Acoustical Society of America).

Tuovila, S.M. (1989) SEARAY sonar simulation model. Nav. Cost. Syst. Ctr., Tech. Note NCSC TN 946–88.

Twersky, V. (1957) On scattering and reflection of sound by rough surfaces. *J. Acoust. Soc. Amer.*, **29**, 209–25.

Untersteiner, N. (1966) Sea ice. In *Encyclopedia of Oceanography*, ed. R.W. Fairbridge. Van Nostrand Reinhold Co., New York, pp. 777–81.

Urick, R.J. (1979) *Sound Propagation in the Sea*. US Government Printing Office, Washington, DC.

Urick, R.J. (1982) *Sound Propagation in the Sea*. Peninsula Publishing, Los Altos Hills, USA.

Urick, R.J. (1983) *Principles of Underwater Sound*, 3rd edn. McGraw-Hill, New York.

Urick, R.J. (1984) *Ambient Noise in the Sea*. Nav. Sea. Syst. Command, Washington, DC.

US General Accounting Office (1979) Guidelines for model evaluation. PAD-79-17.

Vastano, A.C. and Owens, G.E. (1973) On the acoustic characteristics of a Gulf Stream cyclonic ring. *J. Phys. Oceanogr.*, **3**, 470–8.

Vendetti, A., Zeidler, E. and Bartberger, C. (1993a) The NAWC bistatic acoustic model (version 3.0) software modifications and user's guide. Nav. Air Warfare Ctr Rept.

Vendetti, A., Zeidler, E. and Bartberger, C. (1993b) The NAWC monostatic acoustic model (version 1.1) software modifications and user's guide. Nav. Air Warfare Ctr Rept.

Vent, R.J. (1972) Acoustic volume-scattering measurements at 3.5, 5.0, and 12.0 kHz in the eastern Pacific Ocean: Diurnal and season variations. *J. Acoust. Soc. Amer.*, **52**, 373–82.

Von Winkle, W.A. (1963) Vertical directionality of deep ocean noise. Navy Underwater Sound Lab., Rept 600.

Wagstaff, R.A. (1973) RANDI: Research ambient noise directionality model. Nav. Undersea Ctr, Tech. Pub. 349.

Wagstaff, R.A. (1981) Low-frequency ambient noise in the deep sound channel – the missing component. *J. Acoust. Soc. Amer.*, **69**, 1009–14.

Wagstaff, R.A. (1982) Noise field calculation or measurement simulation? Some comments on ambient noise modeling. Proc MTS/IEEE Oceans 82 Conf., pp. 187–91.

Ward, P.D. (1989) A noval method of finding normal modes in shallow water. *J. Sound Vib.*, **128**, 349–54.

Watkins, W.A. and Schevill, W.E. (1977) Sperm whale codas. *J. Acoust. Soc. Amer.*, **62**, 1485–90.

Watkins, W.A., Tyack, P., Moore, K.E. and Bird, J.E. (1987) The 20-Hz signals of Finback whales (Balaenoptera physalus). *J. Acoust. Soc. Amer.*, **82**, 1901–12.

Watson, A.G.D. (1958) The effect of the Earth's sphericity in the propagation of sound in the sea. Admiralty Res. Lab., ARL/N29/L.

Watson, W.H. and McGirr, R. (1975) RAYWAVE II: A propagation loss model for the analysis of complex ocean environments. Nav. Undersea Ctr, Tech. Note 1516.

Weickmann, A.M., Riley, J.P., Georges, T.M. and Jones, R.M. (1989) EIGEN – A Program to compute eigenrays from HARPA/HARPO raysets. Wave Propagation Lab., NOAA Tech. Memo. ERL WPL-160.

Weinberg, H. (1969) CONGRATS I: Ray plotting and eigenray generation. Navy Underwater Sound Lab., Rept 1052.

Weinberg, H. (1971) A continuous-gradient curve-fitting technique for acoustic-ray analysis. *J. Acoust. Soc. Amer.*, **50**, 975–84.

Weinberg, H. (1973) Navy interim surface ship model (NISSM) II. Nav. Underwater Syst. Ctr, Tech. Rept 4527.

Weinberg, H. (1975) Application of ray theory to acoustic propagation in horizontally stratified oceans. *J. Acoust. Soc. Amer.*, **58**, 97–109.

Weinberg, H. (1981) Effective range derivative for acoustic propagation loss in a horizontally stratified ocean. *J. Acoust. Soc. Amer.*, **70**, 1736–42.

Weinberg, H. (1982) Generic sonar model. Proc, MTS/IEEE Oceans 82 Conf., pp. 201–5.

Weinberg, H. (1985a) Multilayer expansion for computing acoustic pressure in a horizontally stratified ocean. Proc 11th IMACS World Congress System Simulation and Scientific Computation (Oslo, Norway), pp. 135–8.

Weinberg, H. (1985b) Generic Sonar Model. Nav. Underwater Syst. Ctr, Tech. Doc. 5971D.

Weinberg, H. and Burridge, R. (1974) Horizontal ray theory for ocean acoustics. *J. Acoust. Soc. Amer.*, **55**, 63–79.

Weinberg, N.L. and Dunderdale, T. (1972) Shallow water ray tracing with nonlinear velocity profiles. *J. Acoust. Soc. Amer.*, **52**, 1000–10.

Weinberg, N.L. and Zabalgogeazcoa, X. (1977) Coherent ray propagation through a Gulf Stream ring. *J. Acoust. Soc. Amer.*, **62**, 888–94.

Wells, D.K. and Wargelin, R.M. (1985) Programs and products of the Naval Oceanographic Office. *Mar. Tech. Soc. J.*, **19**(3), 18–25.

Wenz, G.M. (1962) Acoustic ambient noise in the ocean: Spectra and Sources. *J. Acoust. Soc. Amer.*, **34**, 1936–56.

Weston, D.E. and Rowlands, P.B. (1979) Guided acoustic waves in the ocean. *Rep. Prog. Phys.*, **42**, 347–87.

Westwood, E.K. (1989a) Complex ray methods for acoustic interaction at a fluid-fluid interface. *J. Acoust. Soc. Amer.*, **85**, 1872–84.

Westwood, E.K. (1989b) Ray methods for flat and sloping shallow-water waveguides. *J. Acoust. Soc. Amer.*, **85**, 1885–94.

Westwood, E.K. (1989c) Acoustic propagation modeling in shallow water using ray theory. Appl. Res. Lab., Univ. Texas at Austin, USA, ARL-TR-89-6.

Westwood, E.K. (1990) Ray model solutions to the benchmark wedge problems. *J. Acoust. Soc. Amer.*, **87**, 1539–45.

Westwood, E.K. (1992) Broadband matched-field source localization. *J. Acoust. Soc. Amer.*, **91**, 2777–89.

Westwood, E.K. and Tindle, C.T. (1987) Shallow water time-series simulation using ray theory. *J. Acoust. Soc. Amer.*, **81**, 1752–61.

Westwood, E.K. and Vidmar, P.J. (1987) Eigenray finding and time series simulation in a layered-bottom ocean. *J. Acoust. Soc. Amer.*, **81**, 912–24.

White, D. (1992) Software requirements specification for the ASTRAL model (version 4.1). Sci. Appl. Inter. Corp., OAML-SRS-23.

White, D. and Corley, M. (1992a) Software test description for the ASTRAL model (version 4.1). Sci. Appl. Inter. Corp., OAML-STD-23.

White, D. and Corley, M. (1992b) Software design document for the ASTRAL model (version 4.1). Sci. Appl. Inter. Corp., OAML-STD-23.

White, D., Dozier, L.B. and Pearson, C. (1988) Software product specification for the ASTRAL 2.2 Model, Vols I and II. Sci. Appl. Inter. Corp., OAML-SPS-23.

Williams, A.O. Jr (1976) Hidden depths: Acceptable ignorance about ocean bottoms. *J. Acoust. Soc. Amer.*, **59**, 1175–9.

Willis, C.L. (1992) Oceanographic and atmospheric master library. *Naval Oceanography Command News*, **12**(2), 1–5.

Wilson, W.D. (1960) Equation for the speed of sound in sea water. *J. Acoust. Soc. Amer.*, **32**, 1357.

Wood, D.H. and Papadakis, J.S. (1985) An overview of stochastic signal modeling. Nav. Underwater Syst. Ctr, Tech. Rept 7267.

Wright, E.B., Berman, D.H., Baer, R.N. and Perkins, J.S. (1988) The ocean-refraction bathymetric-scattering (ORBS) model. Nav. Res. Lab., Rept 9123.

Yarger, D.F. (1976) The user's guide for the RAYMODE propagation loss program. Nav. Underwater Syst. Ctr, Tech. Memo. 222-10-76.

Yarger, D.F. (1982) The user's guide for the passive RAYMODE propagation loss program. Nav. Underwater Syst. Ctr, Tech. Memo. 821061.

Yevick, D. and Thomson, D.J. (1994) Split-step/finite-difference and split-step/ Lanczos algorithms for solving alternative higher-order parabolic equations. *J. Acoust. Soc. Amer.*, **96**, 396–405.

Zabal, X., Brill, M.H. and Collins, J.L. (1986) Frequency and angle spreads of acoustic signals reflecting from a fixed rough boundary. *J. Acoust. Soc. Amer.*, **79**, 673–80.

Zhang, R. and Jin, G. (1987) Normal-mode theory of average reverberation intensity in shallow water. *J. Sound Vib.*, **119**, 215–23.

Zhou, J.-X., Zhang, X.-Z. and Rogers, P.H. (1991) Resonant interaction of sound wave with internal solitons in the coastal zone. *J. Acoust. Soc. Amer.*, **90**, 2042–54.

Ziman, J. (1978) *Reliable Knowledge*. Cambridge Univ. Press, New York.

Ziomek, L.J. (1985) *Underwater Acoustics. A Linear Systems Theory Approach*. Academic Press, Orlando, USA.

Zornig, J.G. (1979) Physical modeling of underwater acoustics. In *Ocean Acoustics*, ed. J.A. DeSanto. Springer-Verlag, New York, pp. 159–186.

Author index

Page numbers appearing in *italic* indicate pages on which references are given in full.

Subject index